CALCULATIONS FOR A-LEVEL CHEMISTRY

Other Chemistry books for schools from Stanley Thornes (Publishers) Ltd

MODERN ORGANIC CHEMISTRY by A Atkinson
A COMPLETE O-LEVEL CHEMISTRY by G N Gilmore
A MODERN APPROACH TO COMPREHENSIVE CHEMISTRY by G N Gilmore
REVISION NOTES IN CHEMISTRY by E N Ramsden
A FIRST CHEMISTRY COURSE by E N Ramsden
CALCULATIONS FOR O-LEVEL CHEMISTRY by E N Ramsden

CALCULATIONS FOR A-LEVEL CHEMISTRY

E N Ramsden BSc, PhD, DPhil

Wolfreton School, Hull

Stanley Thornes (Publishers) Ltd

First published in 1982 by Stanley Thornes (Publishers) Ltd,
Educa House, Old Station Drive, Leckhampton,
Cheltenham GL53 0DN

Reprinted 1983 with minor corrections
Reprinted 1984 with minor corrections

British Library in Cataloguing in Publication Data

Ramsden, E.N.
  Calculations for A-Level chemistry.
  1. Chemistry — Mathematics
  I. Title
  540'.1'51   QD39.3.M3

  ISBN 0-85950-309-7

Typeset by Tech-Set, Unit 3, Brewery Lane, Felling, Tyne & Wear.
Printed and bound in Great Britain by Bell and Bain Ltd., Glasgow.

Foreword

It is a common complaint of university and college teachers of physical sciences that many of their incoming students are unable to carry out even simple calculations, although they may appear to have a satisfactory grasp of the underlying subject matter. Moreover, this is by no means a trivial complaint, since inability to solve numerical problems nearly always stems from a failure to understand fundamental principles, rather than from mathematical or computational difficulties. This situation is more likely to arise in chemistry than in physics, since in the latter subject it is much more difficult to avoid quantitative problems and at the same time produce some semblance of understanding.

In attempting to remedy this state of affairs teachers in schools often feel the lack of a single source of well-chosen calculations covering all branches of chemistry. This gap is admirably filled by Dr Ramsden's collection of problems. The brief mathematical introduction serves to remind the student of some general principles, and the remaining sections cover the whole range of chemistry. Each section contains a theoretical introduction, followed by worked examples and a large number of problems, some of them from past examination papers. Since answers are also given, the book will be equally useful in schools and in home study. It should make a real contribution towards improving the facility and understanding of students of chemistry in their last years at school and in the early part of their university or college courses.

R P Bell FRS
Honorary Research Professor, University of Leeds, and
formerly Professor of Chemistry, University of Stirling

Preface

Many topics in Chemistry involve numerical problems. Textbooks are not long enough to include sufficient problems to give students the practice which they need in order to acquire a thorough mastery of calculations. This book aims to fill that need.

Chapter 1 is a quick revision of mathematical techniques, with special reference to the use of the calculator, and some hints on how to tackle chemical calculations. With each topic, a theoretical background is given, leading to worked examples and followed by a large number of problems and a selection of questions from past examination papers. The theoretical section is not intended as a full treatment, to replace a textbook, but is included to make it easier for the student to use the book for individual study as well as for class work. The inclusion of answers is also an aid to private study.

The material will take students up to GCE A- and S-level examinations. It will also serve the needs of students preparing for the Ordinary National Diploma. A few of the topics covered are not in the A-level syllabuses of all the Examination Boards, and it is expected that students will be sufficiently familiar with the syllabus they are following to omit material outside their course if they wish. S-level topics and the more difficult calculations are marked with an asterisk.

E N Ramsden
Hull 1982

Acknowledgements

I thank the following examination boards for permission to print questions from recent A-level papers.

The Associated Examining Board

The Joint Matriculation Board

The Oxford and Cambridge Schools Examination Board

The Oxford Delegacy of Local Examinations

The Southern Universities Joint Board

The University of Cambridge Schools Local Examinations Syndicate

The University of London School Examinations Council

The Welsh Joint Education Committee

Many numerical values have been taken from the *Chemistry Data Book* by J G Stark and H G Wallace (published by John Murray). For help with definitions and physical constants, reference has been made to *Physico-chemical Quantities and Units* by M L McGlashan (published by The Royal Institute of Chemistry).

I have been fortunate in receiving excellent advice from Professor R R Baldwin, Professor R P Bell, FRS, Dr G H Davies, Professor W C E Higginson, Dr K A Holbrook, Dr R B Moyes and Dr J R Shorter. I thank these chemists for the help they have given me. I am indebted to Mr J P D Taylor for checking the answers to the problems and for many valuable comments and corrections.

I thank Stanley Thornes (Publishers) for their collaboration and my family for their encouragement.

E N Ramsden
Hull 1982

Contents

List of Exercises

Questions from A-level papers are on the immediately preceding topic(s). Each question is appended with the name of the Examination Board and the year (80 = 1980 etc). p indicates a part question, S an S-level question and N a Nuffield syllabus. The most difficult (often S-level) questions are also denoted by an asterisk.

ABBREVIATIONS OF EXAMINATION BOARDS

AEB	Associated Examining Board
C	University of Cambridge Schools Local Examinations Syndicate
L	University of London Schools Examinations Council
JMB	Joint Matriculation Board
O	Oxford Delegacy of Local Examinations
O & C	Oxford and Cambridge Schools Examinations Board
SUJB	Southern Universities' Joint Board
WJEC	Welsh Joint Education Committee

1 Basic Mathematics

INTRODUCTION

Calculations are a part of your Chemistry course. To some students, they are a source of distress and dismay. To other students, they give the tremendous satisfaction of knowing that a problem has been solved, a correct answer obtained, and full marks gained for that question – a feat which it is not easy to achieve on other types of question!

Calculations are not just an extra activity: the time you spend on calculations will be richly rewarded. Your perception of Chemistry will become at the same time deeper and more precise. No one can come to an understanding of science without acquiring the sharp, logical approach that is needed for solving numerical problems.

To succeed in solving numerical problems you need two things. The first is an understanding of the Chemistry involved. The second is some facility in simple mathematics. Calculations are a perfectly straightforward matter. A numerical problem gives you some data and asks you to obtain some other numerical values. The connection between the data you are given and the information you are asked for is a chemical relationship. You will need to know your Chemistry to recognise what that relationship is.

This introduction is a reminder of some of the mathematics which you studied earlier in your school career. It is included for the sake of students who are not studying mathematics concurrently with their Chemistry course. A few problems are included to help you to brush up your mathematical skills before you go on to tackle the chemical problems.

WORKING WITH NUMBERS IN STANDARD FORM

You are accustomed to writing numbers in decimal notation, for example $123\,677.54$ and $0.001\,678$. In working with large numbers and small numbers, you will find it convenient to write them in a different way, known as *scientific notation* or *standard form*. This means writing a number as a product of two factors. In the first factor, the decimal point comes after the first digit. The second factor is a multiple of ten. For example, $2123 = 2.123 \times 10^3$ and $0.000\,167 = 1.67 \times 10^{-4}$. 10^3 means $10 \times 10 \times 10$, and 10^{-4} means $1/(10 \times 10 \times 10 \times 10)$. The number 3 or -4 is called the exponent, and the number 10 is the base. 10^3 is referred to as '10 to the power 3' or '10 to the third power'. You will have noticed that, if the exponent is increased by 1, the decimal point must be moved one place to the left.

$$2.5 \times 10^3 = 0.25 \times 10^4 = 25 \times 10^2 = 250 \times 10^1 = 2500 \times 10^0$$

Since $10^0 = 1$, this last factor is normally omitted.

When you multiply numbers in standard form, the exponents are added. The product of 2×10^4 and 6×10^{-2} is given by

$$(2 \times 10^4) \times (6 \times 10^{-2}) = (2 \times 6) \times (10^4 \times 10^{-2})$$
$$= 12 \times 10^2 = 1.2 \times 10^3$$

In division, the exponents are subtracted:

$$\frac{1.44 \times 10^6}{4.50 \times 10^{-2}} = \frac{1.44}{4.50} \times \frac{10^6}{10^{-2}} = 0.320 \times 10^8 = 3.20 \times 10^7$$

In addition and subtraction, it is convenient to express numbers using the same exponents. An example of addition is

$$(6.300 \times 10^2) + (4.00 \times 10^{-1}) = (6.300 \times 10^2) + (0.004\,00 \times 10^2)$$
$$= 6.304 \times 10^2$$

An example of subtraction is

$$(3.60 \times 10^{-3}) - (4.20 \times 10^{-4}) = (3.60 \times 10^{-3}) - (0.420 \times 10^{-3})$$
$$= 3.18 \times 10^{-3}$$

ESTIMATING YOUR ANSWER

One advantage of standard form is that very large and very small numbers can be entered on a calculator. Another advantage is that you can easily estimate the answer to a calculation to the correct order of magnitude (i.e. the correct power of 10).

For example,

$$\frac{2456 \times 0.0123 \times 0.004\,14}{5\,223 \times 60.7 \times 8.51}$$

Putting the numbers into standard form gives

$$\frac{2.456 \times 10^3 \times 1.23 \times 10^{-2} \times 4.14 \times 10^{-3}}{5.223 \times 10^3 \times 6.07 \times 10 \times 8.51}$$

This is approximately

$$\frac{2 \times 1 \times 4}{5 \times 6 \times 8} \times \frac{10^3 \times 10^{-2} \times 10^{-3}}{10^3 \times 10} = \frac{1}{30} \times 10^{-6} = 3 \times 10^{-8}$$

By putting the numbers into standard form, you can estimate the answer very quickly. A complete calculation gives the answer 4.64×10^{-8}. The rough estimate is sufficiently close to this to reassure you that you have not made any slips with exponents of ten.

LOGARITHMS

The logarithm (or 'log') of a number N is the power to which 10 must be raised to give the number.

If $N = 1$, then since $10^0 = 1$, $\lg N = 0$.
If $N = 100$, then since $10^2 = 100$, $\lg N = 2$.

If $N = 0.001$, then since $10^{-3} = 0.001$, $\lg N = -3$.

We say that the logarithm of 100 to the base 10 is 2 or $\lg 100 = 2$.

There is another widely used set of logarithms to the base e. They are called natural logarithms as e is a significant quantity in mathematics. It has the value $2.71828 \ldots$. Natural logarithms are written as $\ln N$. The relationship between the two systems is

$$\ln N = \ln 10 \times \lg N$$

Since $\ln 10 = 2.3026$, for most purposes it is sufficiently accurate to write

$$\ln N = 2.303 \lg N$$

Whenever scientific work gives an equation in which $\ln N$ appears, you can substitute 2.303 times the value of $\lg N$.

You will need to obtain the logs of numbers which are not integral powers of 10 (like 10^3 and 10^{-4}). There are two ways of doing this. One is to enter the number on your calculator and press the log key. The value of the log will appear in the display. This will happen whether you enter the number in standard form or another form. For example, $\lg 12\,345 = 4.0915$, whether you enter the number as $12\,345$ or as 1.2345×10^4. However, there is a limit to the number of digits your calculator will accept, and you need to enter very large and very small numbers in standard notation.

The other way of finding a log is to look it up in a set of log tables. Write the number in standard form, e.g. 2×10^3.

Then $\lg (2 \times 10^3) = \lg 2 + \lg 10^3 = 0.3010 + 3 = 3.3010$

You know that $\lg 10^3$ is 3, and you find out that $\lg 2 = 0.3010$ from the tables. 3 is called the *characteristic* and 0.3010 is called the *mantissa* of the logarithm. The log of 2×10^{-6} has a characteristic of -6 and a mantissa of 0.3010. It is written as $\bar{6}.3010$ or as -5.6990. Your calculator will display it as -5.6990.

Operations on logarithms are:

Multiplication. The logs of the numbers are added:

$$\lg (A \times B) = \lg A + \lg B$$

Division. The logs are subtracted:

$$\lg (P/Q) = \lg P - \lg Q$$

Powers. This is a special case of multiplication.

$$\lg A^2 = \lg A + \lg A = 2 \lg A$$
$$\lg A^{-3} = -3 \lg A$$

Roots. It is easy to show that $\lg \sqrt{B} = \frac{1}{2} \lg B$.

Since
$$B = B^{1/2} \times B^{1/2}$$
$$\lg B = \lg B^{1/2} + \lg B^{1/2}$$
$$\lg B^{1/2} = \frac{1}{2} \lg B$$

Similarly,
$$\lg \sqrt[3]{B} = \frac{1}{3} \lg B$$

ANTILOGARITHMS

To find a number from the logarithm of that number, you look up the mantissa in a table of antilogarithms. The number you obtain must be written with one digit in front of the decimal point. The mantissa 0.5949 gives the number 3.935. The characteristic of the logarithm becomes the exponent of the number. Thus the antilog of 3.5949 is 3.935×10^3.

Your calculator will give you the antilog of a number. You should consult the manual to find out the procedure for your own model of calculator.

Most calculators will give you reciprocals, squares and other powers, square roots and other roots directly. If you have a simpler form of calculator, you can obtain powers and roots by using logarithms.

ROUNDING OFF NUMBERS

Often your calculator will display an answer containing more digits than the numbers you fed into it. Suppose you are given the information that $18.6 \, \text{cm}^3$ of sodium hydroxide solution exactly neutralise $25.0 \, \text{cm}^3$ of a solution of hydrochloric acid of concentration $0.100 \, \text{mol dm}^{-3}$. You want to find the concentration of sodium hydroxide solution, and you put the numbers $(25.0 \times 0.100)/18.6$ into your calculator and obtain a value of $0.134 \, 408 \, 6 \, \text{mol dm}^{-3}$. The concentration of the solution is not known as accurately as this, however, because you cannot read the burette as accurately as this.

Since you read the burette to three figures, you quote your answer to three figures. In the number 0.134 408 6, the figures you are sure of are termed the *significant figures*. The significant figures are retained, and the insignificant figures are dropped. This operation is called *rounding off*. If the first number had been 0.134 708 6, it would have been rounded off to 0.135. If the first of the insignificant figures being dropped is 5 or greater, the last of the significant figures is rounded up to the next digit. If the first of the dropped figures is less than 5, the last significant figure is left unaltered.

Some calculations involve several stages. It is sound practice to give one more significant figure in your answer at each stage than the number of significant figures in the data. Then, in the final stage, the answer is rounded off.

If the calculation were $(25.0 \times 0.100)/26.2 = 0.095\,419\,84\,mol\,dm^{-3}$, would you still round off to 3 significant figures? This would make the answer $0.0954\,mol\,dm^{-3}$. Stated in this way, the answer is claiming an accuracy of 1 part in 954 — about 1 part in 1000. Since the hydrochloric acid concentration is known to about 1 part in 100, the answer cannot be stated to a higher degree of accuracy. You have to use the 3-significant-figure rule sensibly, and say that an error of ± 1 in 95 is about as significant as an error of ± 1 in 134. The answer should therefore be quoted as $0.095\,mol\,dm^{-3}$.

The number of significant figures is the number of figures which is accurately known. The number 123 has 3 significant figures. The number 1.23×10^4 has 3 significant figures, but 12 300 has 5 significant figures because the final zeros mean that each of these digits is known to be zero and not some other digit. The number 0.001 23 has 3 significant figures. The number 25.1 has 3 significant figures, and the number 25.10 has 4 significant figures as the final 0 states that the value of this number is known to an accuracy of 1 part in 2500.

In addition, the sum is known with the accuracy of the least reliable numbers in the sum. For example, the sum of

$$
\begin{array}{r}
1.4167\,g \\
+\,100.5\,g \\
+7.12\,g \\
\hline
\end{array}
$$

is $109.0367\,g$

Since 1 figure is known to only 1 place after the decimal point, the sum also is known to 1 place after decimal point and should be written as 109.0 g. The same guideline is used for subtraction.

In multiplication and division the product or quotient is rounded off to the same number of significant figures as the number with the fewest significant figures. For example, $12\,340 \times 2.7 \times 0.003\,65 = 121.6107$. The product is rounded off to 2 significant figures, 1.2×10^2.

CHOICE OF A CALCULATOR

The functions which you need in a calculator for the problems in this book are:

- Addition, Subtraction, Multiplication and Division
- Squares and other powers (x^2 and x^y keys)
- Square roots and other roots ($\sqrt{x}$ and $x^{1/y}$ keys)
- Reciprocals
- Log_{10} and antilog_{10}(10^x)
- Natural logarithms, $\ln_e$ and antiln_e (e^x)
- Exponent key and $+/-$ key
- Brackets
- Memory

A variety of scientific calculators have these functions and others (such as sin, cos, tan and Σx) which will be useful to you in Physics and Mathematics problems.

UNITS

There are two sets of units currently employed in scientific work. One is the CGS system, based on the centimetre, gram and second. The other is the Système Internationale (SI) which is based on the metre, kilogram, second and ampere. SI units were introduced in 1960, and in 1979 the Association for Science Education published a booklet called *Chemical Nomenclature, Symbols and Terminology for Use in School Science* that recommended that schools and colleges adopt this system.

Listed below are the SI units for the seven fundamental physical quantities on which the system is based and also a number of derived quantities and their units.

Basic SI Units

Physical Quantity	Name of Unit	Symbol
Length	metre	m
Mass	kilogram	kg
Time	second	s
Electric current	ampere	A
Temperature	kelvin	K
Amount of substance	mole	mol
Light intensity	candela	cd

Derived SI Units

Physical Quantity	Name of Unit	Symbol	Definition
Energy	joule	J	$kg\,m^2\,s^{-2}$
Force	newton	N	$J\,m^{-1}$
Electric charge	coulomb	C	$A\,s$
Electric potential difference	volt	V	$J\,A^{-1}\,s^{-1}$
Electric resistance	ohm	Ω	$V\,A^{-1}$
Area	square metre		m^2
Volume	cubic metre		m^3
Density	kilogram per cubic metre		$kg\,m^{-3}$
Pressure	newton per square metre or pascal		$N\,m^{-2}$ or Pa
Molar mass	kilogram per mole		$kg\,mol^{-1}$

With all these units, the following prefixes (and others) may be used:

Prefix	Symbol	Meaning
deci	d	10^{-1}
centi	c	10^{-2}
milli	m	10^{-3}
micro	μ	10^{-6}
nano	n	10^{-9}
kilo	k	10^3
mega	M	10^6
giga	G	10^9
tera	T	10^{12}

Chemists are still using some of the CGS units. You will find mass in g; volume in cm^3 and dm^3; concentrations in $mol\,dm^{-3}$ or $mol\,litre^{-1}$; conductivity in $\Omega^{-1}cm$ as well as $\Omega^{-1}m$. Pressure is sometimes given in mm mercury and temperatures in °C.

It is very important when putting values for physical quantities into an equation to be consistent in the use of units. If you are, then the units can be treated as factors in the same way as numbers. Suppose you are asked to calculate the volume occupied by 0.0110 kg of carbon dioxide at 27 °C and a pressure of $9.80 \times 10^4\,N\,m^{-2}$. You know that the gas constant is $8.31\,J\,mol^{-1}K^{-1}$ and that the molar mass of carbon dioxide is $44.0\,g\,mol^{-1}$. Use the ideal gas equation:

$$PV = nRT$$

The pressure $P = 9.80 \times 10^4\,N\,m^{-2}$

The constant $R = 8.31\,J\,K^{-1}\,mol^{-1}$

The temperature $T = 27 + 273 = 300\,K$

The number of moles

$$n = \text{Mass/Molar mass}$$
$$= 0.0110\,\text{kg}/44.0 \times 10^{-3}\,\text{kg mol}^{-1}$$
$$= 0.250\,\text{mol}$$

Then
$$V = \frac{0.250\,\text{mol} \times 8.31\,\text{J K}^{-1}\text{mol}^{-1} \times 300\,\text{K}}{9.80 \times 10^{4}\,\text{N m}^{-2}}$$
$$= 6.34 \times 10^{-3}\,\text{J N}^{-1}\text{m}^{2}$$

Since
$$J = N\,m \qquad (1\ \text{joule} = 1\ \text{newton metre})$$
$$V = 6.34 \times 10^{-3}\,\text{N m N}^{-1}\text{m}^{2}$$
$$= 6.34 \times 10^{-3}\,\text{m}^{3}$$

Volume has the unit of cubic metre. This calculation illustrates what people mean when they say that SI units form a *coherent system of units.* You can convert from one unit to another by multiplication and division, without introducing any numerical factors.

SOLUTION OF QUADRATIC EQUATIONS

A quadratic equation is the name for an equation where the largest exponent of the unknown quantity is 2. In the equation

$$ax^2 + bx + c = 0$$

x is the unknown quantity, a and b are the coefficients of x, and c is a constant. The solution of this equation is given by

$$x = \frac{-b \pm \sqrt{b^2 - 4ac}}{2a}$$

There are two solutions to the equation. Often you will be able to decide that one solution cannot be allowed. You may be calculating some physical quantity that cannot possibly be negative, so that a negative solution can be disregarded and a positive solution adopted.

DRAWING GRAPHS

Here are some hints for drawing graphs.

a) Whenever possible, data should be plotted in a form that gives a straight line graph. It is easier to draw the best straight line through a set of points than to draw a curve.

If the dimensions x and y are related by the expression $y = ax + b$, then a straight line will result when experimental values of y are plotted against the corresponding values of x. The values of x are plotted against the horizontal axis (the x-axis or abscissa), and the corresponding values of y are plotted along the vertical axis (the y-axis or ordinate). The gradient of the straight line obtained $= a$, and the intercept on the y-axis $= b$.

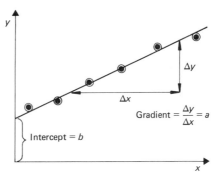

Fig. 1.1   Plotting a graph

b) Choose a scale which will allow the graph to cover as much of the piece of graph paper as possible. There is no need to start at zero. If the points lie between 90 and 100, to start at zero would cramp your graph into a small section at the top of the page.

Choose a scale which will make plotting the data and reading the graph as simple as possible.

c) Label the axes with the dimension and the units. Make the scale units as simple as possible. Instead of plotting as scale units $1 \times 10^{-3}\,\text{mol dm}^{-3}$, $2 \times 10^{-3}\,\text{mol dm}^{-3}$, $3 \times 10^{-3}\,\text{mol dm}^{-3}$, etc., plot 1, 2 and 3, etc., and label the axis as Concentration/$\text{mol dm}^{-3} \times 10^{3}$.

d) When you come to draw a straight line through the points, draw the best straight line you can, to pass through, or close to, as many points as possible. Owing to experimental error, not all the points will fall on to the line. A graph of experimental results gives you a better accuracy than calculating a value from just one point. If you are drawing a curve, draw a smooth curve. Do not join up the points with straight lines. The curve may not pass through every point, but it is more reliable than any one of the points.

A FINAL POINT

Always look critically at your answer. Ask yourself whether it is a reasonable answer. Is it of the right order of magnitude for the data? Is it in the right units? Many errors can be detected by an assessment of this kind.

EXERCISE 1   Practice with Calculations

1. Convert the following numbers into standard form:
    a) 23 678   b) 437.6   c) 0.0169
    d) 0.000 345   e) 672 891

2. Convert each of the following numbers into standard form, enter into your calculator, and multiply by 237. Give the answers in standard form.
    a) 246.8   b) 11 230   c) 267 831   d) 0.051   e) 0.567

3. Find the following quotients:
    a) 2360/0.000 71   b) 28 780/0.106   c) 85.42/460 000
    d) 58/900 670   e) 0.000 88/0.144

4. Find the following sums and differences:
    a) $(2.000 \times 10^4) + (0.10 \times 10^2)$
    b) $48.0 + (5.600 \times 10^3)$
    c) $(1.23 \times 10^5) + (6.00 \times 10^3)$
    d) $(4.80 \times 10^{-4}) - (1.6 \times 10^{-3})$
    e) $(6.300 \times 10^4) - (4.8 \times 10^2)$

5. Make an approximate estimate of the answers to the following:
    a) $\dfrac{4.0 \times 10^3 \times 5.6 \times 10^{-2} \times 7.1 \times 10^6}{8.2 \times 10^{-6} \times 4.9 \times 10^3}$
    b) $567 \times 4183 \times 0.001\,27 \times 0.107$
    c) $\dfrac{496 \times 7124 \times 83\,000 \times 4.7}{7260 \times 41 \times 0.0075}$
    d) $\dfrac{1480 \times 6730 \times 0.173 \times 0.0097}{0.15 \times 0.0088 \times 100\,860 \times 0.10}$
    e) $\dfrac{208 \times 100\,490}{560 \times 0.005\,5 \times 0.000\,49}$

6. Find the logarithms of the following numbers:
    a) 4735   b) 5.072×10^3 c) 0.001 327
    d) 10.076   e) 2.314×10^{-6}

7. Find the antilogarithms of the following:
    a) 3.4567   b) 0.0549   c) 7.8432
    d) -6.4712 e) -2.0571

8. Find the reciprocals of the following:
    a) 234.5   b) 3 488 123   c) 0.002 477
    d) 4.865×10^9 e) 2.645×10^{-5}

9. Solve the quadratic equations:
    a) $(x + 3)^2 = 100$ b) $5x^2 - 8 = 18x$
    c) $(y + 4)^2 = 18y - 9$ d) $(z - 6)^2 = 4z + 8$
    e) $(2x + 4)^2 = 7 - 4x$

2 The Mole

DEFINITIONS

The masses of elements are very small, from 10^{-24} to 10^{-22} grams. Instead of using the actual masses of atoms, we use the *relative atomic masses*. Since hydrogen is the lightest of atoms, it was decided to compare the masses of other atoms with that of hydrogen. Then:

$$\text{Original relative atomic mass} = \frac{\text{Mass of one atom of an element}}{\text{Mass of one atom of hydrogen}}$$

On this scale, the hydrogen atom has a mass of 1 atomic mass unit (m_u) and hydrogen has a relative atomic mass of 1.00000 ($1m_u = 1.6605 \times 10^{-27}$ kg).

Since relative atomic masses are now determined by mass spectrometry, and since volatile carbon compounds are much used in mass spectrometry, the mass of an atom of $^{12}_{6}$C is now taken as the standard of reference ($^{12}_{6}$C is an atom of the isotope of carbon containing 6 protons and 6 neutrons). Thus:

$$\text{Modern relative atomic mass} = \frac{\text{Mass of one atom of an element}}{\frac{1}{12}\text{Mass of one atom of carbon-12}}$$

The difference between the two scales is small. On the carbon-12 scale, the relative atomic mass of carbon is 12.00000, and the relative atomic mass of hydrogen is 1.00897.

You can see from this ratio that, if an atom of carbon is 12 times as heavy as an atom of hydrogen, then 12 g of carbon contain the same number of atoms as 1 g of hydrogen. The same is true of the relative atomic mass expressed in grams for any element: 63.5 g of copper and 27 g of aluminium contain the same number of atoms. This number is 6.022×10^{23}, and is called the Avogadro constant. We call 6.022×10^{23} units of any species a *mole* of that species. Thus, 6.022×10^{23} atoms of copper are a mole of copper; 6.022×10^{23} hydrogen molecules are a mole of hydrogen molecules; and 6.022×10^{23} electrons are a mole of electrons. The symbol for mole is mol.

The mole is defined as the amount of a substance which contains as many elementary entities as there are atoms in 12 grams of carbon-12.

11

The *relative molecular mass* of a compound is defined by the expression:

$$\text{Relative molecular mass} = \frac{\text{Mass of one molecule of the compound}}{\frac{1}{12} \text{ Mass of one atom of carbon-12}}$$

The relative molecular mass of a compound expressed in grams is a mole of the compound; thus 44 g of carbon dioxide is a mole of carbon dioxide, and contains 6.022×10^{23} molecules. In the case of ionic compounds, which do not consist of molecules, you refer to a 'formula unit' of the compound. A formula unit of sodium sulphate is $2Na^+SO_4^{2-}$. Then, the relative formula mass in grams, 142 g, is a mole of sodium sulphate.

The term *relative molar mass* embraces relative molecular mass, relative formula mass and relative atomic mass:

$$\text{Relative molar mass} = \frac{\text{Mass of one mole of substance}}{\frac{1}{12} \text{ Mass of a mole of carbon-12}}$$

Since this equation divides mass by mass, relative molar mass is a ratio and does not have units.

The mass of a mole of an element or compound is referred to as its *molar mass*. The molar mass of sodium hydroxide is 40 g mol^{-1}. The molar mass of copper is 63.5 g mol^{-1}. If you have m grams of a substance which has a molar mass of $M \text{ g mol}^{-1}$, then the number of moles of substance, n, is given by:

$$n \text{ (mol)} = \frac{m \text{ (g)}}{M \text{ (g mol}^{-1}\text{)}}$$

The *number of moles of substance* is referred to simply as the *amount of substance*:

$$\text{Amount of substance (number of moles)} = \frac{\text{Mass}}{\text{Molar mass}}$$

CALCULATION OF MOLAR MASS

The relative molar mass of a compound is the sum of the relative atomic masses of the atoms in a molecule or formula unit of the compound.

EXAMPLE What is the molar mass of glucose?

Formula $= C_6H_{12}O_6$

Relative molar mass $= (6 \times 12) + (12 \times 1) + (6 \times 16) = 180$

ANSWER The relative molar mass is 180, and the molar mass is $180 \, \text{g mol}^{-1}$.

EMPIRICAL FORMULAE

The formula of a compound is composed of the symbols which show which elements are present together with numbers, as subscripts, which show how many atoms of each element are present in a molecule of the compound or in a formula unit of the compound. The empirical formula is the simplest formula which represents the composition of a compound; it shows the ratio of numbers of atoms present. The molecular formula is a simple multiple of the empirical formula. If the empirical formula is CH_2O, the molecular formula may be CH_2O or $C_2H_4O_2$ or $C_3H_6O_3$ and so on. You can tell which molecular formula is correct by finding out which gives the correct molar mass.

To find the empirical formula, you find the ratio of moles of each element present. Remember:

$$\text{Number of moles} = \frac{\text{Mass}}{\text{Molar mass}}$$

EXAMPLE 1 A chloride of iron contains 65.5% chlorine. What is its empirical formula?

METHOD

Elements present	Iron	Chlorine
Percentage by mass	34.5%	65.5%
Relative atomic mass	56	35.5
Number of moles in 100 g of compound	34.5/56 = 0.616	65.5/35.5 = 1.85
Ratio of moles	$\frac{0.616}{0.616}$:	$\frac{1.85}{0.616}$
	= 1 :	3
Ratio of atoms	= 1 :	3

ANSWER The empirical formula is $FeCl_3$.

EXAMPLE 2 Find n in the formula $MgSO_4 \cdot nH_2O$. A sample of 7.38 g of magnesium sulphate crystals lost 3.78 g of water on heating.

METHOD	Compounds present	Magnesium sulphate	Water
	Mass	3.60 g	3.78 g
	Molar mass	$120\,\text{g mol}^{-1}$	$18\,\text{g mol}^{-1}$
	Number of moles	3.60/120	3.78/18
		= 0.030	= 0.21
	Ratio of moles	$\dfrac{0.030}{0.030}$:	$\dfrac{0.21}{0.030}$
		= 1 :	7

ANSWER   The empirical formula is $MgSO_4 \cdot 7H_2O$.

MOLECULAR FORMULAE

The molecular formula can be found from the empirical formula of the compound if the molar mass is known. For methods of finding molar masses, see Chapter 4. The molar mass is a multiple of the empirical formula mass.

EXAMPLE   A compound has the empirical formula CH_2O and molar mass $180\,\text{g mol}^{-1}$. What is its molecular formula?

METHOD   Empirical formula mass = $30\,\text{g mol}^{-1}$

Molar mass = $180\,\text{g mol}^{-1}$

The molar mass is 6 times the empirical formula mass. Therefore the molecular formula is 6 times the empirical formula. Therefore:

ANSWER   The empirical formula is $C_6H_{12}O_6$.

PERCENTAGE COMPOSITION

From the formula of a compound and the relative atomic masses of the elements in it, the percentage of each element in the compound can be calculated. This is called the *percentage composition by mass.*

EXAMPLE 1   Calculate the percentage composition by mass of magnesium oxide.

METHOD   Relative atomic masses are Mg = 24; O = 16

Relative molar mass of MgO = 24 + 16 = 40

Percentage of Mg = $\dfrac{24}{40} \times 100 = 60\%$

Percentage of O = $\dfrac{16}{40} \times 100 = 40\%$

ANSWER   The percentage composition of MgO by mass is 60% Mg, 40% O.

EXAMPLE 2   Calculate the percentage of water of crystallisation in copper(II) sulphate-5-water.

METHOD   Formula is $CuSO_4 \cdot 5H_2O$

Relative atomic masses are Cu, 63.5; S, 32; O, 16; H, 1

Relative molar mass $= 63.5 + 32 + (4 \times 16) + (5 \times 18) = 249.5$

Percentage of water $= \dfrac{90}{249.5} \times 100 = 36.1\%$

ANSWER   The percentage by mass of water of crystallisation is 36.1%.

EXERCISE 2   Problems on Empirical and Molecular Formulae and Percentage Composition

1. Calculate the percentage by mass of the named element in the compound listed:

   a) Na in $Na_2SO_4 \cdot 10H_2O$
   b) Ca in $Ca(CN)_2$
   c) N in $(NH_4)_2CO_3$
   d) U in $UO_2(NO_3)_2 \cdot 6H_2O$
   e) Al in $KAl(SO_4)_2 \cdot 12H_2O$

2. Find the empirical formulae of the compounds formed in the reactions described below:

   a) 10.800 g magnesium form 18.000 g of an oxide
   b) 3.400 g calcium form 9.435 g of a chloride
   c) 3.528 g iron form 10.237 g of a chloride
   d) 2.667 g copper form 4.011 g of a sulphide
   e) 4.662 g lithium form 5.328 g of a hydride.

3. Calculate the empirical formulae of the compounds with the following percentage composition:

   a) 77.7% Fe   22.3% O
   b) 70.0% Fe   30.0% O
   c) 72.4% Fe   27.6% O
   d) 40.2% K   26.9% Cr   32.9% O
   e) 26.6% K   35.4% Cr   38.0% O
   f) 92.3% C   7.6% H
   g) 81.8% C   18.2% H

4. Samples of the following hydrates are weighed, heated to drive off the water of crystallisation, cooled and reweighed. From the results obtained, calculate the values of $a-f$ in the formulae of the hydrates:

   a) 0.869 g of $CuSO_4 \cdot aH_2O$ gave a residue of 0.556 g
   b) 1.173 g of $CoCl_2 \cdot bH_2O$ gave a residue of 0.641 g
   c) 1.886 g of $CaSO_4 \cdot cH_2O$ gave a residue of 1.492 g
   d) 0.904 g of $Pb(C_2H_3O_2)_2 \cdot dH_2O$ gave a residue of 0.774g
   e) 1.144 g of $NiSO_4 \cdot eH_2O$ gave a residue of 0.673 g
   f) 1.175 g of $KAl(SO_4)_2 \cdot fH_2O$ gave a residue of 0.639 g.

5. An organic compound, X, which contains only carbon, hydrogen and oxygen, has a molar mass of about $85\ g\ mol^{-1}$. When 0.43 g of X is burnt in excess oxygen, 1.10 g of carbon dioxide and 0.45 g of water are formed.

   a) What is the empirical formula of X?
   b) What is the molecular formula of X?

6. A liquid, Y, of molar mass $44\,\text{g mol}^{-1}$ contains 54.5% carbon, 36.4% oxygen and 9.1% hydrogen.
   a) Calculate the empirical formula of Y, and
   b) deduce its molecular formula.

7. An organic compound contains 58.8% carbon, 9.8% hydrogen and 31.4% oxygen. The molar mass is $102\,\text{g mol}^{-1}$.
   a) Calculate the empirical formula, and
   b) deduce the molecular formula of the compound.

8. An organic compound has molar mass $150\,\text{g mol}^{-1}$ and contains 72.0% carbon, 6.67% hydrogen and 21.33% oxygen. What is its molecular formula?

CALCULATIONS BASED ON CHEMICAL EQUATIONS

Equations tell us not only what substances react together but also what amounts of substances react together. The equation for the action of heat on sodium hydrogencarbonate

$$2NaHCO_3(s) \longrightarrow Na_2CO_3(s) + CO_2(g) + H_2O(g)$$

tells us that 2 moles of $NaHCO_3$ give 1 mole of Na_2CO_3. Since the molar masses are $NaHCO_3 = 84\,\text{g mol}^{-1}$ and $Na_2CO_3 = 106\,\text{g mol}^{-1}$, it follows that 168 g of $NaHCO_3$ give 106 g of Na_2CO_3.

The amounts of substances undergoing reaction, as given by the balanced chemical equation, are called the *stoichiometric* amounts. *Stoichiometry* is the relationship between the amounts of reactants and products in a chemical reaction. If one reactant is present in excess of the stoichiometric amount required for reaction with another of the reactants, then the excess of one reactant will be left unused at the end of the reaction.

EXAMPLE 1 How many moles of iodine can be obtained from $\frac{1}{6}$ mole of potassium iodate(V)?

METHOD The equation

$$KIO_3(aq) + 5KI(aq) + 6H^+(aq) \longrightarrow 3I_2(aq) + 6K^+(aq) + 3H_2O(l)$$

tells us that 1 mole of KIO_3 gives 3 moles of I_2. Therefore:

ANSWER $\frac{1}{6}$ mole of KIO_3 gives $\frac{1}{6} \times 3$ moles of $I_2 = \frac{1}{2}$ mole of I_2.

EXAMPLE 2   What is the maximum mass of ethyl ethanoate that can be obtained from 0.1 mole of ethanol?

METHOD   Write the equation:

$$C_2H_5OH(l) \ + \ CH_3CO_2H(l) \longrightarrow CH_3CO_2C_2H_5(l) \ + \ H_2O(l)$$

1 mole of C_2H_5OH gives 1 mole $CH_3CO_2C_2H_5$

0.1 mole of C_2H_5OH gives 0.1 mole $CH_3CO_2C_2H_5$

The molar mass of $CH_3CO_2C_2H_5$ is $88 \, g \, mol^{-1}$. Therefore:

ANSWER   0.1 mole of ethanol gives 8.8 g ethyl ethanoate.

EXAMPLE 3   A mixture of 5.00 g of sodium carbonate and sodium hydrogen-carbonate is heated. The loss in mass is 0.31 g. Calculate the percentage by mass of sodium carbonate in the mixture.

METHOD   On heating the mixture, the reaction

$$2NaHCO_3(s) \longrightarrow Na_2CO_3(s) \ + \ CO_2(g) \ + \ H_2O(g)$$

takes place. The loss in mass is due to the decomposition of $NaHCO_3$.

Since 2 mol $NaHCO_3$ form 1 mol CO_2 + 1 mol H_2O

$2 \times 84 \, g \, NaHCO_3$ form $44 \, g \, CO_2$ and $18 \, g \, H_2O$

$168 \, g \, NaHCO_3$ lose $62 \, g$ in mass.

The observed loss in mass of 0.31 g is due to the decomposition of

$$\frac{0.31}{62} \times 168 \, g \, NaHCO_3 \ = \ 0.84 \, g$$

The mixture contains 0.84 g $NaHCO_3$

The difference, $5.00 - 0.84 \ = \ 4.16 \, g \, Na_2CO_3$.

ANSWER   Percentage of $Na_2CO_3 \ = \ \dfrac{4.16}{5.00} \times 100 \ = \ 83.2\%$.

EXAMPLE 4   A mixture of $MgSO_4 \cdot 7H_2O$ and $CuSO_4 \cdot 5H_2O$ is heated at $120\,^{\circ}C$ until a mixture of the anhydrous salts is obtained. If 5.000 g of the mixture give 3.000 g of the anhydrous salts, calculate the percentage by mass of $MgSO_4 \cdot 7H_2O$ in the mixture.

METHOD   Both salts lose water of crystallisation on heating:

$$MgSO_4 \cdot 7H_2O(s) \longrightarrow MgSO_4(s) \ + \ 7H_2O(g)$$

$246 \, g$ of crystals $\longrightarrow$ $120 \, g$ of anhydrous salt $+ 126 \, g$ of steam

$$CuSO_4 \cdot 5H_2O(s) \longrightarrow CuSO_4(s) \ + \ 5H_2O(g)$$

$249.5 \, g$ of crystals $\longrightarrow$ $159.5 \, g$ of anhydrous salt $+ 90.0 \, g$ of steam

Let a g = Mass of $MgSO_4 \cdot 7H_2O$ in the mixture.

Then $(5.000 - a)$ g = Mass of $CuSO_4 \cdot 5H_2O$

a g of $MgSO_4 \cdot 7H_2O$ loses $\dfrac{a}{246} \times 126$ g steam

$(5.000 - a)$ g of $CuSO_4 \cdot 5H_2O$ loses $\dfrac{(5.000 - a)}{249.5} \times 90.0$ g steam

Total loss in mass = 2.000 g

Therefore $\left(\dfrac{a}{246} \times 126 \right) + \dfrac{(5.000 - a) \times 90}{249.5} = 2.000$

Mass of $MgSO_4 \cdot 7H_2O$ = 1.298 g

Mass of $CuSO_4 \cdot 5H_2O$ = 3.702 g

Percentage of $MgSO_4 \cdot 7H_2O$ = $\dfrac{1.298}{5.000} \times 100$ = 25.96%

ANSWER   The mixture contains 26.0% by mass of $MgSO_4 \cdot 7H_2O$.

EXERCISE 3   Problems on Reacting Masses of Solids

1. What mass of glucose must be fermented to give 5.00 kg of ethanol?
$$C_6H_{12}O_6(aq) \longrightarrow 2C_2H_5OH(aq) + 2CO_2(g)$$

2. What mass of sulphuric acid can be obtained from 1000 tonnes of an ore which contains 32.0% of FeS_2?

3. What mass of silver chloride can be precipitated from a solution which contains 1.000×10^{-3} moles of silver ions?

4. The pollutant, sulphur dioxide, can be removed from the air by the reaction
$$2CaCO_3(s) + 2SO_2(g) + O_2(g) \longrightarrow 2CaSO_4(s) + 2CO_2(g)$$
What mass of calcium carbonate is needed to remove 10.0 kg of SO_2?

5. When potassium iodate(V) is allowed to react with acidified potassium iodide, iodine is formed.
$$KIO_3(aq) + KI(aq) + 6H^+(aq) \longrightarrow 3I_2(aq) + 6K^+(aq) + 3H_2O(l)$$
What mass of KIO_3 is required to give 10.00 g of iodine?

6. How many tonnes of iron can be obtained from 10.00 tonnes of
a) Fe_2O_3 and b) Fe_3O_4?

7. What mass of sodium carbonate can be obtained by heating 100 g of sodium hydrogencarbonate?
$$2NaHCO_3(s) \longrightarrow Na_2CO_3(s) + CO_2(g) + H_2O(g)$$

8. What mass of quicklime can be obtained by heating 75.0 g of limestone, which is 86.8% calcium carbonate?

9. What mass of barium sulphate can be precipitated from a solution which contains 4.000 g of barium chloride?

10. Calculate the mass of each of the reactants needed to produce 10.0 g of phosphorus by the reaction

$$2Ca_3(PO_4)_2(s) + 6SiO_2(s) + 5C(s) \longrightarrow P_4(g) + 6CaSiO_3(s) + 5CO_2(g)$$

11. A mixture of calcium and magnesium carbonates weighing 10.0000 g was heated until it reached a constant mass of 5.0960 g. Calculate the percentage composition of the mixture, by mass.

12. A mixture of anhydrous sodium carbonate and sodium hydrogen-carbonate weighing 10.0000 g was heated until it reached a constant mass of 8.7080 g. Calculate the composition of the mixture in grams of each component.

REACTING VOLUMES OF GASES

In stating the volume of a gas, one needs to state the temperature and pressure at which the volume was measured. It is usual to give the volume at 0 °C and 1 atmosphere pressure (273 K and $1.01 \times 10^5 \, N\,m^{-2}$). These conditions are called standard temperature and pressure (s.t.p.). Chapter 5 deals with calculating the volume of a gas at s.t.p. from the volume measured under experimental conditions.

A mole of gas occupies 22.4 dm³ at s.t.p. One can say that *the gas molar volume is 22.4 dm³ at s.t.p.* This makes calculations on reacting volumes of gases very simple. An equation which shows how many moles of different gases react together also shows the ratio of the volumes of the different gases that react together. For example, the equation

$$2NO(g) + O_2(g) \longrightarrow 2NO_2(g)$$

tells us that 2 moles of NO + 1 mole of O_2 form 2 moles of NO_2

∴   44.8 dm³ of NO + 22.4 dm³ of O_2 form 44.8 dm³ of NO_2

In general, 2 volumes of NO + 1 volume of O_2 form 2 volumes of NO_2.

EXAMPLE 1 What is the volume of oxygen needed for the complete combustion of 2 dm³ of propane?

METHOD Write the equation:

$$C_3H_8(g) + 5O_2(g) \longrightarrow 3CO_2(g) + 4H_2O(g)$$

1 mole of C_3H_8 needs 5 moles of O_2

1 volume of C_3H_8 needs 5 volumes of O_2. Therefore:

ANSWER 2 dm³ of propane need 10 dm³ of oxygen.

EXAMPLE 2   What volume of hydrogen is obtained when 3.00 g of zinc react with an excess of dilute sulphuric acid at s.t.p.?

METHOD   Write the equation:

$$Zn(s) \ + \ H_2SO_4(aq) \longrightarrow H_2(g) \ + \ ZnSO_4(aq)$$

1 mole of Zn forms 1 mole of H_2

65 g of Zn form 22.4 dm^3 of H_2 (at s.t.p.)

3.00 g of Zn form $\dfrac{3.00}{65} \times 22.4$ dm^3 $= 1.03$ dm^3 H_2

ANSWER   3.00 g of zinc give 1.03 dm^3 of hydrogen at s.t.p.

EXAMPLE 3   A 250 cm^3 sample of ozonised oxygen is treated with an unsaturated hydrocarbon. A contraction in volume occurs. When a 250 cm^3 sample of ozonised oxygen is heated to 200 °C and then cooled, an expansion occurs, which is half the contraction observed in the previous treatment. Assuming the formula of oxygen is O_2, what is the formula for ozone?

METHOD   Let the formula of ozone be O_x.

Let the contraction on treatment with alkene be a cm^3

Then the volume of ozone in the mixture is a cm^3

When ozone is converted to oxygen an expansion of $a/2$ cm^3 occurs

Thus, a cm^3 of ozone must form $1.5a$ cm^3 of oxygen

and 2 volumes of ozone form 3 volumes of oxygen

and 2 molecules of ozone form 3 molecules of oxygen.

Thus, $2O_x \longrightarrow 3O_2$.

Therefore $x = 3$.

ANSWER   Ozone has the formula O_3.

EXAMPLE 4   10 cm^3 of a hydrocarbon, C_4H_x were allowed to react with an excess of oxygen at 150 °C and 1 atmosphere. There was an expansion of 10 cm^3. Deduce the value of x.

METHOD   Write the equation:

$$C_4H_x(g) \ + \ (4 + x/4)O_2(g) \longrightarrow 4CO_2(g) \ + \ x/2 \ H_2O(g)$$

Volume of hydrocarbon $= 10$ cm^3

Volume of oxygen $= 10(4 + x/4) = (40 + 5x/2)$ cm^3

Volume of carbon dioxide $= 40$ cm^3

Volume of steam $= x/2 \times 10$ cm^3 $= 5x$ cm^3

Let a cm^3 $=$ Volume of unused oxygen.

Since there is an expansion of 10 cm^3,

$$\text{Final volume} = 10 + \text{Initial volume}$$

$$40 + 5x + a = 10 + 10 + 40 + 5x/2 + a$$

and
$$5x/2 = 20$$

$$x = 8$$

ANSWER The formula is C_4H_8.

EXAMPLE 5 10 cm^3 of a hydrocarbon, C_aH_b, are exploded with an excess of oxygen. A contraction of 35 cm^3 occurs, all volumes being measured at room temperature and pressure. On treatment of the products with sodium hydroxide solution, a contraction of 40 cm^3 occurs. Deduce the formula of the hydrocarbon.

METHOD Write the equation:

$$C_aH_b(g) \ + \ (a + b/4)O_2(g) \longrightarrow aCO_2(g) \ + \ b/2 \ H_2O(l)$$

Volume of hydrocarbon $= 10 \text{ cm}^3$

Volume of CO_2 $= a \times$ Volume of C_aH_b

From reaction with NaOH, volume of CO_2 $= 40 \text{ cm}^3$

Therefore $a = 4$

Let the volume of unused oxygen be $c \text{ cm}^3$.

Final volume $=$ Initial volume $- 35 \text{ cm}^3$

Note that $H_2O(l)$ is a liquid at room temperature and pressure, and does not contribute to the final volume of gas.

$$40 + c = 10 + 40 + 5b/2 + c - 35$$

$$25 = 5b/2$$

$$b = 10$$

ANSWER The formula is C_4H_{10}.

EXAMPLE 6 20 cm^3 of ammonia are burned in an excess of oxygen at $110\,^\circ C$. 10 cm^3 of nitrogen and 30 cm^3 of steam are formed. Deduce the formula for ammonia, given that the formula of nitrogen is N_2, and the formula of steam is H_2O.

METHOD Let the formula of ammonia be N_aH_b.

The equation for combustion is

$$N_aH_b \ + \ b/4 \ O_2 \longrightarrow a/2 \ N_2 \ + \ b/2 \ H_2O$$

Volume of N_aH_b $= 20 \text{ cm}^3$

Volume of N_2 $= a/2 \times 20 = 10a \text{ cm}^3 = 10 \text{ cm}^3$ $\therefore a = 1$

Volume of $H_2O(g)$ $= b/2 \times 20 = 10b \text{ cm}^3 = 30 \text{ cm}^3$ $\therefore b = 3$

ANSWER The formula is NH_3.

EXAMPLE 7 25 cm³ of a mixture of methane and ethane were completely oxidised by 72.5 cm³ of oxygen, measured at the same temperature and pressure. What was the composition of the mixture?

METHOD   Let a cm³ = Volume of methane

Then $(25 - a)$ cm³ = Volume of ethane

The equation

$$CH_4(g) \ + \ 2O_2(g) \ \longrightarrow \ CO_2(g) \ + \ 2H_2O(g)$$

tells us that a cm³ of methane need $2a$ cm³ of oxygen.

The equation

$$2C_2H_6(g) \ + \ 7O_2(g) \ \longrightarrow \ 4CO_2(g) \ + \ 6H_2O(g)$$

tells us that $(25 - a)$ cm³ of ethane need $\frac{7}{2}(25 - a)$ cm³ of oxygen.

Thus   Volume of oxygen $= 2a + \frac{7}{2}(25 - a) = 72.5$

$$4a + 175 - 7a = 145$$

$$3a = 30$$

$$a = 10$$

ANSWER   The mixture consists of 10 cm³ methane and 15 cm³ of ethane.

EXERCISE 4   Problems on Reacting Volumes of Gases

1. What volume of oxygen (at s.t.p.) is required to burn exactly:
   a) 1 dm³ of methane, according to the reaction
   $$CH_4(g) \ + \ 2O_2(g) \ \longrightarrow \ CO_2(g) \ + \ 2H_2O(g)$$
   b) 500 cm³ of hydrogen sulphide, according to the reaction
   $$2H_2S(g) \ + \ 3O_2(g) \ \longrightarrow \ 2SO_2(g) \ + \ 2H_2O(g)$$
   c) 250 cm³ of ethyne, according to the equation
   $$2C_2H_2(g) \ + \ 5O_2(g) \ \longrightarrow \ 4CO_2(g) \ + \ 2H_2O(g)$$
   d) 750 cm³ of ammonia, according to the reaction
   $$4NH_3(g) \ + \ 5O_2(g) \ \longrightarrow \ 4NO(g) \ + \ 6H_2O(g)$$
   e) 1 dm³ of phosphine, according to the reaction
   $$PH_3(g) \ + \ 2O_2(g) \ \longrightarrow \ H_3PO_4(s)?$$

2. 1 dm³ of H_2S and 1 dm³ of SO_2 were allowed to react, according to the equation
   $$2H_2S(g) \ + \ SO_2(g) \ \longrightarrow \ 2H_2O(l) \ + \ 3S(s)$$
   What volume of gas will remain after the reaction?

3. 100 cm³ of a mixture of ethane and ethene at s.t.p. were treated with bromine. 0.357 g of bromine was used up. Calculate the percentage by volume of ethene in the mixture.

4. $10\,cm^3$ of a hydrocarbon C_xH_y were exploded with an excess of oxygen. There was a contraction of $30\,cm^3$. When the product was treated with a solution of sodium hydroxide, there was a further contraction of $30\,cm^3$. Deduce the formula of the hydrocarbon. All gas volumes are at s.t.p.

5. $10\,cm^3$ of a hydrocarbon C_aH_b were exploded with excess oxygen. A contraction of $25\,cm^3$ occurred. On treating the product with sodium hydroxide, a further contraction of $40\,cm^3$ occurred. Deduce the values of a and b in the formula of the hydrocarbon. All measurements of gas volumes are at s.t.p.

6. $10\,cm^3$ of a hydrocarbon C_4H_8 were exploded with an excess of oxygen. A contraction of $a\,cm^3$ occurred. On adding sodium hydroxide solution, a further contraction of $b\,cm^3$ occurred. What are the volumes, a and b? All gas volumes are at s.t.p.

7. Hydrogen sulphide burns in oxygen in accordance with the following equation:

$$2H_2S(g) \;+\; 3O_2(g) \longrightarrow 2H_2O(g) \;+\; 2SO_2(g)$$

If $4\,dm^3$ of H_2S are burned in $10\,dm^3$ of oxygen at 1 atmosphere pressure and $120\,°C$, what is the final volume of the mixture?

a $6\,dm^3$ b $8\,dm^3$ c $10\,dm^3$ d $12\,dm^3$ e $14\,dm^3$

EXERCISE 5   Problems on Reactions Involving Solids and Gases

1. In the Solvay process

$$NaCl(aq) + NH_3(g) + H_2O(l) + CO_2(g) \longrightarrow NaHCO_3(s) + NH_4Cl(aq)$$

what volume of carbon dioxide (at s.t.p.) is required to produce $1.00\,kg$ of sodium hydrogencarbonate?

2. What volume of ethyne (at s.t.p.) can be prepared from $10.0\,g$ of calcium carbide by the reaction

$$CaC_2(s) \;+\; 2H_2O(l) \longrightarrow Ca(OH)_2(aq) \;+\; C_2H_2(g)?$$

3. What mass of phosphorus is required for the preparation of $200\,cm^3$ of phosphine (at s.t.p.) by the reaction

$$P_4(s) + 3NaOH(aq) + 3H_2O(l) \longrightarrow 3NaH_2PO_4(aq) + PH_3(g)?$$

4. Calculate the mass of ammonium chloride required to produce $1.00\,dm^3$ of ammonia (at s.t.p.) in the reaction

$$2NH_4Cl(s) \;+\; Ca(OH)_2(s) \longrightarrow 2NH_3(g) \;+\; CaCl_2(s) \;+\; 2H_2O(g)$$

5. What mass of potassium chlorate(V) must be heated to give $1.00\,dm^3$ of oxygen at s.t.p.? The reaction is

$$2KClO_3(s) \longrightarrow 2KCl(s) \;+\; 3O_2(g)$$

6. What volume of chlorine (at s.t.p.) can be obtained from the electrolysis of a solution containing 60.0 g of sodium chloride?

7. What volume of oxygen (at s.t.p.) is needed for the complete combustion of 1.00 kg of octane? The reaction is

$$2C_8H_{18}(l) \ + \ 25O_2(g) \longrightarrow 16CO_2(g) \ + \ 18H_2O(g)$$

PERCENTAGE YIELD

There are many reactions which do not go to completion. Reactions between organic compounds do not often give a 100% yield of product. The actual yield is compared with the yield calculated from the molar masses of the reactants. The equation

$$\text{Percentage yield} \ = \ \frac{\text{Actual mass of product}}{\text{Calculated mass of product}} \times 100$$

is used to give the percentage yield.

EXAMPLE   From 23 g of ethanol are obtained 44 g of ethyl ethanoate by esterification with ethanoic acid in the presence of concentrated sulphuric acid. What is the percentage yield of the reaction?

METHOD   Write the equation:

$$CH_3CO_2H(l) \ + \ C_2H_5OH(l) \longrightarrow CH_3CO_2C_2H_5(l) \ + \ H_2O(l)$$

46 g of C_2H_5OH forms 108 g of $CH_3CO_2C_2H_5$

23 g of C_2H_5OH should give $\dfrac{23}{46} \times 108\,g \ = \ 54\,g$ of $CH_3CO_2C_2H_5$

Actual mass obtained $= \ 44\,g$

$$\text{Percentage yield} \ = \ \frac{\text{Actual mass of product}}{\text{Calculated mass of product}} \times 100$$

ANSWER $\text{Percentage yield} \ = \ \dfrac{44}{54} \times 100 \ = \ 81.5\%.$

EXERCISE 6   Problems on Percentage Yield

1. Phenol, C_6H_5OH, is converted to trichlorophenol, $C_6H_2Cl_3OH$. If 488 g of product are obtained from 250 g of phenol, calculate the percentage yield.

2. 29.5 g of ethanoic acid, CH_3CO_2H, are obtained from the oxidation of 25.0 g of ethanol, C_2H_5OH. What percentage yield does this represent?

3. 0.8500 g of hexanone, $C_6H_{12}O$, is converted to its 2,4-dinitrophenylhydrazone. After isolation and purification, 2.1180 g of product, $C_{12}H_{18}N_4O_4$, are obtained. What percentage yield does this represent?

4. Benzaldehyde, C_7H_6O, forms a hydrogensulphite compound of formula $C_7H_7SO_4Na$. From 1.210 g of benzaldehyde, a yield of 2.181 g of the product was obtained. Calculate the percentage yield.

5. 100 cm^3 of barium chloride solution of concentration 0.0500 mol dm^{-3} were treated with an excess of sulphate ions in solution. The precipitate of barium sulphate formed was dried and weighed. A mass of 1.1558 g was recorded. What percentage yield does this represent?

LIMITING REACTANT

In a chemical reaction, the reactants are often added in amounts which are not stoichiometric. One or more of the reactants is in excess and is not completely used up in the reaction. The amount of product is determined by the amount of the reactant that is not in excess and is used up completely in the reaction. This is called the *limiting reactant*. You first have to decide which is the limiting reactant before you can calculate the amount of product formed.

EXAMPLE   5.00 g of iron and 5.00 g of sulphur are heated together to form iron(II) sulphide. Which reactant is present in excess? What mass of product is formed?

METHOD   Write the equation:

$$Fe(s) \ + \ S(s) \longrightarrow FeS(s)$$

1 mole of Fe + 1 mole of S form 1 mole of FeS

∴   56 g Fe and 32 g S form 88 g FeS

5.00 g Fe is 5/56 mol $=$ 0.0893 mol Fe

5.00 g S is 5/32 mol $=$ 0.156 mol S

There is insufficient Fe to react with 0.156 mol S; iron is the limiting reactant.

0.0893 mol Fe forms 0.0893 mol FeS $=$ 0.0893 $\times$ 88 g $=$ 7.86 g

ANSWER   Mass formed $=$ 7.86 g.

EXERCISE 7   Problems on Limiting Reactant

1. In the blast furnace, the overall reaction is

$$2Fe_2O_3(s) \ + \ 3C(s) \longrightarrow 3CO_2(g) \ + \ 4Fe(s)$$

What is the maximum mass of iron that can be obtained from 700 tonnes of iron(III) oxide and 70 tonnes of coke? (1 tonne $=$ 1 000 kg.)

2. In the manufacture of calcium carbide

$$CaO(s) \ + \ 3C(s) \longrightarrow CaC_2(s) \ + \ CO(g)$$

What is the maximum mass of calcium carbide that can be obtained from 40 kg of quicklime and 40 kg of coke?

3. In the manufacture of the fertiliser ammonium sulphate
$$H_2SO_4(aq) + 2NH_3(g) \longrightarrow (NH_4)_2SO_4(aq)$$
What is the maximum mass of ammonium sulphate that can be obtained from 2.0 kg of sulphuric acid and 1.0 kg of ammonia?

4. In the Solvay process, ammonia is recovered by the reaction
$$2NH_4Cl(s) + CaO(s) \longrightarrow CaCl_2(s) + H_2O(g) + 2NH_3(g)$$
What is the maximum mass of ammonia that can be recovered from 2.00×10^3 kg of ammonium chloride and 500 kg of quicklime?

5. In the Thermit reaction
$$2Al(s) + Cr_2O_3(s) \longrightarrow 2Cr(s) + Al_2O_3(s)$$
Calculate the percentage yield when 180 g of chromium are obtained from a reaction between 100 g of aluminium and 400 g of chromium(III) oxide.

DERIVING THE EQUATION FOR A REACTION

If you know the mass of each solid or the volume of each gas taking part in a reaction, you can calculate the number of moles of each substance taking part, and this will tell you the equation for the reaction.

EXAMPLE   When 100 cm^3 of a hydrocarbon X burn in 500 cm^3 of oxygen, 50 cm^3 of oxygen are unused, 300 cm^3 of carbon dioxide are formed, and 300 cm^3 of steam are formed. Deduce the equation for the reaction and the formula of the hydrocarbon.

METHOD
$$X + O_2(g) \longrightarrow CO_2(g) + H_2O(g)$$

100 cm^3 450 cm^3 300 cm^3 300 cm^3

The volumes of gases reacting tell us that
$$X + 4\tfrac{1}{2}O_2(g) \longrightarrow 3CO_2(g) + 3H_2O(g)$$
To balance the equation, X must be C_3H_6. Then,
$$C_3H_6(g) + 4\tfrac{1}{2}O_2(g) \longrightarrow 3CO_2(g) + 3H_2O(g)$$

ANSWER
$$2C_3H_6(g) + 9O_2(g) \longrightarrow 6CO_2(g) + 6H_2O(g)$$

EXERCISE 8   Problems on Deriving Equations

1. To a solution containing 2.975 g of sodium persulphate, $Na_2S_2O_8$, is added an excess of potassium iodide solution. A reaction occurs, in which sulphate ions are formed and 3.175 g of iodine are formed. Deduce the equation for the reaction.

2. Amidosulphuric acid, H_2NSO_3H, reacts with warm sodium hydroxide solution to give ammonia and a solution which contains sulphate ions. 0.540 g of the acid, when treated with an excess of alkali, gave 153 cm³ of ammonia at 60 °C and 1 atm. Deduce the equation for the reaction.

3. A solution containing 5.00×10^{-3} mol of sodium thiosulphate was shaken with 1 g of silver chloride. 0.717 5 g of silver chloride dissolved, and analysis showed that 5.00×10^{-3} mol of chloride ions were present in the resulting solution. Derive an equation for the reaction.

4. An unsaturated hydrocarbon of molar mass 80 g mol⁻¹ reacts with bromine. If 0.250 g of hydrocarbon reacts with 1.00 g of bromine, what is the equation for the reaction?

5. Given that 1.00 g of phenylamine, $C_6H_5NH_2$, reacts with 5.16 g of bromine, derive an equation for the reaction.

EXERCISE 9   Questions from A-level Papers

1. a) When 0.203 g of hydrated magnesium chloride, $MgCl_m \cdot nH_2O$, was dissolved in water and titrated with 0.1 M silver nitrate ($AgNO_3$) solution, 20.0 cm³ of the latter were required. A sample of the hydrated chloride lost 53.2% of its mass when heated in a stream of hydrogen chloride, leaving a residue of anhydrous magnesium chloride. From these figures calculate the values of m and n.

   b) When the hydrated chloride was heated in air instead of hydrogen chloride, the loss in mass was greater than 53.2% and both HCl and H_2O were evolved. What reaction do you think might have occurred and what was the solid product?   (L80)

2. A hydrated aluminium sulphate, $Al_2(SO_4)_3 \cdot xH_2O$, contains 8.1% of aluminium by mass. Find the value of x.   (L78, p)

3. A hydrated iron sulphate, $Fe_2(SO_4)_3 \cdot xH_2O$, contains 19.9% of iron by mass. Find the value of x.   (L78, p)

4. A study of germanium chemistry is reported below. Use the data to calculate the formulae of the products A and B. Also draw diagrams of possible structural formulae for A and B.

   a) 13 g of germanium(IV) bromide were dissolved in dry benzene and treated with 5 g of magnesium and 22 g of bromoethane, previously reacted together in dry ethoxyethane (ether). After heating for eight hours, the ethoxyethane and benzene were removed by distillation and the product A collected as an oily liquid, b.p. 164 °C. The germanium content was determined by nitric acid oxidation to germanium(IV) oxide, 0.155 g of A gave GeO_2, 0.086 g; the carbon and hydrogen content was determined by combustion, 0.174 g of A gave CO_2, 0.325 g and H_2O, 0.166 g.

b) 1.9 g of the product A was treated with bromine in an inert solvent for six days at 40 °C. An oily product B was obtained in almost 100% yield, b.p. 190 °C, yield 2.4 g. On analysis by nitric acid oxidation 0.540 g of B gave GeO_2, 0.237 g. The bromine content of B was determined as silver bromide, 0.384 g of B gave AgBr, 0.301 g.   (L78, N)

5. If $\frac{1}{10}$ mol of ethanol, $\frac{1}{8}$ mol of potassium bromide and $\frac{1}{6}$ mol of sulphuric acid were used to prepare bromoethane, the maximum mass of bromoethane (relative molecular mass 109), which could be formed is:

a $\frac{1}{10} \times 109$ g   b $\frac{1}{6} \times 109$ g   c $\frac{1}{8} \times 109$ g

d $\frac{1}{10} \times \frac{109}{46}$ g   e $\frac{1}{8} \times \frac{109}{46}$ g   (C79)

6. Tin(IV) iodide (stannic iodide) is prepared by direct combination of the elements. Add 2 g of granulated tin to a solution of 6.35 g of iodine in 25 cm^3 of tetrachloromethane (carbon tetrachloride) in a 100 cm^3 flask fitted with a reflux condenser. Reflux gently until reaction is complete. Filter through pre-heated funnel, wash residue with 10 cm^3 of hot tetrachloromethane and add washings to the filtrate. Cool in ice until orange crystals of tin(IV) iodide separate. Filter off the crystals and dry. The melting-point of tin(IV) iodide is 144 °C. (Sn = 119, I = 127.)

a) Write an equation for the reaction.
b) Suggest reasons why tetrachloromethane is used.
c) Which reactant is in excess?
d) Calculate the maximum theoretical yield of product.   (SUJB79, p)

7. When 10 cm^3 of the hydrocarbon but-1-ene ($CH_3CH_2CH{=}CH_2$) were exploded with an excess of oxygen, there was a contraction of x cm^3. A further contraction of y cm^3 took place on addition of aqueous sodium hydroxide solution (all measurements being made at 0 °C and at atmospheric pressure). Deduce the values of x and y and explain your reasoning fully.   (O79)

8. a) Give the equations for the combustion of: i) ethane, and ii) propane in an excess of oxygen.

b) What change in volume (+ for expansion and − for contraction) takes place when one volume of each of these two hydrocarbons is burnt in an excess of oxygen, the reacting gases and products being measured at the same room temperature and pressure: i) ethane, and ii) propane?

c) Use the above information to calculate the volume of ethane in 100 cm^3 of a mixture of this gas with propane, given that when this volume was combusted under the conditions specified in b) above, there was a change in volume of 265 cm^3. You must decide for yourself whether this is an expansion or contraction.   (O78)

9. $60 \, cm^3$ of oxygen was added to $10 \, cm^3$ of a gaseous unsaturated hydrocarbon. After explosion and cooling to room temperature the gases occupied $45 \, cm^3$ and after absorption by aqueous potassium hydroxide solution $15 \, cm^3$ of oxygen remained. Calculate the molecular formula of the hydrocarbon (all measurements were taken under the same conditions of temperature and pressure). (AEB79, p)

10. When $40 \, cm^3$ of a gaseous hydrocarbon were exploded with $400 \, cm^3$ of oxygen, the volume of gas remaining, after cooling to room temperature was $320 \, cm^3$. This was reduced to $40 \, cm^3$ by the addition of aqueous potassium hydroxide. Assume that all volumes are recorded at the same temperature and pressure.
    a) What is the volume of carbon dioxide produced in the explosion?
    b) What is the volume of oxygen used in the explosion?
    c) What is the molecular formula of the hydrocarbon?
    d) State one chemical principle which has been used in c) in deriving the molecular formula of the hydrocarbon from the experimental data. (JMB79, p)

11. $50.0 \, cm^3$ of a mixture of carbon monoxide, carbon dioxide and hydrogen were exploded with $25.0 \, cm^3$ of oxygen. After explosion, the volume measured at the original room temperature and pressure was $37.0 \, cm^3$. After treatment with potassium hydroxide solution the volume was reduced to $5.0 \, cm^3$. Calculate the percentage composition by volume of the original mixture. (AEB79)

12. Under like conditions, carbon monoxide competes 200 times more effectively than oxygen for haemoglobin available in blood. The effects in the body of carboxy-haemoglobin are detectable when it reaches 5% of the concentration of oxy-haemoglobin in the blood. Air contains 20% of oxygen by volume. By using these data, it can be deduced that the minimum concentration (parts per million, p.p.m.) of carbon monoxide in the atmosphere that will result in detectable effects in the body is:

    a 10 p.p.m.   b 50 p.p.m.   c 250 p.p.m.
    d 500 p.p.m.   e 1000 p.p.m.   (C78)

13. On repeated sparking $10 \, cm^3$ of a mixture of carbon monoxide and nitrogen required $3 \, cm^3$ oxygen for combustion. What was the volume of nitrogen in the mixture? (All volumes were measured at room temperature and pressure.)

    a $3\frac{1}{3} \, cm^3$ b $4 \, cm^3$ c $5 \, cm^3$
    d $7 \, cm^3$ e $8\frac{1}{2} \, cm^3$ (C80)

3 Volumetric Analysis

CONCENTRATION

Chemical reactions are often carried out between substances in solution. The concentration of a solution is measured in terms of the number of moles of solute contained in a cubic decimetre of solution.

$$\text{Concentration (mol dm}^{-3}) = \frac{\text{Number of moles of solute}}{\text{Volume (dm}^3) \text{ of solution}}$$

Another way of writing this is:

$$\left(\begin{array}{l}\text{Number of moles} \\ \text{of solute}\end{array}\right) = \text{Concentration (mol dm}^{-3}) \times \text{Volume (dm}^3)$$

The concentration in mol dm^{-3} used to be referred to as the *molarity* of a solution. (In strict SI units, concentration is expressed in mol m^{-3}.)

A solution of known concentration is called a *standard solution*. Such a solution can be used to find the concentrations of solutions of other reagents.

In *volumetric analysis*, the concentration of a solution is found by measuring the volume of solution that will react with a known volume of a standard solution. The procedure of adding one solution to another in a measured way until the reaction is complete is called *titration*. Volumetric analysis is often referred to as *titrimetric analysis* or *titrimetry*.

ACID-BASE TITRATIONS

A standard solution of acid can be used to find the concentration of a solution of alkali. A known volume of alkali is taken by pipette, a suitable indicator is added, and the alkali is titrated against the standard acid until the neutral point is reached. The number of moles of acid used can be calculated and the equation used to give the number of moles of alkali neutralised.

EXAMPLE 1 *Standardising sodium hydroxide solution*
What is the concentration of a solution of sodium hydroxide, 25.0 cm^3 of which requires 20.0 cm^3 of hydrochloric acid of concentration $0.100 \text{ mol dm}^{-3}$ for neutralisation?

METHOD   a) Write the equation:

$$NaOH(aq) + HCl(aq) \longrightarrow NaCl(aq) + H_2O(l)$$

1 mole of NaOH needs 1 mole of HCl for neutralisation.

b) Find the number of moles of the reagent of known concentration, in this case HCl.

$$\text{No. of moles of HCl} = \text{Volume (dm}^3) \times \text{Concn (mol dm}^{-3})$$

$$= 20.0 \times 10^{-3} \times 0.100 = 2.00 \times 10^{-3}\, mol$$

From equation: No. of moles of NaOH = No. of moles of HCl

$$= 2.00 \times 10^{-3}\, mol$$

But: No. of moles of NaOH = Volume (dm^3) $\times$ Concn (mol dm^{-3})

$$= 25.0 \times 10^{-3} \times c$$

(where c = concn)

Equate these two values: $2.00 \times 10^{-3} = 25.0 \times 10^{-3} \times c$

$$c = (2.00 \times 10^{-3})/(25.0 \times 10^{-3})$$

$$= 0.080\, mol\, dm^{-3}$$

ANSWER   The concentration of sodium hydroxide is 0.080 mol dm^{-3}.

EXAMPLE 2   *Standardising hydrochloric acid*
Sodium carbonate (anhydrous) is used as a primary standard in volumetric analysis. A solution of sodium carbonate of concentration 0.100 mol dm^{-3} is used to standardise a solution of hydrochloric acid. 25.0 cm^3 of the standard solution of sodium carbonate require 35.0 cm^3 of the acid for neutralisation. Calculate the concentration of the acid.

METHOD   a) Write the equation:

$$Na_2CO_3(aq) + 2HCl(aq) \longrightarrow 2NaCl(aq) + CO_2(g) + H_2O(l)$$

1 mole of Na$_2$CO$_3$ neutralises 2 moles of HCl.

b) Find the number of moles of the standard reagent used.

$$\text{No. of moles of Na}_2CO_3(aq) = \text{Volume (dm}^3) \times \text{Concn (mol dm}^{-3})$$

$$= 25.0 \times 10^{-3} \times 0.100$$

$$= 2.50 \times 10^{-3}\, mol$$

From equation: No. of moles of HCl = 2 $\times$ No. of moles of Na$_2$CO$_3$

$$= 5.00 \times 10^{-3}\, mol$$

But: No. of moles of HCl(aq) $=$ Volume $(dm^3) \times$ Concn $(mol\,dm^{-3})$

$$= 35.0 \times 10^{-3} \times c$$

(where $c =$ concn)

Equate these two values: $5.00 \times 10^{-3} = 35.0 \times 10^{-3} \times c$

$$c = (5.00 \times 10^{-3})/(35.0 \times 10^{-3})$$

$$= 0.143\,mol\,dm^{-3}$$

ANSWER   The concentration of hydrochloric acid is $0.143\,mol\,dm^{-3}$.

EXAMPLE 3   *Calculating the percentage of sodium carbonate in washing soda crystals*
5.125 g of washing soda crystals are dissolved and made up to $250\,cm^3$ of solution. A $25.0\,cm^3$ aliquot of the solution requires $35.8\,cm^3$ of $0.0500\,mol\,dm^{-3}$ sulphuric acid for neutralisation. Calculate the percentage of sodium carbonate in the crystals.

METHOD   a) Write the equation:

$$Na_2CO_3(aq) + H_2SO_4(aq) \longrightarrow Na_2SO_4(aq) + CO_2(g) + H_2O(l)$$

1 mole of Na_2CO_3 neutralises 1 mole of H_2SO_4.

b) Calculate the number of moles of the standard reagent.

No. of moles of $H_2SO_4 = 35.8 \times 10^{-3} \times 0.0500 = 1.79 \times 10^{-3}\,mol$

No. of moles of $Na_2CO_3 = 1.79 \times 10^{-3}\,mol$

But: No. of moles of $Na_2CO_3 = 25.0 \times 10^{-3} \times c\,mol$

(where $c =$ concn)

Equate these two values: $1.79 \times 10^{-3} = 25.0 \times 10^{-3} \times c$

$$c = (1.79 \times 10^{-3})/(25.0 \times 10^{-3})$$

$$= 0.0716\,mol\,dm^{-3}$$

No. of moles of Na_2CO_3 in whole solution $=$ Volume $\times$ Concn

$$= 250 \times 10^{-3} \times 0.0716$$

$$= 0.0179\,mol$$

Mass of $Na_2CO_3 =$ No. of moles $\times$ Molar mass $= 0.0179 \times 106\,g$

$$= 1.90\,g$$

% of $Na_2CO_3 = \dfrac{\text{Mass of sodium carbonate}}{\text{Mass of crystals}} \times 100$

$$= \frac{1.90}{5.125} \times 100 = 37.1\%$$

ANSWER   Washing soda crystals are 37.1% sodium carbonate.

EXAMPLE 4 *Estimating ammonium salts*

A sample containing ammonium sulphate was warmed with $250 \, cm^3$ of $0.800 \, mol \, dm^{-3}$ sodium hydroxide solution. After the evolution of ammonia had ceased, the excess of sodium hydroxide solution was neutralised by $85.0 \, cm^3$ of hydrochloric acid of concentration $0.500 \, mol \, dm^{-3}$. What mass of ammonium sulphate did the sample contain?

METHOD

a) There are two reactions taking place:

i) the reaction between the ammonium salt and the alkali:

$$(NH_4)_2SO_4(s) + 2NaOH(aq) \longrightarrow 2NH_3(g) + Na_2SO_4(aq) + 2H_2O(l)$$

ii) the reaction between the excess alkali and the hydrochloric acid:

$$NaOH(aq) \ + \ HCl(aq) \longrightarrow NaCl(aq) \ + \ H_2O(l)$$

b) Pick out the substance for which you have the information you need to calculate the number of moles. As you know its volume and concentration, you can calculate the number of moles of HCl. This will tell you the number of moles of NaOH left over after reaction i). Subtract this from the number of moles of NaOH added to the ammonium salt to obtain the number of moles of NaOH used in reaction i). This will give you the number of moles of $(NH_4)_2SO_4$ with which it reacted.

No. of moles of HCl $= 85.0 \times 10^{-3} \times 0.500 = 0.0425 \, mol$

No. of moles of NaOH left over from reaction i) $= 0.0425 \, mol$

No. of moles of NaOH added $= 250 \times 10^{-3} \times 0.800 = 0.200 \, mol$

No. of moles of NaOH used in reaction i) $= 0.200 - 0.0425$
$$= 0.1575 \, mol$$

No. of moles of $(NH_4)_2SO_4 = 0.5 \times$ No. of moles of NaOH
$$= 0.0788 \, mol$$

Molar mass of $(NH_4)_2SO_4 = 132 \, g \, mol^{-1}$

Mass of ammonium sulphate $= 0.0788 \times 132 = 10.4 \, g$

ANSWER

The sample contained $10.4 \, g$ of ammonium sulphate.

EXAMPLE 5 *Determination of the amounts of sodium hydroxide and sodium carbonate in a mixture*

Different indicators change colour at different values of pH. When a solution of sodium carbonate is titrated against hydrochloric acid, two reactions occur:

$$Na_2CO_3(aq) \ + \ HCl(aq) \longrightarrow NaHCO_3 aq) \ + \ NaCl(aq)$$
$$NaHCO_3(aq) \ + \ HCl(aq) \longrightarrow NaCl(aq) \ + \ CO_2(g) \ + \ H_2O(l)$$

The indicator phenolphthalein changes from pink to colourless at the end of the first reaction, when the pH of the solution is 9. The indicator methyl orange changes from yellow to orange at the end of the second reaction when the pH is 7. This two-stage titration can be used to estimate sodium hydroxide and sodium carbonate in a mixture of both.

$25.0 \, cm^3$ of a solution containing sodium hydroxide and sodium carbonate were titrated against $0.100 \, mol \, dm^{-3}$ hydrochloric acid, using phenolphthalein as indicator. After $18.50 \, cm^3$ of acid had been added, the indicator was decolourised. Methyl orange was added, and a further $7.50 \, cm^3$ of hydrochloric acid were needed to turn the indicator orange. Calculate the concentrations of sodium hydroxide and sodium carbonate in the solution.

METHOD   The second stage, using methyl orange as indicator, gives

$$HCO_3^-(aq) + H^+(aq) \longrightarrow CO_2(g) + H_2O(l)$$

No. of moles of HCl(aq) used = $7.50 \times 10^{-3} \times 0.100$

$= 0.750 \times 10^{-3} \, mol$

$=$ No. of moles of $HCO_3^-(aq)$ formed in first stage,

$=$ No. of moles of $CO_3^{2-}(aq)$ in the $25.0 \, cm^3$ of solution.

$\therefore$ $25.0 \, cm^3$ solution contains 0.750×10^{-3} moles of $CO_3^{2-}(aq)$.

$\therefore$ Concentration of $CO_3^{2-}(aq) = \dfrac{0.750 \times 10^{-3}}{25.0 \times 10^{-3}} = 0.0300 \, mol \, dm^{-3}$

Volume of acid needed by NaOH(aq) = $11.00 \times 10^{-3} \times 0.100$

$= 1.10 \times 10^{-3} \, mol$

No. of moles of NaOH(aq) in $25 \, cm^3$ of solution = $1.10 \times 10^{-3} \, mol$

Concn of NaOH(aq) = $\dfrac{1.10 \times 10^{-3}}{25.0 \times 10^{-3}} = 0.0440 \, mol \, dm^{-3}$

ANSWER   The concentrations are $0.0300 \, mol \, dm^{-3}$ sodium carbonate and $0.0440 \, mol \, dm^{-3}$ sodium hydroxide.

EXAMPLE 6   *Estimation of sodium carbonate and sodium hydrogencarbonate in a solution*
The same technique, employing two indicators, can be used to estimate a mixture of sodium carbonate and sodium hydrogencarbonate. Using phenolphthalein as indicator, the reaction measured is

$$CO_3^{2-}(aq) + H^+(aq) \longrightarrow HCO_3^-(aq)$$

When methyl orange is added, and titration is continued, the reaction measured is

$$HCO_3^-(aq) + H^+(aq) \longrightarrow CO_2(g) + H_2O(l)$$

A 25 cm^3 aliquot of a solution containing sodium carbonate and sodium hydrogencarbonate needed 27.5 cm^3 of a solution of hydrochloric acid of concentration 0.0800 mol dm^{-3} to decolourise phenolphthalein, which had been added as indicator. On addition of methyl orange, a further 32.5 cm^3 of the acid were needed to turn this indicator to its neutral colour.

METHOD   In the first stage, the CO_3^{2-}(aq) in the mixture is converted to HCO_3^-(aq).

No. of moles of HCl(aq) needed to convert CO_3^{2-}(aq) to HCO_3^-(aq)
$$= 27.5 \times 10^{-3} \times 0.0800 = 2.20 \times 10^{-3} \, mol$$
No. of moles of CO_3^{2-}(aq) $= 2.20 \times 10^{-3} \, mol$
Concentration of CO_3^{2-}(aq) $= (2.20 \times 10^{-3})/(25.0 \times 10^{-3})$
$$= 8.80 \times 10^{-2} \, mol \, dm^{-3}$$
The volume of acid needed to neutralise the total HCO_3^-(aq) is 32.5 cm^3.

Of this, 27.5 cm^3 are needed to neutralise the HCO_3^-(aq) formed from CO_3^{2-}(aq) in the first stage. The remaining 5.00 cm^3 neutralise the HCO_3^-(aq) present in the original solution.

No. of moles of HCl(aq) needed for HCO_3^-(aq)
$$= 5.00 \times 10^{-3} \times 0.0800$$
$$= 0.0400 \times 10^{-3} \, mol$$
No. of moles of HCO_3^-(aq) $= 0.0400 \times 10^{-3} \, mol$
Concentration of HCO_3^-(aq) $= (0.0400 \times 10^{-3})/(25.0 \times 10^{-3})$
$$= 1.60 \times 10^{-3} \, mol \, dm^{-3}$$

ANSWER   The concentrations are 8.80×10^{-2} mol dm^{-3} sodium carbonate and 1.60×10^{-3} mol dm^{-3} sodium hydrogencarbonate.

EXERCISE 10   Problems on Neutralisation

1. What mass of the solute must be used in order to prepare the required solutions listed below?
   a) 500 cm^3 of 0.100 mol dm^{-3} $H_6C_4O_4$(aq) from $H_6C_4O_4$(s)
   b) 250 cm^3 of 0.200 mol dm^{-3} Na_2CO_3(aq) from Na_2CO_3(s)
   c) 750 cm^3 of 0.100 mol dm^{-3} $H_2C_2O_4$(aq) from $H_2C_2O_4 \cdot 2H_2O$(s)
   d) 2.50 dm^3 of 0.200 mol dm^{-3} $NaHCO_3$(aq) from $NaHCO_3$(s)
   e) 500 cm^3 of 0.100 mol dm^{-3} $Na_2B_4O_7$(aq) from $Na_2B_4O_7 \cdot 10H_2O$(s)

2. What volumes of the following concentrated solutions are required to give the stated volumes of the more dilute solutions?
   a) 2.00 dm^3 of 0.500 mol dm^{-3} H_2SO_4(aq) from 2.00 mol dm^{-3} H_2SO_4(aq)
   b) 1.00 dm^3 of 0.750 mol dm^{-3} HCl(aq) from 10.0 mol dm^{-3} HCl(aq)

c) 250 cm^3 of 0.250 mol dm^{-3} NaOH(aq) from 5.50 mol dm^{-3} NaOH(aq)

d) 500 cm^3 of 1.25 mol dm^{-3} HNO$_3$(aq) from 3.25 mol dm^{-3} HNO$_3$(aq)

e) 250 cm^3 of 2.00 mol dm^{-3} KOH(aq) from 2.60 mol dm^{-3} KOH(aq)

3. a) What mass of the following will react completely with 50.0 cm^3 of 0.100 mol dm^{-3} sulphuric acid?
   i) zinc
   ii) solid potassium hydroxide
   iii) solid copper(II) carbonate
   iv) sodium carbonate-10-water crystals

   b) What volumes of the following solutions will react with 50.0 cm^3 of 0.100 mol dm^{-3} sulphuric acid?
   i) 0.200 mol dm^{-3} potassium hydroxide solution
   ii) 0.0500 mol dm^{-3} sodium carbonate solution
   iii) 0.500 mol dm^{-3} lead nitrate solution
   iv) 0.250 mol dm^{-3} sodium hydrogencarbonate solution

4. What volume of 0.125 mol dm^{-3} sodium hydroxide solution is needed to titrate 25.0 cm^3 of 0.085 mol dm^{-3} sulphuric acid?

5. Arsenic(V) acid, H$_3$AsO$_4$, is a tribasic acid. 25.0 cm^3 of a solution of the acid require 35.7 cm^3 of a solution of sodium hydroxide of concentration 0.100 mol dm^{-3} for neutralisation. What is the concentration of the acid?

6. Soda lime is 85.0% NaOH and 15.0% CaO. What volume of 0.500 mol dm^{-3} nitric acid is needed to neutralise 2.50 g of soda lime?

7. Carbon dioxide is prepared by the action of 2.00 mol dm^{-3} hydrochloric acid on marble. What volume of acid is required to give 10.0 dm^3 of carbon dioxide, measured at 9.9 × 10^4 N m^{-2} and 26 °C?

8. 0.500 g of impure ammonium chloride is warmed with an excess of sodium hydroxide solution. The ammonia liberated is absorbed in 25.0 cm^3 of 0.200 mol dm^{-3} sulphuric acid. The excess of sulphuric acid requires 5.64 cm^3 of 0.200 mol dm^{-3} sodium hydroxide solution for titration. Calculate the percentage of ammonium chloride in the original sample.

9. A 1.00 g sample of limestone is allowed to react with 100 cm^3 of 0.200 mol dm^{-3} hydrochloric acid. The excess acid required 24.8 cm^3 of 0.100 mol dm^{-3} sodium hydroxide solution. Calculate the percentage of calcium carbonate in the limestone.

10. An impure sample of barium hydroxide of mass 1.6524 g was allowed to react with 100 cm³ of hydrochloric acid of concentration 0.200 mol dm⁻³. When the excess of acid was titrated against sodium hydroxide, 10.9 cm³ of sodium hydroxide solution were required. 25.0 cm³ of the sodium hydroxide required 28.5 cm³ of the hydrochloric acid in a separate titration. Calculate the percentage purity of the sample of barium hydroxide.

11. A solution contains sodium carbonate and sodium hydrogencarbonate. After the addition of a few drops of phenolphthalein, 25.0 cm³ of the solution were titrated against a 0.200 mol dm⁻³ solution of hydrochloric acid, until the indicator turned colourless. Methyl orange was added, and titration was continued until the indicator changed colour. Fig. 3.1 shows how the pH changed during the titration and the pH ranges of the two indicators. Calculate the concentrations of sodium carbonate and sodium hydrogencarbonate in the solution.

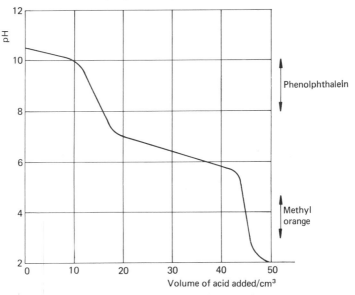

Fig. 3.1   pH changes in Question **11**

12. A household cleaner contains ammonia. A 25.37 g sample of the cleaner is dissolved in water and made up to 250 cm³. A 25.0 cm³ portion of this solution requires 37.3 cm³ of 0.360 mol dm⁻³ sulphuric acid for neutralisation. What is the percentage by mass of ammonia in the cleaner?

13. A fertiliser contains ammonium sulphate and potassium sulphate. A sample of 0.225 g of fertiliser was warmed with sodium hydroxide solution. The ammonia evolved required 15.7 cm³ of 0.100 mol dm⁻³ hydrochloric acid for neutralisation. Calculate the percentage of ammonium sulphate in the sample.

14. Calculate the number of carboxyl groups in the compound $C_6H_8O_6$, given that 0.440 g of it neutralised 37.5 cm^3 of sodium hydroxide of concentration 0.200 mol dm^{-3}.

15. 0.500 g of a mixture of sodium hydrogencarbonate and anhydrous sodium carbonate was dissolved in water. Phenolphthalein was used as indicator when the solution was titrated against hydrochloric acid of concentration 0.100 mol dm^{-3}, 15.0 cm^3 of acid were required. Calculate the percentage by mass of sodium hydrogencarbonate in the mixture.

16. Sodium carbonate crystals (27.823 0 g) were dissolved in water and made up to 1.00 dm^3. 25.0 cm^3 of the solution were neutralised by 48.8 cm^3 of hydrochloric acid of concentration 0.100 mol dm^{-3}. Find n in the formula $Na_2CO_3 \cdot nH_2O$.

OXIDATION–REDUCTION REACTIONS

Oxidation–reduction (or 'redox') reactions involve a transfer of electrons. The oxidising agent accepts electrons, and the reducing agent gives electrons. In working out the equation for a redox reaction, a good method is to work out the 'half-reaction equation' for the oxidising agent and the 'half-reaction equation' for the reducing agent, and then add them together.

Examples of half-reaction equations

a) Iron(III) salts are reduced to iron(II) salts. The equation is

$$Fe^{3+} \longrightarrow Fe^{2+}$$

For the equation to balance, the charge on the right-hand side (RHS) must equal the charge on the left-hand side (LHS). This can be accomplished by inserting an electron on the LHS:

$$Fe^{3+} + e^- \longrightarrow Fe^{2+}$$

b) When chlorine acts as an oxidising agent, it is reduced to chloride ions:

$$Cl_2 \longrightarrow 2Cl^-$$

To obtain a balanced half-reaction equation, 2e$^-$ must be inserted on the LHS:

$$Cl_2 + 2e^- \longrightarrow 2Cl^-$$

c) Sulphites can be oxidised to sulphates:

$$SO_3^{2-} \longrightarrow SO_4^{2-}$$

To balance the equation with respect to mass, an extra oxygen atom is needed on the LHS. If H_2O is introduced on the LHS to supply this oxygen, the equation becomes

$$SO_3^{2-} + H_2O \longrightarrow SO_4^{2-} + 2H^+$$

To balance the equation with respect to charge, $2e^-$ are needed on the RHS:

$$SO_3^{2-} + H_2O \longrightarrow SO_4^{2-} + 2H^+ + 2e^-$$

d) Potassium manganate(VII) is an oxidising agent. In acid solution, it is reduced to a manganese(II) salt:

$$MnO_4^- + H^+ \longrightarrow Mn^{2+}$$

To balance the equation with respect to mass, $8H^+$ are needed to combine with 4 oxygen atoms:

$$MnO_4^- + 8H^+ \longrightarrow Mn^{2+} + 4H_2O$$

To balance the equation with respect to charge, $5e^-$ are needed on the LHS:

$$MnO_4^- + 8H^+ + 5e^- \longrightarrow Mn^{2+} + 4H_2O$$

It is a good idea to make a final check. Charge on LHS $= -1 + 8 - 5 = +2$. Charge on RHS $= +2$. The equation is balanced.

e) Potassium dichromate(VI) is an oxidising agent in acid solution, being reduced to a chromium(III) salt:

$$Cr_2O_7^{2-} + H^+ \longrightarrow Cr^{3+}$$

To balance the equation for mass, $14H^+$ are needed:

$$Cr_2O_7^{2-} + 14H^+ \longrightarrow 2Cr^{3+} + 7H_2O$$

To balance the equation for charge, $6e^-$ are needed on the LHS:

$$Cr_2O_7^{2-} + 14H^+ + 6e^- \longrightarrow 2Cr^{3+} + 7H_2O$$

A final check shows that the charge on the LHS $= -2 + 14 - 6 = +6$.

Charge on RHS $= 2(+3) = +6$. The equation is balanced.

You may like to practise with the half-reaction equations on p. 54.

Using half-reaction equations to obtain the equation for a redox reaction

a) In the reaction between iodine and thiosulphate ions, the two half-reaction equations are

$$I_2 + 2e^- \longrightarrow 2I^- \qquad [1]$$
$$2S_2O_3^{2-} \longrightarrow S_4O_6^{2-} + 2e^- \qquad [2]$$

Adding.[1] and [2] gives

$$I_2 + 2e^- + 2S_2O_3^{2-} \longrightarrow 2I^- + S_4O_6^{2-} + 2e^-$$

Deleting the $2e^-$ term from both sides of the equation gives

$$I_2 + 2S_2O_3^{2-} \longrightarrow 2I^- + S_4O_6^{2-}$$

A check shows that the charges on the LHS and the RHS are both -4.

b) When potassium manganate(VII) oxidises an iron(II) salt to an iron (III) salt, the equations for the half-reactions are

$$MnO_4^- + 8H^+ + 5e^- \longrightarrow Mn^{2+} + 4H_2O \qquad [3]$$

$$Fe^{2+} \longrightarrow Fe^{3+} + e^- \qquad [4]$$

One manganate(VII) ion needs 5 electrons, and one iron(II) ion gives only one. Equation [4] must therefore be multiplied by 5:

$$5Fe^{2+} \longrightarrow 5Fe^{3+} + 5e^- \qquad [5]$$

Equations [3] and [5] can now be added to give

$$MnO_4^- + 8H^+ + 5Fe^{2+} \longrightarrow Mn^{2+} + 4H_2O + 5Fe^{3+}$$

c) When potassium manganate(VII) oxidises sodium ethanedioate, the equation for the manganate(VII) half-reaction is [3] as in Example 2, and the equation for the reduction of ethanediaote is

$$C_2O_4^{2-} \longrightarrow 2CO_2 + 2e^- \qquad [6]$$

d) One manganate(VII) ion needs $5e^-$, and one ethanedioate ion gives $2e^-$. Multiplying equation [3] by 2 and equation [6] by 5 and adding gives

$$2MnO_4^- + 16H^+ + 5C_2O_4^{2-} \longrightarrow 2Mn^{2+} + 10CO_2 + 8H_2O$$

e) Potassium dichromate(VI) oxidises iron(II) salt to iron(III) salts. The equations for the two half-reactions are

$$Cr_2O_7^{2-} + 14H^+ + 6e^- \longrightarrow 2Cr^{3+} + 7H_2O \qquad [7]$$

$$Fe^{2+} \longrightarrow Fe^{3+} + e^- \qquad [8]$$

One dichromate ion will oxidise six iron(II) ions:

$$Cr_2O_7^{2-} + 14H^+ + 6Fe^{2+} \longrightarrow 2Cr^{3+} + 6Fe^{3+} + 7H_2O$$

You may like to try the problems on balancing equations on p. 54 before going on to tackle the numerical problems.

There is another method of balancing redox equations. It is explained in the following section on oxidation numbers.

Oxidation numbers

It is helpful to discuss oxidation–reduction reactions in terms of the change in the *oxidation number* of each reactant. In the reaction

$$Cu(s) \ + \ \tfrac{1}{2}O_2(g) \longrightarrow Cu^{2+}O^{2-}(s)$$

copper is oxidised and oxygen is reduced. It is said that the oxidation number of copper increases from zero to $+2$, and the oxidation number of oxygen decreases from zero to -2. The following rules are followed in assigning oxidation numbers:

a) The oxidation number of an uncombined element is zero.

b) In ionic compounds, the oxidation number of each element is the charge on its ion. In NaCl, the oxidation number of Na $= +1$, and that of Cl $= -1$.

c) The sum of the oxidation numbers of all the elements in a compound is zero. In $AlCl_3$, the oxidation numbers are: Al $= +3$; Cl $= -1$, so that the sum of the oxidation numbers is $+3 + 3(-1) = 0$.

d) The sum of the oxidation numbers of all the elements in an ion is equal to the charge on the ion. In SO_4^{2-}, the oxidation numbers are S $= +6$, O $= -2$. The sum of the oxidation numbers for all the atoms is $+6 + 4(-2) = -2$, the same as the charge of the SO_4^{2-} ion.

e) In a covalent compound, one element must be given a positive oxidation number and the other a negative oxidation number, such that the sum of the oxidation numbers for all the atoms is zero. The following elements always have the same oxidation numbers in all their compounds. A knowledge of their oxidation numbers helps one to assign oxidation numbers to the other elements combined with them:

Na, K $+1$ H $+1$, except in metal hydrides

Mg, Ca $+2$ F -1

Al $+3$ Cl -1, except in compounds with O and F

O -2, except in peroxides and compounds with F

The oxidation number method

A consideration of the changes in oxidation numbers which occur during a redox reaction helps you to decide which reactants have been oxidised and which have been reduced. It can also be very helpful when you need to balance the equation for the reaction. The following two points cover what is involved when you use the oxidation number method to balance the equation for a redox reaction:

a) When an element is oxidised, its oxidation number increases; when an element is reduced, its oxidation number decreases. If x atoms (or ions) of an element A react with y atoms (or ions) of an element B, i.e.

$$x \, A \ + \ y \, B \ \longrightarrow$$

then, if the oxidation number of A changes by a units, and the oxidation number of B changes by b units, you can see that

$$xa \ = \ yb$$

For example, in the reaction between tin(II) and iron(III) ions,

$$Sn^{2+}(aq) \ + \ 2Fe^{3+}(aq) \ \longrightarrow \ Sn^{4+}(aq) \ + \ 2Fe^{2+}$$

For Sn, no. of ions = 1, change in ox. no. = 2, and product = 2

For Fe, no. of ions = 2, change in ox. no. = 1, and product = 2

b) In a balanced equation

LHS sum of ox. nos. of elements = RHS sum of ox. nos. of elements

In the reaction

$$KIO_3(aq) \ + \ 2Na_2SO_3(aq) \ \longrightarrow \ KIO(aq) \ + \ 2Na_2SO_4(aq)$$

the elements K, Na and O keep the same oxidation states during the reaction, while I and S change.

Ox. no. of I in KIO_3 = $+5$; in KIO = $+1$

Ox. no. of S in Na_2SO_3 = $+4$; in Na_2SO_4 = $+6$

Sum of ox. nos. on LHS = $(+5) + 2(+4)$ = $+13$

Sum of ox. nos. on RHS = $(+1) + 2(+6)$ = $+13$

When applying the oxidation number method to a reaction between A and B, remember:

$$\left(\begin{array}{c} \text{No. of atoms of A} \times \text{Change} \\ \text{in oxidation number of A} \end{array} \right) = \left(\begin{array}{c} \text{No. of atoms of B} \times \text{Change} \\ \text{in oxidation number of B} \end{array} \right)$$

Sum of ox. nos. on LHS = Sum of ox. nos. on RHS

EXAMPLE 1   What is the oxidation number of germanium in $GeCl_4$?

METHOD   Chlorine is one of the elements with a constant oxidation number of -1.

(Oxidation number of Ge) + $4(-1)$ = 0.

ANSWER   Oxidation number of Ge = $+4$.

EXAMPLE 2   What is the oxidation number of manganese in Mn_2O_7?

METHOD   Oxygen is one of the elements with a constant oxidation number of -2.

$2(\text{Oxidation number of Mn}) + 7(-2) = 0.$

ANSWER   Oxidation number of Mn $= +7$.

EXAMPLE 3   What is the oxidation number of iron in $Fe(CN)_6{}^{3-}$?

METHOD   Since the cyanide ion is CN^-, it has an oxidation number of -1.
(Oxidation number of Fe) $+ 6(-1) = -3$.

ANSWER   Oxidation number of Fe $= +3$.

EXAMPLE 4   Use the oxidation number method to balance the equation

$$MnO_4{}^-(aq) + H^+(aq) + Fe^{2+}(aq) \longrightarrow Mn^{2+}(aq) + Fe^{3+}(aq) + H_2O(l)$$

METHOD   Hydrogen and oxygen have the same oxidation numbers on both sides of the equation; only manganese and iron need be considered.

In $MnO_4{}^-$, the oxidation number of Mn $= +7$

In Mn^{2+}, the oxidation number of Mn $= +2$

Thus, manganese decreases its oxidation number by 5 units, and iron must increase its oxidation number by 5 units.

From Fe^{2+} to Fe^{3+} is an increase of 1 unit; therefore the equation needs $5Fe^{2+} \longrightarrow 5Fe^{3+}$. This makes the equation

$$MnO_4{}^-(aq) + H^+(aq) + 5Fe^{2+}(aq) \longrightarrow Mn^{2+}(aq) + 5Fe^{3+}(aq) + H_2O(l)$$

To combine with 4 oxygen atoms, $8H^+$ are needed:

$$MnO_4{}^-(aq) + 8H^+(aq) + 5Fe^{2+}(aq) \longrightarrow Mn^{2+}(aq) + 5Fe^{3+}(aq) + 4H_2O(l)$$

EXERCISE 11   Problems on Oxidation Numbers

1. What is the oxidation number of the named element in the following compounds?

    a) Ba in $BaCl_2$
    b) Fe in $Fe(CN)_6{}^{4-}$
    c) Cl in Cl_2
    d) Li in Li_2O
    e) Fe in $Fe(CN)_6{}^{3-}$
    f) Cl in ClO^-
    g) P in P_2O_3
    h) Br in $BrO_3{}^-$
    i) Cl in $ClO_3{}^-$
    j) C in CCl_4
    k) I in I_2
    l) Cl in Cl_2O_7
    m) C in CO
    n) I in I^-
    o) Cl in Cl_2O_3
    p) Cr in CrO_3
    q) I in $IO_3{}^-$
    r) O in H_2O_2
    s) Cr in $CrO_4{}^{2-}$
    t) N in NO_2
    u) H in LiH
    v) Cr in $Cr_2O_7{}^{2-}$
    w) N in N_2O_4
    x) H in HBr
    y) S in $SO_3{}^{2-}$
    z) P in $PO_4{}^{3-}$

2. a) Calculate the oxidation numbers of tin and lead on each side of the equation

   $$PbO_2(s) + 4H^+(aq) + Sn^{2+}(aq) \longrightarrow Pb^{2+}(aq) + Sn^{4+}(aq) + 2H_2O(l)$$

   and state which element has been oxidised and which has been reduced.

   b) In the redox reaction

   $$2Mn^{2+}(aq) + 5BiO_3^-(aq) + 14H^+(aq) \longrightarrow 2MnO_4^-(aq) + 5Bi^{3+}(aq) + 7H_2O(l)$$

   calculate the oxidation numbers of all the elements, and state which have been oxidised and which have been reduced.

   c) Calculate the oxidation numbers of arsenic and manganese in each of the species in the reaction:

   $$5As_2O_3(s) + 4MnO_4^-(aq) + 12H^+(aq) \longrightarrow 5As_2O_5(s) + 4Mn^{2+}(aq) + 6H_2O(l)$$

   State which element has been oxidised and which has been reduced.

3. In each of the following equations, one element is underlined. Calculate its oxidation number in each species, and state whether an oxidation or a reduction has occurred.

   a) $2\underline{F}_2(g) + 2OH^-(aq) \longrightarrow \underline{F}_2O(g) + 2\underline{F}^-(aq) + H_2O(l)$
   b) $3\underline{Cl}_2(g) + 6OH^-(aq) \longrightarrow \underline{Cl}O_3^-(aq) + 5\underline{Cl}^-(aq) + 3H_2O(l)$
   c) $\underline{N}H_4^+\underline{N}O_3^-(s) \longrightarrow \underline{N}_2O(g) + 2H_2O(l)$
   d) $\underline{Cr}_2O_7^{2-}(aq) + 14H^+(aq) + 6e^- \longrightarrow 2\underline{Cr}^{3+}(aq) + 7H_2O(l)$
   e) $\underline{C}_2O_4^{2-}(aq) \longrightarrow 2\underline{C}O_2(g) + 2e^-$

4. a) Only N and I alter in oxidation number in the reaction

   $$N_2H_6O(aq) + IO_3^-(aq) + 2H^+(aq) + Cl^-(aq) \longrightarrow N_2(g) + ICl(aq) + 4H_2O(l)$$

   Calculate the oxidation number of N in N_2H_6O.

   b) In the reaction below, only S and Br change in oxidation number.

   $$Na_2H_{10}S_2O_8(aq) + 4Br_2(aq) \longrightarrow 2H_2SO_4(aq) + 2NaBr(aq) + 6HBr(aq)$$

   Calculate the oxidation number of S in $Na_2H_{10}S_2O_8$.

5. Use the oxidation number method to balance the equations

   a) $IO_4^-(aq) + I^-(aq) + H^+(aq) \longrightarrow I_2(aq) + 4H_2O(l)$
   b) $BrO_3^-(aq) + I^-(aq) + H^+(aq) \longrightarrow Br^-(aq) + I_2(aq) + H_2O(l)$
   c) $V^{3+}(aq) + H_2O_2(aq) \longrightarrow VO^{2+}(aq) + H^+(aq)$
   d) $SO_2(g) + H_2O(l) + Br_2(aq) \longrightarrow H^+(aq) + SO_4^{2-}(aq) + Br^-(aq)$
   e) $NH_3(g) + O_2(g) \longrightarrow N_2(g) + H_2O(g)$
   f) $NH_3(g) + O_2(g) \longrightarrow N_2O(g) + H_2O(g)$
   g) $NH_3(g) + O_2(g) \longrightarrow NO(g) + H_2O(g)$
   h) $Fe^{2+}C_2O_4^{2-}(aq) + Ce^{3+}(aq) \longrightarrow CO_2(g) + Ce^{2+}(aq) + Fe^{3+}(aq)$

6. When potassium dichromate solution reacts with acidified potassium iodide solution, titration shows that 1 mole of potassium dichromate produces 3 moles of iodine. Use the oxidation number method to complete and balance the equation

$$Cr_2O_7^{2-}(aq) + I^-(aq) + H^+(aq) \longrightarrow 3I_2(aq)$$

POTASSIUM MANGANATE(VII) TITRATIONS

When potassium manganate(VII) acts as an oxidising agent in acid solution, it is reduced to a manganese(II) salt:

$$MnO_4^-(aq) + 8H^+(aq) + 5e^- \longrightarrow Mn^{2+}(aq) + 4H_2O(l)$$

Potassium manganate(VII) is not sufficiently pure to be used as a primary standard, and solutions of the oxidant are standardised by titration against a primary standard such as sodium ethanedioate. This reductant can be obtained in a high state of purity as crystals of formula $Na_2C_2O_4 \cdot 2H_2O$, which are neither deliquescent nor efflorescent, and can be weighed out exactly to make a standard solution.

Once it has been standardised, a solution of potassium manganate(VII) can be used to estimate reducing agents such as iron(II) salts. No indicator is needed as the oxidant changes from purple to colourless at the end point.

EXAMPLE 1 *Standardising potassium manganate(VII) against the primary standard, sodium ethanedioate*
A 25.0 cm^3 portion of sodium ethanedioate solution of concentration $0.200 \text{ mol dm}^{-3}$ is warmed and titrated against a solution of potassium manganate(VII). If 17.2 cm^3 of potassium manganate(VII) are required, what is its concentration?

METHOD Let M = concentration of the manganate(VII) solution.
Number of moles of ethanedioate = $25.0 \times 10^{-3} \times 0.200$ mol
Number of moles of manganate(VII) = $17.2 \times 10^{-3} \times M$ mol

The equations for the half-reactions are

$$MnO_4^-(aq) + 8H^+(aq) + 5e^- \longrightarrow Mn^{2+}(aq) + 4H_2O(l) \quad [1]$$
$$C_2O_4^{2-}(aq) \longrightarrow 2CO_2(g) + 2e^- \quad [2]$$

Multiplying [1] by 2 and [2] by 5, and adding the two equations gives

$$2MnO_4^-(aq) + 16H^+(aq) + 5C_2O_4^{2-}(aq) \longrightarrow 2Mn^{2+}(aq) + 8H_2O(l) + 10CO_2(g)$$

No. of moles of MnO_4^- = $\frac{2}{5} \times$ No. of moles of $C_2O_4^{2-}$.

$$\therefore \quad 17.2 \times 10^{-3} \times M = \tfrac{2}{5} \times 25.0 \times 10^{-3} \times 0.200$$

$$M = \frac{2 \times 25.0 \times 10^{-3} \times 0.200}{5 \times 17.2 \times 10^{-3}} = 0.116 \, \text{mol dm}^{-3}$$

ANSWER   The potassium manganate(VII) solution has a concentration of $0.116 \, \text{mol dm}^{-3}$.

EXAMPLE 2   *Oxidising iron(II) compounds*
Ammonium iron(II) sulphate crystals have the following formula: $(NH_4)_2SO_4 \cdot FeSO_4 \cdot nH_2O$. In an experiment to determine n, $8.492 \, \text{g}$ of the salt were dissolved and made up to $250 \, \text{cm}^3$ of solution with distilled water and dilute sulphuric acid. A $25.0 \, \text{cm}^3$ portion of the solution was further acidified and titrated against potassium manganate(VII) solution of concentration $0.0150 \, \text{mol dm}^{-3}$. A volume of $22.5 \, \text{cm}^3$ was required.

METHOD   The equations for the two half-reactions are

$$MnO_4^-(aq) + 8H^+(aq) + 5e^- \longrightarrow Mn^{2+}(aq) + 4H_2O(l) \quad [1]$$
$$Fe^{2+}(aq) \longrightarrow Fe^{3+}(aq) + e^- \quad [2]$$

Multiplying [2] by 5 and then adding it to [1] gives

$$MnO_4^-(aq) + 8H^+(aq) + 5Fe^{2+}(aq) \longrightarrow Mn^{2+}(aq) + 5Fe^{3+}(aq) + 4H_2O(l)$$

$$\text{No. of moles of manganate(VII)} = 22.5 \times 10^{-3} \times 0.0150$$
$$= 0.338 \times 10^{-3} \, \text{mol}$$

$$\text{No. of moles of iron(II)} = 5 \times \text{No. of moles of manganate(VII)}$$
$$= 5 \times 0.338 \times 10^{-3} = 1.69 \times 10^{-3} \, \text{mol}$$

$$\text{Concn of iron(II)} = \frac{1.69 \times 10^{-3}}{25.0 \times 10^{-3}} = 0.0674 \, \text{mol dm}^{-3}$$

$$\text{Concn of } (NH_4)_2SO_4 \cdot FeSO_4 \cdot nH_2O = \frac{\text{Mass in 1 dm}^3 \text{ of solution}}{\text{Molar mass}}$$

$$= \frac{4 \times 8.492}{\text{Molar mass}}$$

$$0.0674 = \frac{4 \times 8.492}{\text{Molar mass}}$$

$$\text{Molar mass} = 503.9 \, \text{g mol}^{-1}$$

Molar mass of $(NH_4)_2SO_4 \cdot FeSO_4 \cdot nH_2O = 284 + 18n = 504 \, \text{g mol}^{-1}$

$$\therefore \quad n = 12$$

ANSWER   The formula of the crystals is $(NH_4)_2SO_4 \cdot 12H_2O$

EXAMPLE 3   *Oxidising hydrogen peroxide*

A solution of hydrogen peroxide was diluted 20.0 times. A 25.0 cm^3 portion of the diluted solution was acidified and titrated against 0.0150 mol dm^{-3} potassium manganate(VII) solution. 45.7 cm^3 of the oxidant were required. Calculate the concentration of the hydrogen peroxide solution a) in mol dm^{-3} and b) the 'volume concentration'. (This means the number of volumes of oxygen obtained from one volume of the solution.)

METHOD   The equations for the half-reactions are

$$MnO_4^-(aq) + 8H^+(aq) + 5e^- \longrightarrow Mn^{2+}(aq) + 4H_2O(l) \qquad [1]$$

$$H_2O_2(aq) \longrightarrow O_2(g) + 2H^+(aq) + 2e^- \qquad [2]$$

Multiplying [1] by 2 and [2] by 5, and adding the two equations gives

$$2MnO_4^-(aq) + 6H^+(aq) + 5H_2O_2(aq) \longrightarrow 2Mn^{2+}(aq) + 8H_2O(l) + 5O_2(g)$$

No. of moles of $MnO_4^-(aq)$ = 45.7 × 10^{-3} × 0.0150

= 0.685 × 10^{-3} mol

No. of moles of H_2O_2 = $\frac{5}{2}$ × No. of moles of MnO_4^-

= $\frac{5}{2}$ × 0.685 × 10^{-3} = 1.71 × 10^{-3} mol

Concn of H_2O_2 = (1.71 × 10^{-3})/(25.0 × 10^{-3}) = 0.0684 mol dm^{-3}

Concn of original solution = 20.0 × 0.0684 = 1.37 mol dm^{-3}.

When hydrogen peroxide decomposes,

$$2H_2O_2(aq) \longrightarrow 2H_2O(l) + O_2(g)$$

2 moles of hydrogen peroxide form 1 mole of oxygen. Therefore a solution of hydrogen peroxide of concentration 2 mol dm^{-3} is a 22.4 volume solution (the volume of 1 mole of oxygen).

A solution of H_2O_2 of concentration 1.37 mol dm^{-3} is a 22.4 × 1.37/2 = 15.4 volume solution.

ANSWER   The concentration of hydrogen peroxide is: a) 1.37 mol dm^{-3}, and b) 15.4 volume.

EXAMPLE 4   *Finding the percentage of iron in ammonium iron(III) sulphate*

Iron(III) ions can be estimated by first reducing them to iron(II) ions, and then, after destroying the excess of reducing agent, oxidising them to iron(III) ions with a standard solution of potassium manganate(VII). Zinc amalgam and sulphuric acid are used as the reducing agent. Note that hydrochloric acid cannot be used, and the reducing agent tin(II) chloride cannot be used as potassium manganate(VII) oxidises chloride ions to chlorine.

7.418 g of ammonium iron(III) sulphate are dissolved and made up to 250 cm^3 after the addition of dilute sulphuric acid, 25.0 cm^3 of the solution are pipetted into a bottle containing zinc amalgam, and shaken until a drop of the solution gives no colour when tested with a solution of a thiocyanate (which turns deep red in the presence of iron(III) ions). The aqueous solution is then separated by decantation from the zinc amalgam. On addition of more dilute sulphuric acid and titration against standard potassium manganate(VII) solution, 18.7 cm^3 of 0.0165 mol dm^{-3} solution are required. Calculate the percentage of iron in ammonium iron(III) sulphate.

METHOD

No. of moles of manganate(VII) in volume used $= 18.7 \times 10^{-3} \times 0.0165$

$$= 0.0309 \times 10^{-3} \, mol$$

From the equation

$$MnO_4^-(aq) + 8H^+(aq) + 5Fe^{2+}(aq) \longrightarrow Mn^{2+}(aq) + 5Fe^{3+}(aq) + 4H_2O(l)$$

No. of moles of Fe^{2+}(aq) in 25.0 cm^3 $= 5 \times 0.309 \times 10^{-3}$

$$= 1.55 \times 10^{-3} \, mol$$

No. of moles of Fe^{2+}(aq) in whole solution $= 1.55 \times 10^{-2} \, mol$

Mass of iron in sample $=$ No. of moles $\times$ Relative atomic mass

$$= 1.55 \times 10^{-2} \times 55.8 = 0.865 \, g$$

ANSWER

Percentage of iron $= \dfrac{0.865}{7.418} \times 100 = 11.7\%$.

POTASSIUM DICHROMATE(VI) TITRATIONS

Potassium dichromate(VI) can be obtained in a high state of purity, and its aqueous solutions are stable. It is used as a primary standard. The colour change when chromium(VI) changes to chromium(III) in the reaction

$$Cr_2O_7^{2-}(aq) + 14H^+(aq) + 6e^- \longrightarrow 2Cr^{3+}(aq) + 7H_2O(l)$$

is from orange to green. As it is not possible to see a sharp change in colour, an indicator is used. Barium N-phenylphenylamine–4–sulphonate gives a sharp colour change, from blue-green to violet, when a slight excess of potassium dichromate has been added. Phosphoric(V) acid must be present to form a complex with the Fe^{3+} ions formed during the oxidation reaction; otherwise Fe^{3+} ions affect the colour change of the indicator.

Since dichromate(VI) has a slightly lower redox potential than manganate(VII), it can be used in the presence of chloride ions, without oxidising them to chlorine.

EXAMPLE *Determination of the percentage of iron in iron wire*

A piece of iron wire of mass 2.225 g was put into a conical flask containing dilute sulphuric acid. The flask was fitted with a bung carrying a Bunsen valve, to allow the hydrogen generated to escape but prevent air from entering. The mixture was warmed to speed up reaction. When all the iron had reacted, the solution was cooled to room temperature and made up to 250 cm^3 in a graduated flask. With all these precautions, iron is converted to Fe^{2+} ions only, and no Fe^{3+} ions are formed. 25.0 cm^3 of the solution were acidified and titrated against a 0.0185 mol dm^{-3} solution of potassium dichromate(VI). The volume required was 31.0 cm^3. Calculate the percentage of iron in the iron wire.

METHOD Number of moles of $Cr_2O_7{}^{2-}$(aq) used $= 31.0 \times 10^{-3} \times 0.0185$

$$= 0.574 \times 10^{-3} \, mol$$

The equations for the two half-reactions are

$$Cr_2O_7{}^{2-}(aq) + 14H^+(aq) + 6e^- \longrightarrow 2Cr^{3+}(aq) + 7H_2O(l) \quad [1]$$
$$Fe^{2+}(aq) \longrightarrow Fe^{3+}(aq) + e^- \qquad [2]$$

Multiplying [2] by 6 and adding gives

$$Cr_2O_7{}^{2-}(aq) + 14H^+(aq) + 6Fe^{2+}(aq) \longrightarrow 2Cr^{3+}(aq) + 6Fe^{3+}(aq) + 7H_2O(l)$$

No. of moles of Fe^{2+} in 25.0 cm^3 $= 6 \times 0.574 \times 10^{-3}$

$$= 3.45 \times 10^{-3} \, mol$$

No. of moles of Fe^{2+} in the whole solution $= 3.45 \times 10^{-2} \, mol$

Mass of Fe in the whole solution $= 3.45 \times 10^{-2} \times 55.8 = 1.93 \, g$

Percentage of Fe in wire $= \dfrac{1.93}{2.225} \times 100 = 86.7\%$

ANSWER The wire is 86.7% iron.

SODIUM THIOSULPHATE TITRATIONS

Sodium thiosulphate reduces iodine to iodide ions, and forms sodium tetrathionate, $Na_2S_4O_6$:

$$2S_2O_3{}^{2-}(aq) + I_2(aq) \longrightarrow 2I^-(aq) + S_4O_6{}^{2-}(aq)$$

Sodium thiosulphate, $Na_2S_2O_3 \cdot 5H_2O$, is not used as a primary standard as the water content of the crystals is variable. A solution of sodium thiosulphate can be standardised against a solution of iodine, or a solution of potassium iodate(V) or potassium dichromate or potassium manganate(VII).

EXAMPLE 1   *Standardisation of a sodium thiosulphate solution, using iodine*
Iodine has a limited solubility in water. It dissolves in a solution of
potassium iodide because it forms the very soluble complex ion, I_3^-.

$$I_2(s) \ + \ I^-(aq) \ \rightleftharpoons \ I_3^-(aq)$$

An equilibrium is set up between iodine and tri-iodide ions, and if
iodine molecules are removed from solution by a reaction, tri-iodide
ions dissociate to form more iodine molecules. A solution of iodine
in potassium iodide can thus be titrated as though it were a solution
of iodine in water.

When sufficient of a solution of thiosulphate is added to a solution of
iodine, the colour of iodine fades to a pale yellow. Then $2 \, cm^3$ of
starch solution are added to give a blue colour with the iodine.
Addition of thiosulphate is continued drop by drop, until the blue
colour disappears.

$2.835 \, g$ of iodine and $6 \, g$ of potassium iodide are dissolved in distilled
water and made up to $250 \, cm^3$. A $25.0 \, cm^3$ portion titrated against
sodium thiosulphate solution required $17.7 \, cm^3$ of the solution.
Calculate the concentration of the thiosulphate solution.

METHOD   Molar mass of iodine $= \ 2 \times 127 \ = \ 254 \, g \, mol^{-1}$
Concn of iodine solution $= \ 2.835 \times 4/254 \ = \ 0.0446 \, mol \, dm^{-3}$
No. of moles of I_2 in $25.0 \, cm^3 \ = \ 25.0 \times 10^{-3} \times 0.0446$
$$= \ 1.115 \times 10^{-3} \, mol$$

From the equation

$$2S_2O_3{}^{2-}(aq) \ + \ I_2(aq) \ \longrightarrow \ 2I^-(aq) \ + \ S_4O_6{}^{2-}(aq)$$

No. of moles of 'thio' $= \ 2 \times$ No. of moles of I_2
No. of moles of 'thio' in volume used $= \ 2.23 \times 10^{-3} \, mol$
$$Concn \ of \ 'thio' \ = \ \frac{2.23 \times 10^{-3}}{17.7 \times 10^{-3}} = 0.126 \, mol \, dm^{-3}$$

ANSWER   The concentration of the thiosulphate solution is $0.126 \, mol \, dm^{-3}$.

EXAMPLE 2   *Standardisation of thiosulphate against potassium iodate (V)*
Potassium iodate(V) is a primary standard. It reacts with iodide ions
in the presence of acid to form iodine:

$$IO_3^-(aq) \ + \ 5I^-(aq) \ + \ 6H^+(aq) \ \longrightarrow \ 3I_2(aq) \ + \ 3H_2O(l)$$

A standard solution of iodine can be prepared by weighing out the
necessary quantity of potassium iodate(V) and making up to a known
volume of solution. When a portion of this solution is added to an
excess of potassium iodide in acid solution, a calculated amount
of iodine is liberated.

1.105 g of potassium iodate(V) are dissolved and made up to 250 cm³. To a 25.0 cm³ portion are added an excess of potassium iodide and dilute sulphuric acid. The solution is titrated with a solution of sodium thiosulphate, starch solution being added near the end-point. 29.8 cm³ of thiosulphate solution are required. Calculate the concentration of the thiosulphate solution.

METHOD

Molar mass of $KIO_3 = 39.1 + 127 + (3 \times 16.0) = 214 \, g \, mol^{-1}$

Concn of KIO_3 solution $= 1.015 \times 4/214 = 0.0189 \, mol \, dm^{-3}$

No. of moles of KIO_3 in 25 cm³ $= 25.0 \times 10^{-3} \times 0.0189$
$$= 0.473 \times 10^{-3} \, mol$$

Since

$$IO_3^-(aq) + 5I^-(aq) + 6H^+(aq) \longrightarrow 3I_2(aq) + 3H_2O(l)$$

and
$$2S_2O_3^{2-}(aq) + I_2(aq) \longrightarrow 2I^-(aq) + S_4O_6^{2-}(aq)$$

No. of moles of 'thio' $= 6 \times$ No. of moles of IO_3^-
$$= 6 \times 0.473 \times 10^{-3} = 2.84 \times 10^{-3} \, mol$$

Concn of 'thio' $= (2.84 \times 10^{-3})/(29.8 \times 10^{-3}) = 0.0950 \, mol \, dm^{-3}$

ANSWER

The sodium thiosulphate solution has a concentration $0.0950 \, mol \, dm^{-3}$.

EXAMPLE 3 *Standardisation of thiosulphate solution with potassium dichromate(VI)*
A standard solution is made by dissolving 1.015 g of potassium dichromate(VI) and making up to 250 cm³. A 25.0 cm³ portion is added to an excess of potassium iodide and dilute sulphuric acid, and the iodine liberated is titrated with sodium thiosulphate solution. 19.2 cm³ of this solution are needed. Find the concentration of the thiosulphate solution.

METHOD

Molar mass of $K_2Cr_2O_7 = 294 \, g \, mol^{-1}$

Concn of dichromate solution $= 1.015 \times 4/294 = 0.0138 \, mol \, dm^{-3}$

No. of moles of dichromate in 25 cm³ $= 25.0 \times 10^{-3} \times 0.138 \, mol$
$$= 0.345 \times 10^{-3} \, mol$$

The equations for the half-reactions are

$$Cr_2O_7^{2-}(aq) + 14H^+(aq) + 6e^- \longrightarrow 2Cr^{3+}(aq) + 7H_2O(l) \quad [1]$$
$$2I^-(aq) \longrightarrow I_2(aq) + 2e^- \quad [2]$$

Multiplying [2] by 3, and adding to [1] gives the equation

$$Cr_2O_7^{2-}(aq) + 14H^+(aq) + 6I^-(aq) \longrightarrow 2Cr^{3+}(aq) + 7H_2O(l) + 3I_2(aq)$$

No. of moles of $I_2 = 3 \times$ No. of moles of $Cr_2O_7^{2-}$

No. of moles of I_2 in 25 cm^3 = 3 X 0.345 X 10^{-3} = 1.035 X 10^{-3} mol
No. of moles of 'thio' = 2 X No. of moles of I_2 (see Example 1)
No. of moles of 'thio' in volume used = 2.07 X 10^{-3} mol
Concn. of 'thio' = (2.07 X 10^{-3})/(19.2 X 10^{-3}) = 0.108 mol dm^{-3}

ANSWER   The concentration of the thiosulphate solution is 0.108 mol dm^{-3}.

EXAMPLE 4   *Estimation of chlorine*
Chlorine displaces iodine from iodides. The iodine formed can be determined by titration with a standard thiosulphate solution. Chlorate(I) solutions are often used as a source of chlorine as they liberate chlorine readily on reaction with acid:

$$ClO^-(aq) + 2H^+(aq) + Cl^-(aq) \longrightarrow Cl_2(aq) + H_2O(l)$$

The amount of chlorine available in a domestic bleach which contains sodium chlorate(I) can be found by allowing the bleach to react with an iodide solution to form iodine, and then titrating with thiosulphate solution:

$$ClO^-(aq) + 2H^+(aq) + 2I^-(aq) \longrightarrow I_2(aq) + Cl^-(aq) + H_2O(l)$$

A domestic bleach in solution is diluted by pipetting 10.0 cm^3 and making this volume up to 250 cm^3. A 25.0 cm^3 portion of the solution is added to an excess of potassium iodide and ethanoic acid and titrated against sodium thiosulphate solution of concentration 0.0950 mol dm^{-3}, using starch as an indicator. The volume required is 21.3 cm^3. Calculate the percentage of available chlorine in the bleach.

METHOD   No. of moles of 'thio' = 21.3 X 10^{-3} X 0.0950 = 2.03 X 10^{-3} mol
Since $2S_2O_3^{2-}(aq) + I_2(aq) \longrightarrow S_4O_6^{2-}(aq) + 2I^-(aq)$
No. of moles of I_2 = 1.015 mol
Since iodine is produced in the reaction

$$ClO^-(aq) + 2I^-(aq) + 2H^+(aq) \longrightarrow I_2(aq) + Cl^-(aq) + H_2O(l)$$

No. of moles of ClO^- in 25 cm^3 of solution = 1.015 X 10^{-3} mol
Since chlorate(I) liberates chlorine in the reaction

$$ClO^-(aq) + 2H^+(aq) + Cl^-(aq) \longrightarrow Cl_2(aq) + H_2O(l)$$

No. of moles of Cl_2 = No. of moles of chlorate(I)
= 1.015 X 10^{-3} mol
Mass of chlorine = 71.0 X 1.015 X 10^{-3} = 0.0720 g

This is the mass of chlorine available in $25 \, cm^3$ of solution

$$\text{Percentage of available Cl}_2 = \frac{\text{Mass of Cl}_2 \text{ from } 250 \, cm^3 \text{ solution}}{\text{Mass of bleach solution used}} \times 100$$

$$= \frac{0.0720 \times 10}{10} \times 100 = 7.2\%$$

ANSWER The percentage of available chlorine in bleach is 7.2%.

EXAMPLE 5 *Estimation of copper(II) salts*
Copper(II) ions oxidise iodide ions to iodine. The iodine produced can be titrated with standard thiosulphate solution, and, from the amount of iodine produced, the concentration of copper(II) ions in the solution can be calculated.

A sample of $4.256 \, g$ of copper(II) sulphate-5-water is dissolved and made up to $250 \, cm^3$. A $25.0 \, cm^3$ portion is added to an excess of potassium iodide. The iodine formed required $18.0 \, cm^3$ of a 0.0950 mol dm^{-3} solution of sodium thiosulphate for reduction. Calculate the percentage of copper in the crystals.

METHOD No. of moles of 'thio' $= 18.0 \times 10^{-3} \times 0.0950 = 1.71 \times 10^{-3} \, mol$
No. of moles of $I_2 = \frac{1}{2} \times$ No. of moles of 'thio' $= 0.855 \times 10^{-3} \, mol$

Since $\quad 2Cu^{2+}(aq) + 4I^-(aq) \longrightarrow Cu_2I_2(s) + I_2(aq)$

No. of moles of Cu $= 2 \times$ No. of moles of $I_2 = 1.71 \times 10^{-3} \, mol$
Mass of Cu $= 63.5 \times 1.71 \times 10^{-3} = 0.109 \, g$
Mass of Cu in whole solution $= 1.09 \, g$

$$\text{Percentage of Cu} = \frac{1.09}{4.256} \times 100 = 25.6\%.$$

ANSWER The percentage of copper in the crystals is 25.6%.

*EXAMPLE 6 *Deriving an equation for the reaction between bromine and thiosulphate ions*
A solution of bromine was prepared and two titrations were performed:

a) $25.0 \, cm^3$ of the solution were added to an excess of potassium iodide. The iodine liberated required $19.5 \, cm^3$ of a 0.120 mol dm^{-3} solution of sodium thiosulphate.

b) $25.0 \, cm^3$ of the bromine solution were titrated directly against the thiosulphate solution. $2.45 \, cm^3$ of thiosulphate solution were required.

c) The solution from titration b) was tested for the presence of various anions. Sulphate ions were detected.

Derive an equation for the reaction.

METHOD   In titration a)

No. of moles of 'thio' $= 19.5 \times 10^{-3} \times 0.120 = 2.35 \times 10^{-3}$ mol

No. of moles of $I_2 = \frac{1}{2} \times$ No. of moles of 'thio' $= 1.18 \times 10^{-3}$ mol

Since the reaction is

$$Br_2(aq) \ + \ 2I^-(aq) \longrightarrow 2Br^-(aq) \ + \ I_2(aq)$$

No. of moles of $Br_2 =$ No. of moles of I_2

No. of moles of Br_2 in 25.0 cm^3 $= 1.18 \times 10^{-3}$ mol

In titration b)

No. of moles of 'thio' in volume used $= 2.45 \times 10^{-3} \times 0.120$
$$= 0.294 \text{ mol}$$

Moles of Br_2/Moles of $S_2O_3{}^{2-}$ $= 1.18 \times 10^{-3}/0.294 \times 10^{-3} = 4/1$

Bromine is reduced to bromide ions:

$$4Br_2(aq) \ + \ 8e^- \longrightarrow 8Br^-(aq)$$

Thiosulphate ions form sulphate ions:

$$S_2O_3{}^{2-}(aq) \longrightarrow 2SO_4{}^{2-}(aq)$$

To balance the equation, H_2O is needed to supply the extra oxygen:

$$S_2O_3{}^{2-}(aq) \ + \ 5H_2O(l) \longrightarrow 2SO_4{}^{2-}(aq) \ + \ 10H^+(aq) \ + \ 8e^-$$

Putting the two half-reactions together gives

ANSWER $4Br_2(aq) + S_2O_3{}^{2-}(aq) + 5H_2O(l) \longrightarrow 8Br^-(aq) + 2SO_4{}^{2-}(aq) + 10H^+(aq)$

EXERCISE 12   Problems on Redox Reactions

1. Write balanced half-reaction equations for the oxidation of each of the following:
   a) NO_2^- to NO_3^- b) $AsO_3{}^{3-}$ to $AsO_4{}^{3-}$ c) $Hg_2{}^{2+}$ to Hg^{2+}
   d) H_2O_2 to O_2 e) V^{3+} to VO^{2+}

2. Write balanced half-reaction equations for the reduction of each of the following in acid solution:
   a) NO_3^- to NO_2 b) NO_3^- to NO c) NO_3^- to NH_4^+
   d) BrO_3^- to Br_2 e) PbO_2 to Pb^{2+}

3. Complete and balance the following ionic equations:
   a) $MnO_4^-(aq) \ + \ H_2O_2(aq) \ + \ H^+(aq) \longrightarrow$
   b) $MnO_2(s) \ + \ H^+(aq) \ + \ Cl^-(aq) \longrightarrow$
   c) $MnO_4^-(aq) \ + \ C_2O_4{}^{2-}(aq) \ + \ H^+ \longrightarrow$
   d) $Cr_2O_7{}^{2-}(aq) \ + \ C_2O_4{}^{2-}(aq) \ + \ H^+(aq) \longrightarrow$
   e) $Cr_2O_7{}^{2-}(aq) \ + \ I^-(aq) \ + \ H^+(aq) \longrightarrow$
   f) $H_2O_2(aq) \ + \ NO_2^-(aq) \longrightarrow$

4. How many moles of the following reductants will be oxidised by 3.0×10^{-3} mol of potassium manganate(VII) in acid solution?

   a) Fe^{2+} b) Sn^{2+} c) $(CO_2^-)_2$ d) H_2O_2 e) I^-

5. How many moles of the following will be oxidised by 1.0×10^{-4} mol of potassium dichromate(VI)?

   a) Fe^{2+} b) SO_3^{2-} c) Br^- d) $(CO_2^-)_2$ e) Hg_2^{2+}?

6. How many moles of the following will be reduced by 2.0×10^{-3} moles of Sn^{2+}?

   a) $Fe(CN)_6^{3-}$ b) Cl_2 c) Mn^{4+} (to Mn^{2+})

   d) Ce^{4+} (to Ce^{3+})   e) BrO_3^- (to Br^-)?

7. What volumes of the following solutions will be oxidised by $25.0\,cm^3$ of $0.0200\,mol\,dm^{-3}$ potassium manganate(VII) in acid solution?

   a) $0.0200\,mol\,dm^{-3}$ tin(II) nitrate

   b) $0.0100\,mol\,dm^{-3}$ iron(II) sulphate

   c) $0.250\,mol\,dm^{-3}$ hydrogen peroxide

   d) $0.200\,mol\,dm^{-3}$ chromium(II) nitrate

   e) $0.150\,mol\,dm^{-3}$ sodium ethanedioate

8. What volumes of the following solutions will be oxidised by $20.0\,cm^3$ of $0.0150\,mol\,dm^{-3}$ potassium dichromate(VI) in acid solution?

   a) $0.0200\,mol\,dm^{-3}$ tin(II) chloride

   b) $0.150\,mol\,dm^{-3}$ iron(II) chloride

   c) $0.125\,mol\,dm^{-3}$ sodium ethanedioate

   d) $0.300\,mol\,dm^{-3}$ sodium sulphite (sulphate(IV))

   e) $0.100\,mol\,dm^{-3}$ mercury(I) nitrate, $Hg_2(NO_3)_2$

9. $25.0\,cm^3$ of a sodium sulphite solution require $45.0\,cm^3$ of $0.0200\,mol\,dm^{-3}$ potassium manganate(VII) solution for oxidation. What is the concentration of the sodium sulphite solution?

10. $35.0\,cm^3$ of potassium manganate(VII) solution are required to oxidise a $0.2145\,g$ sample of ethanedioic acid-2-water, $H_2C_2O_4 \cdot 2H_2O$. What is the concentration of the potassium manganate(VII) solution?

11. $37.5\,cm^3$ of cerium(IV) sulphate solution are required to titrate a $0.2245\,g$ sample of sodium ethanedioate, $Na_2C_2O_4$. What is the concentration of the cerium(IV) sulphate solution?

12. A piece of iron wire weighs $0.2756\,g$. It is dissolved in acid, reduced to the Fe^{2+} state, and titrated with $40.8\,cm^3$ of $0.0200\,mol\,dm^{-3}$ potassium dichromate solution. What is the percentage purity of the iron wire?

13. A piece of limestone weighing $0.1965\,g$ was allowed to react with an excess of hydrochloric acid. The calcium in it was precipitated as calcium ethanedioate. The precipitate was dissolved in sulphuric acid, and the ethanedioate in the solution needed $35.6\,cm^3$ of a $0.0200\,mol\,dm^{-3}$ solution of potassium manganate(VII) for titration. Calculate the percentage of $CaCO_3$ in the limestone.

14. A solution of potassium dichromate is standardised by titration with sodium ethanedioate solution. If $47.0\,cm^3$ of the dichromate solution were needed to oxidise $25.0\,cm^3$ of ethanedioate solution of concentration $0.0925\,mol\,dm^{-3}$, what is the concentration of the potassium dichromate solution?

15. $2.4680\,g$ of sodium ethanedioate are dissolved in water and made up to $250\,cm^3$ of solution. When a $25.0\,cm^3$ portion of the solution is titrated against cerium(IV) sulphate, $35.7\,cm^3$ of the cerium(IV) sulphate solution are required. What is its concentration?

16. A $25.0\,cm^3$ aliquot of a solution containing Fe^{2+} ions and Fe^{3+} ions was acidified and titrated against potassium manganate(VII) solution. $15.0\,cm^3$ of a $0.0200\,mol\,dm^{-3}$ solution of potassium manganate(VII) were required. A second $25.0\,cm^3$ aliquot was reduced with zinc and titrated against the same manganate(VII) solution. $19.0\,cm^3$ of the oxidant solution were required. Calculate the concentrations of a) Fe^{2+}, and b) Fe^{3+} in the solution.

17. a) What volume of acidified potassium manganate(VII) of concentration $0.0200\,mol\,dm^{-3}$ is decolourised by $100\,cm^3$ of hydrogen peroxide of concentration $0.0100\,mol\,dm^{-3}$?

    b) What volume of oxygen is evolved at s.t.p.?

18. A $0.6125\,g$ sample of potassium iodate(V), KIO_3, is dissolved in water and made up to $250\,cm^3$. A $25.0\,cm^3$ aliquot of the solution is added to an excess of potassium iodide in acid solution. The iodine formed requires $22.5\,cm^3$ of sodium thiosulphate solution for titration. What is the concentration of the thiosulphate solution?

19. $25.0\,cm^3$ of a solution of X_2O_5 of concentration $0.100\,mol\,dm^{-3}$ is reduced by sulphur dioxide to a lower oxidation state. To reoxidise X to its original oxidation number required $50.0\,cm^3$ of $0.0200\,mol\,dm^{-3}$ potassium manganate(VII) solution. To what oxidation number was X reduced by sulphur dioxide?

20. Manganese(II) sulphate is oxidised to manganese(IV) oxide by potassium manganate(VII) in acid solution. A flocculant is added to settle the solid MnO_2 so that it does not obscure the colour of the manganate(VII). If $25.0\,cm^3$ of manganese(II) sulphate solution require $22.5\,cm^3$ of $0.0200\,mol\,dm^{-3}$ potassium manganate(VII) solution, what is the concentration of $MnSO_4$?

*21. A solution of hydroxylamine hydrochloride contains $0.1240\,g$ of $NH_2OH \cdot HCl$. On boiling, it is oxidised by an excess of acidified iron(III) sulphate. The iron formed is titrated against potassium manganate(VII) solution of concentration $0.0160\,mol\,dm^{-3}$. A volume of $44.6\,cm^3$ of the oxidant is required.

    a) Find the ratio of moles NH_2OH : moles Fe^{3+}.
    b) State the change in oxidation number of Fe.
    c) State the oxidation number of N in NH_2OH.

d) Deduce the oxidation number of N in the product of the reaction.
e) Decide what compound of nitrogen in this oxidation state is likely to be formed in the reaction.
f) Write the equation for the reaction.

22. A piece of impure copper was allowed to react with dilute nitric acid. The copper(II) nitrate solution formed liberated iodine from an excess of potassium iodide solution. The iodine was estimated by titration with a solution of sodium thiosulphate. If a 0.877 g sample of copper was used, and the volume required was 23.7 cm^3 of 0.480 mol dm^{-3} thiosulphate solution, what is the percentage of copper in the sample?

23. A household bleach contains sodium chlorate(I), NaOCl. The chlorate(I) ion will react with potassium iodide to give iodine, which can be estimated with a standard thiosulphate solution.

a) Write the equations for the reaction of ClO$^-$ and I$^-$ to give I$_2$ and for the reaction of iodine and thiosulphate ions.

b) A 25.0 cm^3 sample of household bleach is diluted to 250 cm^3. A 25.0 cm^3 portion of the solution is added to an excess of potassium iodide solution and titrated against 0.200 mol dm^{-3} sodium thiosulphate solution. The volume required is 18.5 cm^3. What is the concentration of sodium chlorate(I) in the bleach?

COMPLEXOMETRIC TITRATIONS

The complexes formed by a number of metal ions with

bis[bis(carboxymethyl)amino]ethane,
$(HO_2CCH_2)_2NCH_2CH_2N(CH_2CO_2H)_2$,

which is usually referred to as edta (short for its old name) are very stable, and can be used for the estimation of metal ions by titration. The end-point in the titration is shown by an indicator which forms a coloured complex with the metal ion being titrated. If Eriochrome Black T is used as indicator, the metal-indicator colour of red is seen at the beginning of the titration. As the titrant is added, the metal ions are removed from the indicator and complex with edta. At the end-point, the colour of the free indicator, blue, is seen:

Metal-indicator (red) + edta $\longrightarrow$ Metal-edta + Indicator (blue)

EXAMPLE *Determination of the hardness of tap water*
Hardness in water is caused by the presence of calcium ions and magnesium ions. Both these ions complex strongly with edta. The amounts of temporary hardness and permanent hardness can be determined separately by performing complexometric titrations on tap water and boiled tap water. 100 cm^3 of tap water are measured into a flask. An alkaline buffer and Eriochrome Black T are added, and the solution is titrated against 0.100 mol dm^{-3} edta solution. The volume required is 2.10 cm^3.

A second $100\,cm^3$ of tap water are measured into a $250\,cm^3$ beaker, and boiled for 30 minutes. After cooling, the water is filtered into a $100\,cm^3$ graduated flask, and made up to the mark by the addition of distilled water. On titration as before, the volume of edta needed is $1.25\,cm^3$. Calculate the concentration of calcium and magnesium present as permanent hardness and the concentration of calcium and magnesium present as temporary hardness.

METHOD   Total hardness requires $2.10\,cm^3$ of $0.100\,mol\,dm^{-3}$ edta

Permanent hardness requires $1.25\,cm^3$ of $0.100\,mol\,dm^{-3}$ edta

Temporary hardness requires $0.85\,cm^3$ of $0.100\,mol\,dm^{-3}$ edta

Moles of metal as permanent hardness

$$= 1.25 \times 10^{-3} \times 0.100\,mol$$

$$= 0.125 \times 10^{-3}\,mol\ in\ 100\,cm^3\ water$$

$$= 1.25 \times 10^{-3}\,mol\,dm^{-3}$$

Moles of metal as temporary hardness

$$= 0.85 \times 10^{-3} \times 0.100\,mol$$

$$= 0.085 \times 10^{-3}\,mol\ in\ 100\,cm^3\ water$$

$$= 0.85 \times 10^{-3}\,mol\,dm^{-3}$$

ANSWER   The concentration of calcium and magnesium present as temporary hardness is $8.5 \times 10^{-4}\,mol\,dm^{-3}$; the concentration of calcium and magnesium present as permanent hardness is $1.25 \times 10^{-3}\,mol\,dm^{-3}$.

EXERCISE 13   Problems on Complexometric Titrations

1. Calculate the concentration of a solution of zinc sulphate from the following data. $25.0\,cm^3$ of the solution, when added to an alkaline buffer and Eriochrome Black T indicator, required $22.3\,cm^3$ of a $1.05 \times 10^{-2}\,mol\,dm^{-3}$ solution of edta for titration. The equation for the reaction can be represented as

$$Zn^{2+}(aq)\ +\ edta^{4-}(aq)\ \longrightarrow\ Znedta^{2-}(aq)$$

2. To a $50.0\,cm^3$ sample of tap water were added a buffer and a few drops of Eriochrome Black T. On titration against a $0.0100\,mol\,dm^{-3}$ solution of edta, the indicator turned blue after the addition of $9.80\,cm^3$ of the titrant. Calculate the hardness of water in parts per million of calcium, assuming that the hardness is entirely due to the presence of calcium salts. (1 p.p.m. $=$ 1 g in $10^6\,g$ water.)

3. A $0.2500\,g$ sample of a mixture of magnesium oxide and calcium oxide was dissolved in dilute nitric acid and made up to $1.00\,dm^3$ of solution with distilled water. A $50.0\,cm^3$ portion was buffered and, after addition of indicator, was titrated against $0.0100\,mol\,dm^{-3}$ edta solution. $25.8\,cm^3$ of the titrant were required. Find the percentage by mass of calcium oxide and magnesium oxide in the mixture.

4. Find n in the formula $Al_2(SO_4)_3 \cdot nH_2O$ from the following analysis. 2.000 g of aluminium sulphate hydrate were weighed out and made up to $250\ cm^3$. A $25.0\ cm^3$ portion was allowed to complex with edta by being boiled with $50.0\ cm^3$ of edta solution of concentration $1.00 \times 10^{-2}\ mol\ dm^{-3}$. The excess of edta was determined by adding Eriochrome Black T and titrating against a solution of 1.115×10^{-2} $mol\ dm^{-3}$ solution of zinc sulphate. $17.9\ cm^3$ of zinc sulphate solution were required to turn the indicator from blue to red. The reactions taking place are

$$Al^{3+}(aq)\ +\ edta^{4-}(aq)\ \longrightarrow\ Aledta^{-}(aq)$$
$$Zn^{2+}(aq)\ +\ edta^{4-}(aq)\ \longrightarrow\ Znedta^{2-}(aq)$$

PRECIPITATION TITRATIONS

In a precipitation titration, the two solutions react to form a precipitate of an insoluble salt. A solution of a chloride can be estimated by finding the volume of a standard solution of silver nitrate that will precipitate all the chloride ions as insoluble silver chloride:

$$Ag^{+}(aq)\ +\ Cl^{-}(aq)\ \longrightarrow\ AgCl(s)$$

Bromides and iodides and thiocyanates can be titrated in the same way. There are various ways of finding out when the end-point has been reached.

EXAMPLE 1 *Determination of chlorides*
$25.0\ cm^3$ of a sodium chloride solution required $18.7\ cm^3$ of $0.100\ mol$ dm^{-3} silver nitrate solution for complete precipitation. Calculate the concentration of the sodium chloride solution.

METHOD Since $\qquad Ag^{+}(aq)\ +\ Cl^{-}(aq)\ \longrightarrow\ AgCl(s)$

No. of moles of $AgNO_3$ = No. of moles of Cl^-

No. of moles of $AgNO_3$ in volume used = $18.7 \times 10^{-3} \times 0.100$
$$= 1.87 \times 10^{-3}\ mol$$

No. of moles of Cl^- in $25.0\ cm^3$ = $1.87 \times 10^{-3}\ mol$

Concn of Cl^- = $(1.87 \times 10^{-3})/(25.0 \times 10^{-3})$ = $7.50 \times 10^{-2}\ mol\ dm^{-3}$

ANSWER The concentration of sodium chloride is $7.50 \times 10^{-2}\ mol\ dm^{-3}$.

EXAMPLE 2 *Determination of a mixture of halides*
2.95 g of a mixture of potassium chloride and potassium bromide is dissolved in water, and the solution is made up to $250\ cm^3$. $25.0\ cm^3$ of this solution required $31.5\ cm^3$ of $0.100\ mol\ dm^{-3}$ silver nitrate solution. Calculate the percentages of potassium chloride and potassium bromide in the mixture.

METHOD   Let x grams = Mass of potassium chloride

Then $(2.95 - x)$ grams = Mass of potassium bromide

No. of moles of KCl in $25.0 \, cm^3$ = $\dfrac{x}{74.5} \times \dfrac{25.0}{250} = \dfrac{0.1x}{74.5}$ mol

The KCl in $25.0 \, cm^3$ requires $\dfrac{x}{74.5}$ dm^3 of $0.100 \, mol \, dm^{-3} \, AgNO_3(aq)$

No. of moles of KBr in $25.0 \, cm^3$ = $\dfrac{(2.95 - x)}{119} \times \dfrac{25.0}{250}$

$$= \dfrac{0.100(2.95 - x)}{119}$$

The KBr in $25.0 \, cm^3$ requires $\dfrac{(2.95 - x)}{119}$ dm^3 of $0.100 \, mol \, dm^{-3} \, AgNO_3(aq)$

Therefore, $\dfrac{x}{74.5} + \dfrac{(2.95 - x)}{119} = 31.5 \times 10^{-5}$

$$x = 1.34$$

ANSWER   The mass of potassium chloride is $1.34 \, g$; the mass of potassium bromide is $1.61 \, g$.

EXERCISE 14   Problems on Precipitation Reactions

1. A $25.0 \, cm^3$ portion of a solution of potassium chloride required $18.5 \, cm^3$ of a silver nitrate solution of concentration $0.0200 \, mol \, dm^{-3}$ for titration. What is the concentration of the KCl solution?

2. A solid mixture contains sodium chloride and sodium nitrate. A $0.5800 \, g$ sample of the mixture was dissolved in water and made up to $250 \, cm^3$. A $25.0 \, cm^3$ portion was titrated against silver nitrate solution of concentration $0.0180 \, mol \, dm^{-3}$. The volume required was $27.4 \, cm^3$. Calculate the percentage by mass of sodium chloride in the mixture.

3. $1.2400 \, g$ of a mixture of sodium chloride and sodium bromide was dissolved and made up to $1.00 \, dm^3$. A $25.0 \, cm^3$ portion of this solution required $21.7 \, cm^3$ of a silver nitrate solution of concentration $0.0175 \, mol \, dm^{-3}$ for titration. Calculate the percentage composition by mass of the mixture.

4. $25.0 \, cm^3$ of a solution of potassium cyanide required $19.8 \, cm^3$ of a solution of silver nitrate of concentration $0.0350 \, mol \, dm^{-3}$ for titration. Calculate the concentration of the KCN solution.

5. A solution contains sodium chloride and hydrochloric acid. A 25.0 cm^3 aliquot required 38.2 cm^3 of a 0.0325 mol dm^{-3} solution of silver nitrate for titration. A second 25.0 cm^3 aliquot required 7.2 cm^3 of a 0.0550 mol dm^{-3} solution of sodium hydroxide for neutralisation. Calculate the concentrations of a) sodium chloride, and b) hydrochloric acid in the solution.

6. Find the percentage by mass of silver in an alloy from the following information. A sample of 1.245 g of the alloy was dissolved in dilute nitric acid and made up to 250 cm^3. A 25.0 cm^3 portion required 29.8 cm^3 of a 0.0214 mol dm^{-3} solution of potassium thiocyanate for titration.

EXERCISE 15   Questions from A-level Papers

1. A sample of xenon tetrafluoride reacts with iodide ion as shown in the equation

$$XeF_4 + 4I^- \longrightarrow 2I_2 + Xe + 4F^-$$

The liberated iodine from such a reaction was reduced by 40.0 cm^3 of sodium thiosulphate(VI) solution of concentration 0.01 mol/l. Calculate the volume at s.t.p. of xenon which was liberated in the reaction.   (AEB78, p)

2. The following are the results of titrating weighed aliquots of a metal chloride with 0.1 M silver nitrate solution:

	Mass taken/g	Titre/cm^3
First titration	0.190	27.0
Second titration	0.200	30.0
Third titration	0.180	27.0

a) Which of the titrations are concordant?
b) What is the percentage of chloride ion in the salt?   (JMB77, p)

3. Sulphur dichloride dioxide, SO_2Cl_2, reacts with water to give a mixture of sulphuric acid and hydrochloric acid. How many moles of sodium hydroxide, NaOH, would be needed to neutralise the solution formed by adding 1 mol of SO_2Cl_2 to an excess of water?

a 1   b 2   c 3   d 4   e 6   (C79)

4. Briefly describe the preparation of a) chlorine, and b) iodine.

Suggest a reason why the same method is not used for the preparation of both halogens.

When an excess of liquid chlorine is added to iodine a reaction occurs. When the unreacted chlorine is allowed to evaporate an orange solid of formula ICl_x remains. One mole of ICl_x reacts with excess potassium iodide solution to liberate two moles of iodine (I_2).

Write an equation for the reaction between ICl_x and iodide ions.

What is the oxidation state of the iodine in ICl_x?   (O & C79)

5. White phosphorus reacts with dilute aqueous solutions of copper(II) sulphate to deposit metallic copper and produce a strongly acidic solution.

   In an experiment to investigate this reaction, 0.31 g of white phosphorus reacted in excess aqueous copper(II) sulphate giving 1.60 g of metallic copper (P = 31, Cu = 64).

   a) i) Calculate the number of moles of phosphorus atoms used.
      ii) Calculate the number of moles of copper produced.
     iii) Hence calculate the number of moles of copper deposited by one mole of phosphorus atoms.

   b) i) State the change in oxidation number of the copper in this reaction.
      ii) Calculate the new oxidation number of the phosphorus after the reaction.
     iii) In the reaction, the phosphorus forms an acid, HPO_n. What is the value of n?

   c) Now write a balanced equation showing the action of white phosphorus on copper(II) sulphate in the presence of water.

   d) i) In performing this experiment, what practical difficulty would you expect in weighing the piece of phosphorus?
      ii) Describe how you would try to overcome this difficulty.

   (L76, N)

6. The reaction of uranyl(VI) methanoate (uranyl(VI) formate) with excess ethanedioic acid (oxalic acid) produces the salt uranyl(IV) ethanedioate hexahydrate, $UO(C_2O_4) \cdot 6H_2O$.

   a) When 585 mg of the salt $UO(C_2O_4) \cdot 6H_2O$ was left in a vacuum desiccator for forty-eight hours, the mass changed to 535 mg. What formula would you predict for the resulting substance?

   b) When 300 mg of the salt $UO(C_2O_4) \cdot 6H_2O$ was dissolved in hot dilute sulphuric acid and titrated with 0.05 M potassium manganate(VII) ($KMnO_4$) solution, the end point was obtained after the addition of 10.7 cm³ of the $KMnO_4$ solution. What change, if any, is there in the oxidation number of the uranium?

      The half-equations for the reduction of manganate(VII) ion and the oxidation of the ethanedioate ion are

   $$MnO_4^-(aq) + 8H^+(aq) + 5e^- \longrightarrow Mn^{2+}(aq) + 4H_2O(l)$$
   $$C_2O_4^{2-}(aq) \longrightarrow 2CO_2(g) + 2e^-$$

   (L77, N)

7. Manganate(VII) ions, MnO_4^-, are reduced in acid conditions to manganese(II) ions, Mn^{2+}, whereas they are reduced in neutral conditions to manganese(IV) oxide, MnO_2. The oxidation of 25 cm³ of

a solution X containing iron(II) ions required, in acid conditions, $20 \, cm^3$ of a solution Y containing manganate(VII) ions. What volume of solution Y would be required to oxidise $25 \, cm^3$ of solution X in neutral conditions?

a $20 \times 5/3 \, cm^3$ b $20 \times 7/4 \, cm^3$ c $20 \times 5/4 \, cm^3$

d $20 \times 3/5 \, cm^3$ e $20 \times 4/7 \, cm^3$ (C79)

*8. The following is a method by which the reaction between iron(III) ions and hydroxylammonium chloride, $NH_3OH^+Cl^-$, may be investigated.

25.0 cm^3 of a solution containing $3.60 \, g \, dm^{-3}$ of hydroxylammonium chloride is added to a solution containing an excess of Fe^{3+} ions and about $25 \, cm^3$ of 1 M sulphuric acid, and the mixture boiled. It is then diluted with water, allowed to cool, and the Fe^{2+} ions titrated with 0.02 M potassium manganate(VII) (potassium permanganate) of which $25.9 \, cm^3$ were required.

a) Calculate the molar ratio Fe^{3+}/NH_3OH^+ in the reaction, and hence determine the oxidation number of nitrogen in the product.

b) Using the oxidation number concept, or otherwise, deduce the equation for the reaction (H $= 1$, N $= 14$, O $= 16$, Cl $= 35.5$).
                                                                                        (L80, S)

9. a) Explain what is meant by the italicised terms in the statement: 'the *oxidation state* of chromium is three in the *6-coordinated complex ion* $[Cr(OH_2)_6]^{3+}$'.

b) A compound, Y, of chromium(III) is believed to contain the complex species $[Cr(OH_2)_x Cl_{6-x}]^z$. A sample of Y was boiled with sodium hydroxide solution and the resulting liquid was filtered. After acidification with nitric acid, the filtrate required $10.0 \, cm^3$ of silver nitrate solution of concentration $0.100 \, mol \, dm^{-3}$ to precipitate all of the chloride present.

The precipitate of chromium(III) hydroxide, left after the reaction with sodium hydroxide solution, dissolved in a boiling alkaline solution of hydrogen peroxide to give a yellow solution; after removal of the excess of hydrogen peroxide, the yellow solution became orange in colour when it was acidified. The iodine liberated when potassium iodide was added to the acidified solution required $30.0 \, cm^3$ of sodium thiosulphate(VI) solution of concentration $0.100 \, mol \, dm^{-3}$ for titration.

i) What reactions occurred with alkaline hydrogen peroxide and subsequently with the acid?

ii) What is the molar ratio chloride : chromium in the complex?

iii) What is the charge, z, on the complex species?

$[I_2 + 2S_2O_3^{2-} \longrightarrow S_4O_6^{2-} + 2I^-$; any other ions present in Y may be assumed not to affect the titrations.]   (O & C78)

*10. Sulphur dioxide was passed through $100 \, cm^3$ of solution A of potassium chlorate(V) until reduction was complete according to the equation

$$KClO_3 + 3SO_2 + 3H_2O \longrightarrow KCl + 3H_2SO_4$$

The solution was then boiled and treated with a slight excess of silver nitrate solution. The precipitate of silver chloride was collected and found to weigh $0.3010 \, g$. Exactly $25 \, cm^3$ of a solution B of iron(II) sulphate in dilute sulphuric acid were titrated with a solution of $0.016 \, 67 \, mol \, 1^{-1}$ potassium dichromate(VI); $44.00 \, cm^3$ of dichromate(VI) were required.

$$K_2Cr_2O_7 + 6FeSO_4 + 7H_2SO_4 \longrightarrow K_2SO_4 + Cr_2(SO_4)_3 + 3Fe_2(SO_4)_3 + 7H_2O$$

Calculate the concentrations in $mol \, 1^{-1}$ of solutions A and B.

$25 \, cm^3$ of solution A were then mixed with $25 \, cm^3$ of solution B and boiled for ten minutes.

To find the iron(II) sulphate remaining the mixture was then titrated with the $0.016 \, 67 \, mol \, 1^{-1}$ potassium dichromate(VI) solution used above. $12.50 \, cm^3$ were required.

Calculate the number of moles of Fe^{2+} which have reacted with one mole of ClO_3^-.

Hence suggest a balanced ionic equation for the reaction of the type

$$ClO_3^- + xFe^{2+} + yH^+ \longrightarrow \qquad \text{(JMB80, S)}$$

4   The Atom

MASS SPECTROMETRY

In a mass spectrometer, an element or compound is vaporised and then ionised. The ions are accelerated, collimated into a beam and deflected by a magnetic field. The amount of the deflection depends on the ratio of mass/charge of the ions, as well as the values of the accelerating voltage and the magnetic field. The magnetic field is kept constant while the accelerating voltage is varied continuously to focus each species in turn into the ion detector. The detector records each species as a peak on a trace. From the value of the voltage associated with a particular peak the ratio of mass/charge for that ionic species can be found. Since each ion has a charge of $+1$, the ratio mass/charge is equal to the mass of the ion. The mass spectrometer can be calibrated to read out ionic masses directly. The heights of the peaks are proportional to the relative abundance of the different ions.

EXAMPLE 1   The mass spectrum of boron shows two peaks, one at $10.0\,m_u$, and the other at $11.0\,m_u$. The heights of the peaks are in the ratio $18.7\% : 81.3\%$. Calculate the relative atomic mass of boron.

METHOD   The relative heights of the peaks show that the relative abundance of ^{10}B and ^{11}B is 18.7% $^{10}B : 81.3\%$ ^{11}B.

In 1000 atoms, there are 187 of mass $10.0\,m_u$ $=$ $1870\,m_u$

and 813 of mass $11.0\,m_u$ $=$ $8943\,m_u$

The mass of 1000 atoms $=$ $10\,813\,m_u$

The average atomic mass $=$ $10.8\,m_u$

ANSWER   The relative atomic mass of boron is 10.8.

EXAMPLE 2   The mass spectrum of neon shows three peaks, corresponding to masses of 22, 21 and $20\,m_u$. The heights of the peaks are in the ratio $11.2 : 0.2 : 114$. Calculate the average atomic mass of neon.

METHOD   Multiplying the relative abundance (the height of the peak) by the mass to find the total mass of each isotope present gives

Mass of neon-22 $=$ 11.2×22.0 $=$ $246.4\,m_u$

Mass of neon-21 $=$ 0.2×21.0 $=$ $4.2\,m_u$

Mass of neon-20 $=$ 114×20.0 $=$ $2280\,m_u$

Totals are 125.4 $=$ $2530.6\,m_u$

Average mass of neon atom $=$ $2530.6/125.4$ $=$ $20.18\,m_u$.

ANSWER   The average atomic mass of neon is $20.2\,m_u$.

EXERCISE 16   Problems on Mass Spectrometry

1. The mass spectrum of rubidium consists of a peak at mass 85 and a peak at mass $87 m_u$. The relative abundance of the isotopes is 72 : 28. Calculate the mean atomic mass of rubidium.

2. If ^{69}Ga and ^{71}Ga occur in the proportions 60 : 40, calculate the average atomic mass of gallium.

3. Fig. 4.1 shows the mass spectrum of magnesium. The heights of the three peaks and the mass numbers of the isotopes are shown in Fig. 4.1. Calculate the relative atomic mass of magnesium.

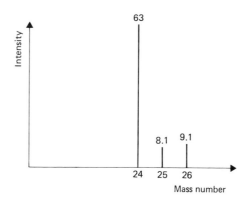

Fig. 4.1   Mass spectrum of magnesium

4. The mass spectrum of chlorine shows peaks at masses 70, 72 and $74 m_u$. The heights of the peaks are in the ratio of 9 to 6 to 1. What is the relative abundance of ^{35}Cl and ^{37}Cl? What is the average atomic mass of chlorine?

5. Calculate the relative atomic mass of lithium, which consists of 7.4% of ^6Li and 92.6% of ^7Li.

6. A sample of water containing ^1H, ^2H and ^{16}O was analysed in a mass spectrometer. The trace showed peaks at mass numbers 1, 2, 3, 4, 17, 18, 19 and 20. Suggest which ions are responsible for these peaks.

7. Calculate the average atomic mass of potassium, which consists of 93% ^{39}K and 7.0% ^{41}K.

8. Fig. 4.2 shows a mass spectrometer trace for copper nitrate. Each of the eight peaks is produced by a different species of ion. Suggest what these ions are.

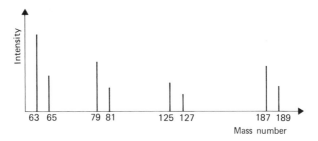

Fig. 4.2 The mass spectrum of copper nitrate

EXERCISE 17   Questions from A-level Papers

1. a) The mass spectrum of an element, which is diatomic in the vapour state, gave peaks which corresponded to masses of 158, 160 and 162. What deduction may be made about this element if there were no other peaks in the vicinity of the above numbers?

   b) The heights of the peaks in a) were in the ratio of 1 : 2 : 1. What further deductions can be made?   (O76)

2. Calculate, showing your reasoning, the relative atomic mass of chlorine to three significant figures given that the element consists of two isotopes with relative atomic masses of 34.98 and 36.98 and relative abundances of 75.4% and 24.6%.   (O75)

3. Copper (atomic number 29) has two isotopes, the first of relative atomic mass 62.9 and abundance 65%, the second of relative atomic mass 64.9 and abundance 35%.

   a) What do you understand by: i) atomic number, and ii) isotope?

   b) Calculate the mean relative atomic mass of naturally-occurring copper.   (WJEC78, p)

4. The mass spectrum of neon consists of two major peaks, one at mass 20 and the other at mass 22, of relative abundance 10 : 1. Show how these data can be used to determine the relative atomic mass of neon.

   Interpret as fully as you can the mass spectrum of sulphur vapour shown in the figure below.   (C80, p)

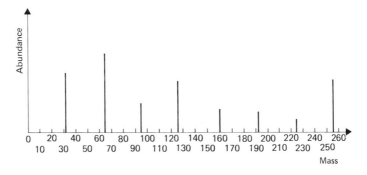

NUCLEAR REACTIONS

In a nuclear reaction, a rearrangement of the protons and neutrons in the nuclei of the atoms takes place, and new elements are formed.

The atomic number or proton number, Z, of an element is the number of protons in the nucleus of an atom of the element. The mass number or nucleon number, A, is the number of protons and neutrons in the nucleus of an atom. Isotopes of an element differ in mass number but have the same atomic number. Isotopes are represented as ^A_Z Symbol, e.g. $^{12}_6C$. Protons are represented as ^1_1H, electrons (β-particles) as $^0_{-1}e$, neutrons as ^1_0n, and α-particles as ^4_2He. In the equation for a nuclear reaction, the sum of the mass numbers is the same on both sides, and the sum of the atomic numbers is the same on both sides of the equation.

For practice in balancing nuclear equations, study the following examples.

EXAMPLE 1 Complete the equation

$$^{16}_7N \longrightarrow {}^a_bO + {}^0_{-1}e$$

METHOD Consider mass numbers: $16 = a + 0 \quad \therefore a = 16$

Consider atomic numbers: $7 = b + (-1) \quad \therefore b = 8$

ANSWER $^{16}_7N \longrightarrow {}^{16}_8O + {}^0_{-1}e$

EXAMPLE 2 Find the values of a and b in the equation

$$^{27}_{13}Al + {}^1_0n \longrightarrow {}^4_2He + {}^a_bX$$

METHOD Consider mass numbers: $27 + 1 = 4 + a \quad \therefore a = 24$

Consider atomic numbers: $13 + 0 = 2 + b \quad \therefore b = 11$

ANSWER $a = 24$ and $b = 11$.

EXERCISE 18   Problems on Nuclear Reactions

Complete the following equations, supplying values for the missing mass numbers (nucleon numbers) and atomic numbers (proton numbers).

1. a) $^9_4Be + \gamma \longrightarrow {}^8_4Be + {}^a_bX$

   b) $^{14}_7N + {}^4_2He \longrightarrow {}^1_1H + {}^a_bY$

   c) $^9_4Be + {}^1_1H \longrightarrow {}^6_3Li + {}^a_bZ$

   d) $^{209}_{83}Bi + {}^2_1D \longrightarrow {}^a_bX + {}^1_1H$

   e) $^{16}_8O + {}^1_0n \longrightarrow {}^{13}_6C + {}^a_bY$

   f) $^{10}_5B + {}^a_bY \longrightarrow {}^{13}_7N + {}^1_0n$

   g) $^{14}_7N + {}^1_0n \longrightarrow {}^a_bQ + {}^1_1H$

h) $^{19}_{9}F + ^{1}_{0}n \longrightarrow ^{a}_{b}Z + ^{4}_{2}He$

i) $^{207}_{82}Pb \longrightarrow ^{a}_{b}X + ^{0}_{-1}e$

j) $^{27}_{13}Al + ^{1}_{0}n \longrightarrow ^{24}_{b}Y + ^{a}_{2}Z$

k) $^{35}_{17}Cl + ^{p}_{q}X \longrightarrow ^{r}_{16}S + ^{1}_{1}H$

l) $^{6}_{3}Li + ^{1}_{0}n \longrightarrow ^{4}_{2}He + ^{c}_{d}X$

EXERCISE 19   Questions from A-level Papers

1. What are a, b, and Z in the following nuclear equation?

$$^{27}_{13}Al + ^{1}_{0}n = ^{4}_{2}He + ^{a}_{b}Z$$ (O79)

2. Complete the following nuclear equations, adding all the missing mass numbers, atomic number and symbols:

   a) $^{27}Al + ^{1}_{1}H \longrightarrow ^{24}Mg + $ _____

   b) $^{32}S + ^{1}_{0}n \longrightarrow ^{1}_{1}H + $ _____   (C78)

3. a) What are:
   i)   α-radiation,
   ii) β-radiation,
   iii) the half-life of a radioactive element?

   b) In the radioactive series

$$^{212}_{82}Pb \xrightarrow{(1)} ^{212}_{83}X \xrightarrow{(2)} ^{208}_{81}Y \xrightarrow[\text{emission}]{\beta} Z$$

   identify the emitted particles (1) and (2) and give the periodic groups in which Pb, X, Y and Z occur. (X, Y and Z are not chemical symbols.) Give the action of water on the chlorides of Pb and X.   (WJEC78)

4. a) What are α, β and γ emissions and how do they differ in their penetrating power and their behaviour in a magnetic field? Explain the meaning of the two numbers before the symbol for uranium and identify P, Q, R and S in the following equations.

$$^{234}_{92}U \longrightarrow \alpha + P$$
$$^{239}_{92}U \longrightarrow \beta + Q$$
$$^{235}_{92}U \longrightarrow \gamma + R$$
$$^{238}_{92}U + ^{2}_{1}H \longrightarrow ^{239}_{92}U + S$$

   b) Deduce the nature of X in the following reaction.

$$^{235}_{92}U + ^{1}_{0}n \longrightarrow ^{95}_{42}Mo + ^{139}_{57}La + 2^{1}_{0}n + 7X$$

   There is approximately 0.1 per cent less mass on the right-hand side of the equation than on the left-hand side. What is the significance of this?   (C77)

5　Gases

THE GAS LAWS

The behaviour of gases is described by the Gas Laws: Boyle's law, Charles' law, the equation of state for an ideal gas, Graham's law of diffusion, Gay-Lussac's law, Avogadro's law, Dalton's law of partial pressures, Henry's law of the solubility of gases and the ideal gas equation. We look at each of these in turn.

BOYLE'S LAW

Boyle's law states that the pressure of a fixed mass of gas at a constant temperature is inversely proportional to its volume:

$$PV = \text{Constant}$$

where P = pressure, V = volume.

CHARLES' LAW

Charles' law states that the volume of a fixed mass of gas at constant pressure is directly proportional to its temperature on the Kelvin scale:

$$\frac{V}{T} = \text{Constant}$$

where T = temperature in kelvins.

Temperature on the Kelvin scale (called the absolute temperature) is obtained by adding 273 to the temperature on the Celsius scale.

$$\text{Temperature (K)} = \text{Temperature (}^{\circ}\text{C)} + 273$$

$$273\,\text{K} = 0\,^{\circ}\text{C}$$

THE EQUATION OF STATE FOR AN IDEAL GAS

Gases which obey Boyle's law and Charles' law are called *ideal* gases. By combining these two laws, the following equation can be obtained. It is called the *equation of state for an ideal gas*:

$$\frac{P_1 V_1}{T_1} = \frac{P_2 V_2}{T_2}$$

A gas has a volume of V_1 at a temperature T_1 and pressure P_1. If the conditions are changed to a pressure P_2 and a temperature T_2, the new volume can be calculated from the equation. It is usual to compare gas volumes at $0\,°C$ and 1 atmosphere (abbreviated as 1 atm). These conditions are referred to as standard temperature and pressure (s.t.p.) or normal temperature and pressure (n.t.p.). Some authors calculate volumes at room temperature ($20\,°C$) and 1 atm. The SI unit of pressure is the pascal.

$$1\,\text{atm} \; = \; 1.01 \times 10^5\,\text{pascals (Pa)} \; = \; 101\,\text{kilopascals (kPa)}$$

$$= \; 1.01 \times 10^5\,\text{newtons per square metre (N m}^{-2})$$

$$= \; 760\,\text{mm mercury}$$

Volumes can be measured in the SI unit, the cubic metre (m^3) or in cubic decimetres (dm^3) or cubic centimetres (cm^3).

$$10^3\,\text{cm}^3 \; = \; 1\,\text{dm}^3 \; = \; 10^{-3}\,\text{m}^3$$

Temperatures must be in kelvins.

EXAMPLE A volume of gas, $265\,\text{cm}^3$, is collected at $70\,°C$ and $1.05 \times 10^5\,\text{N m}^{-2}$. What volume would the gas occupy at s.t.p.?

METHOD 1 The experimental conditions are

$$P_1 \; = \; 1.05 \times 10^5\,\text{N m}^{-2}$$

$$T_1 \; = \; 273 + 70 \; = \; 343\,\text{K}$$

$$V_1 \; = \; 265\,\text{cm}^3$$

Standard conditions are
$$P_2 \; = \; 1.01 \times 10^5\,\text{N m}^{-2}$$

$$T_2 \; = \; 273\,\text{K}$$

$$\frac{P_1 V_1}{T_1} \; = \; \frac{P_2 V_2}{T_2}$$

$$\frac{1.05 \times 10^5 \times 265}{343} \; = \; \frac{1.01 \times 10^5 \times V_2}{273}$$

$$V_2 \; = \; 219\,\text{cm}^3$$

ANSWER The volume of gas at s.t.p. would be $219\,\text{cm}^3$.

(*Note* that the pressure and volume are in the same units on both sides of the equation.)

METHOD 2 Some students prefer to tackle this type of calculation in a slightly different manner.

First, consider the effect of the change in pressure.

Since the pressure decreases from $1.05 \times 10^5 \, \text{N m}^{-2}$ to $1.01 \times 10^5 \, \text{N m}^{-2}$, the volume will increase in the same ratio.

$$\text{New volume} = 265 \times \frac{1.05 \times 10^5}{1.01 \times 10^5}$$

Now consider the effect of the change in temperature.

Since the temperature drops from 343 to 273, the volume will decrease by the same ratio.

$$\text{New volume} = 265 \times \frac{1.05 \times 10^5}{1.01 \times 10^5} \times \frac{273}{343} = 219 \, \text{cm}^3 \quad \text{(as before).}$$

EXERCISE 20   Problems on Gas Volumes

1. Correct the following gas volumes to s.t.p.:
   a) $205 \, \text{cm}^3$ at $27 \, ^\circ\text{C}$ and $1.01 \times 10^5 \, \text{N m}^{-2}$
   b) $355 \, \text{cm}^3$ at $310 \, \text{K}$ and $1.25 \times 10^5 \, \text{N m}^{-2}$
   c) $5.60 \, \text{dm}^3$ at $425 \, \text{K}$ and $1.75 \times 10^5 \, \text{N m}^{-2}$
   d) $750 \, \text{cm}^3$ at $308 \, \text{K}$ and $2.00 \times 10^4 \, \text{N m}^{-2}$
   e) $1.25 \, \text{dm}^3$ at $25 \, ^\circ\text{C}$ and $2.14 \times 10^5 \, \text{N m}^{-2}$

2. A certain mass of an ideal gas has a volume of $3.25 \, \text{dm}^3$ at $25 \, ^\circ\text{C}$ and $1.01 \times 10^5 \, \text{N m}^{-2}$. What pressure is required to compress it to $1.88 \, \text{dm}^3$ at the same temperature?

3. An ideal gas occupies a volume $2.00 \, \text{dm}^3$ at $25 \, ^\circ\text{C}$ and $1.01 \times 10^5 \, \text{N m}^{-2}$. What will the volume of gas become at $40 \, ^\circ\text{C}$ and $2.25 \times 10^5 \, \text{N m}^{-2}$?

4. An ideal gas occupies $2.75 \, \text{dm}^3$ at $290 \, \text{K}$ and $8.70 \times 10^4 \, \text{N m}^{-2}$. At what temperature will it occupy $3.95 \, \text{dm}^3$ at $1.01 \times 10^5 \, \text{N m}^{-2}$?

5. An ideal gas occupies $365 \, \text{cm}^3$ at $298 \, \text{K}$ and $1.56 \times 10^5 \, \text{N m}^{-2}$. What will be its volume at $310 \, \text{K}$ and $1.01 \times 10^5 \, \text{N m}^{-2}$?

6. Correct the following gas volumes to s.t.p.:
   a) $256 \, \text{cm}^3$ of an ideal gas measured at $50 \, ^\circ\text{C}$ and $650 \, \text{mm Hg}$
   b) $47.2 \, \text{cm}^3$ of an ideal gas measured at $62 \, ^\circ\text{C}$ and $726 \, \text{mm Hg}$
   c) $10.0 \, \text{dm}^3$ of an ideal gas measured at $200 \, ^\circ\text{C}$ and $850 \, \text{mm Hg}$
   d) $4.25 \, \text{dm}^3$ of an ideal gas measured at $370 \, ^\circ\text{C}$ and $2.12 \, \text{atm}$
   e) $600 \, \text{cm}^3$ of an ideal gas measured at $95 \, ^\circ\text{C}$ and $0.98 \, \text{atm}$

GRAHAM'S LAW OF DIFFUSION

At constant temperature and pressure, the rate of diffusion of a gas is inversely proportional to the square root of its density:

$$r \propto \frac{1}{\sqrt{\rho}}$$

where r = rate of diffusion and ρ = density.

Comparing the rates of diffusion of two gases A and B gives

$$\frac{r_A}{r_B} = \sqrt{\frac{\rho_B}{\rho_A}}$$

This expression applies to rates of effusion (passage through a small aperture) as well as to diffusion (passage from a region of high concentration to a region of low concentration). It provides a method of measuring molar masses. The molar mass of a gas is proportional to its density (see p. 81: Density = Molar mass/Gas molar volume).

Graham's law can therefore be written as

$$\frac{r_A}{r_B} = \sqrt{\frac{M_B}{M_A}}$$

where M_A and M_B are the molar masses of A and B.

EXAMPLE 1 A gas, A, diffuses through a porous plug at a rate of $1.43\,\text{cm}^3\,\text{s}^{-1}$. Carbon dioxide diffuses through the plug at a rate of $0.43\,\text{cm}^3\,\text{s}^{-1}$. Calculate the molar mass of A.

METHOD Molar mass of carbon dioxide $= 44.0\,\text{g mol}^{-1}$

$$\frac{r_{CO_2}}{r_A} = \sqrt{\frac{M_A}{M_{CO_2}}}$$

$$\frac{0.43}{1.43} = \sqrt{\frac{M_A}{44.0}}$$

$$M_A = 4.0$$

ANSWER The molar mass of A is $4.0\,\text{g mol}^{-1}$.

EXAMPLE 2 It takes 54.4 seconds for $100\,\text{cm}^3$ of a gas, X, to effuse through an aperture, and 36.5 seconds for $100\,\text{cm}^3$ of oxygen to effuse through the same aperture. What is the molar mass of X?

METHOD Since

$$\frac{r_{O_2}}{r_X} = \sqrt{\frac{M_X}{M_{O_2}}}$$

$$\frac{100/36.5}{100/54.4} = \sqrt{\frac{M_X}{32}}$$

$$M_X = 71$$

ANSWER The molar mass of X is $71\,\text{g mol}^{-1}$.

EXERCISE 21 Problems on Diffusion and Effusion

1. A certain volume of hydrogen takes 2 min 10 s to diffuse through a porous plug, and an oxide of nitrogen takes 10 min 23 s. What is: a) the molar mass, b) the formula of the oxide of nitrogen?

2. Plugs of cotton wool, one soaked in concentrated ammonia solution and the other soaked in concentrated hydrochloric acid, are inserted into opposite ends of a horizontal glass tube. A disc of solid ammonium chloride forms in the tube. If the tube is 1 m long, how far from the ammonia plug is the solid deposit?

3. A certain volume of sulphur dioxide diffuses through a porous plug in 10.0 min, and the same volume of a second gas takes 15.8 min. Calculate the relative molecular mass of the second gas.

4. Nickel forms a carbonyl, $Ni(CO)_n$. Deduce the value of n from the fact that carbon monoxide diffuses 2.46 times faster than the carbonyl compound.

5. A certain volume of oxygen diffuses through an apparatus in 60.0 seconds. The same volumes of gases A and B, in the same apparatus under the same conditions, diffuse in 15.0 and 73.5 seconds respectively. Gas A is flammable and gas B turns starch-iodide paper blue. Identify A and B.

6. $25 \, cm^3$ of ethane effuses through a small aperture in 40 s. What time is taken by $25 \, cm^3$ of carbon dioxide?

7. $100 \, cm^3$ of oxygen effused in 42 s through a small hole. $100 \, cm^3$ of nitrogen dioxide took 60 s. Calculate the molar mass of nitrogen dioxide. How does your answer compare with the molar mass calculated from the formula? Explain the difference.

8. $200 \, cm^3$ of chlorine diffuse out of a porous container in 2 min 14 s. How long will it take for the same volume of argon to diffuse?

9. Xenon diffuses through a pin-hole at a rate of $2.00 \, cm^3 \, min^{-1}$. At what rate will hydrogen effuse through the same hole at the same temperature and pressure?

10. In 3.00 minutes, $7.50 \, cm^3$ of carbon dioxide effuse through a pinhole. What volume of helium would effuse through the same hole under the same conditions in the same time?

11. In 5.00 minutes, $12.6 \, cm^3$ of a gas, X, diffuse through a porous partition. In the same time, $14.8 \, cm^3$ of oxygen diffuse through the same partition. Calculate the molar mass of X.

12. A mixture of carbon monoxide and carbon dioxide diffuses through a porous diaphragm in one half of the time taken for the same volume of bromine vapour. What is the composition by volume of the mixture?

13. In 4.00 minutes, $16.2 \, cm^3$ of water vapour effuse through a small hole. In the same time, $8.1 \, cm^3$ of a mixture of NO_2 and N_2O_4 effuse through the same hole. Calculate the percentage by volume of NO_2 in the mixture.

GAY-LUSSAC'S LAW

Gay-Lussac's law states that in a reaction between gases, the volumes of the reacting gases, measured at the same temperature and pressure, are in simple ratio to one another and to the volumes of any gaseous products.

The law was stated as a result of observations such as the fact that one volume of oxygen reacts with twice its volume of hydrogen to form two volumes of steam. An explanation of Gay-Lussac's observations was provided by Avogadro.

AVOGADRO'S LAW

Avogadro's law states that equal volumes of gases, measured at the same temperature and pressure, contain the same number of molecules.

Experimental results on reacting volumes of gases can be interpreted using Avogadro's law to give the molecular formulae of gases.

Since

$$\left(\begin{array}{c}\text{1 volume of}\\ \text{hydrogen}\end{array}\right) + \left(\begin{array}{c}\text{1 volume of}\\ \text{chlorine}\end{array}\right) \longrightarrow \left(\begin{array}{c}\text{2 volumes of}\\ \text{hydrogen chloride}\end{array}\right)$$

then

$$\left(\begin{array}{c}\text{1 molecule of}\\ \text{hydrogen}\end{array}\right) + \left(\begin{array}{c}\text{1 molecule of}\\ \text{chlorine}\end{array}\right) \longrightarrow \left(\begin{array}{c}\text{2 molecules of}\\ \text{hydrogen chloride}\end{array}\right)$$

The formation of one molecule of hydrogen chloride can occur if the molecules of hydrogen and chlorine consist of two atoms, so that

$$\tfrac{1}{2}(H_2) + \tfrac{1}{2}(Cl_2) \longrightarrow \text{1 molecule of hydrogen chloride}$$

Hydrogen chloride must therefore have the formula HCl.

EXAMPLE Cyanogen is a compound of carbon and nitrogen. On combustion in excess oxygen, $250\,cm^3$ of cyanogen form $500\,cm^3$ of carbon dioxide and $250\,cm^3$ of nitrogen (measured at the same temperature and pressure). What is the formula of cyanogen?

METHOD Let the formula of cyanogen be C_xN_y.

Then
$$C_xN_y + O_2 \longrightarrow N_2 + CO_2$$
$$250\,cm^3 \qquad\qquad 250\,cm^3 \quad 500\,cm^3$$
$$\text{1 volume} \qquad\qquad \text{1 volume} \quad \text{2 volumes}$$

Therefore,

$$C_xN_y + nO_2 \longrightarrow N_2 + 2CO_2$$

Balancing the C atoms gives $x = 2$, and balancing the N atoms gives $y = 2$.

ANSWER   The formula is C_2N_2.

More examples and problems are given in the section on Reacting Volumes of Gases in Chapter 2, on p. 19 onwards.

THE GAS MOLAR VOLUME

Avogadro's law states that equal volumes of gases, measured at the same temperature and pressure, contain equal numbers of molecules. It follows that the volume occupied by a mole of gas is the same for all gases. It is called the *gas molar volume* and measures $22.4 \, dm^3$ at s.t.p. ($24.0 \, dm^3$ at $20 \, ^\circ C$ and 1 atm).

If the volume occupied by a known mass of gas is known, the molar mass of the gas can be calculated.

EXAMPLE $11.0 \, g$ of a gas occupy $5.60 \, dm^3$ at s.t.p. What is the molar mass of the gas?

METHOD   Mass of $5.60 \, dm^3$ of gas $= 11.0 \, g$
Mass of $22.4 \, dm^3$ of gas $= 11.0 \times 22.4/5.60 = 44.0 \, g$

ANSWER   The molar mass of the gas is $44.0 \, g \, mol^{-1}$.

EXERCISE 22   Problems on Gas Molar Volume

Use $R = 8.314 \, J \, K^{-1} mol^{-1}$; GMV $= 22.41 \, dm^3$ at s.t.p.

1. Calculate the molar mass of a gas which has a density of $1.798 \, g \, dm^{-3}$ at 298 K and $101 \, kN \, m^{-2}$.

2. At 273 K and $1.01 \times 10^5 \, N \, m^{-2}$, $2.965 \, g$ of argon occupy $1.67 \, dm^3$. Calculate the molar mass of the gas.

3. Calculate the volume occupied by $0.250 \, mol$ of an ideal gas at $1.01 \times 10^5 \, N \, m^{-2}$ and $20 \, ^\circ C$.

4. A volume, $500 \, cm^3$ of krypton, measured at $0 \, ^\circ C$ and $9.8 \times 10^4 \, N \, m^{-2}$, has a mass of $1.809 \, g$. Calculate the molar mass of krypton.

5. What amount (number of moles) of an ideal gas occupies $5.80 \, dm^3$ at $2.50 \times 10^5 \, N \, m^{-2}$ and 300 K?

6. Propane has a density of $1.655 \, g \, dm^{-3}$ at 323 K and $1.01 \times 10^5 \, N \, m^{-2}$. Calculate its molar mass.

7. What volume is occupied by $0.250 \, mole$ of an ideal gas at 373 K and $1.25 \times 10^5 \, N \, m^{-2}$?

8. An ideal gas occupies $1.50 \, dm^3$ at $300 \, K$ and $1.25 \times 10^5 \, N \, m^{-2}$. What is the amount (in moles) of gas present?

DALTON'S LAW OF PARTIAL PRESSURES

In a mixture of gases, the total pressure is the sum of the pressures that each of the gases would exert if it alone occupied the same volume as the mixture. The contribution that each gas makes to the total pressure is called the *partial pressure.*

EXAMPLE $3.0 \, dm^3$ of carbon dioxide at a pressure of $200 \, kPa$ and $1.0 \, dm^3$ of nitrogen at a pressure of $300 \, kPa$ are introduced into a $1.5 \, dm^3$ vessel. What is the total pressure in the vessel?

METHOD When the carbon dioxide contracts from $3.0 \, dm^3$ to $1.5 \, dm^3$, the pressure increases from 200 to $200 \times \dfrac{3.0}{1.5} \, kPa$, i.e. $400 \, kPa$. The partial pressure of carbon dioxide in the vessel is $400 \, kPa$.

When the nitrogen expands from $1.0 \, dm^3$ to $1.5 \, dm^3$, the pressure decreases from 300 to $300 \times 1.0/1.5 = 200 \, kPa$. The partial pressure of nitrogen is $200 \, kPa$.

$$\text{Total pressure} = P_{CO_2} + P_{N_2}$$
$$= 400 + 200 = 600 \, kPa$$

ANSWER The total pressure is $600 \, kPa$.

EXERCISE 23   Problems on Partial Pressures of Gases

1. Use the following values of the vapour pressure of water at various temperatures.

Temperature	Vapour Pressure/$N \, m^{-2}$
$15 \, °C$	1.70×10^3
$20 \, °C$	2.33×10^3
$25 \, °C$	3.16×10^3
$30 \, °C$	4.23×10^3

a) $200 \, cm^3$ of oxygen are collected over water at an atmospheric pressure of $9.80 \times 10^4 \, N \, m^{-2}$ and a temperature of $20 \, °C$. What is the partial pressure of the oxygen? What will be its volume at s.t.p.?

b) $250 \, cm^3$ of gas are collected over water at an atmospheric pressure of $9.70 \times 10^4 \, N \, m^{-2}$ and a temperature of $30 \, °C$. What is the partial pressure of the gas? Correct its volume to s.t.p.

c) What is the volume of 1 mole of nitrogen measured over water at an atmospheric pressure of $9.70 \times 10^4 \, N \, m^{-2}$ and a temperature of $25 \, °C$?

2. $2.00 \, dm^3$ of nitrogen at a pressure of $1.01 \times 10^5 \, N \, m^{-2}$ and $5.00 \, dm^3$ of hydrogen at a pressure of $5.05 \times 10^5 \, N \, m^{-2}$ are injected into a $10.0 \, dm^3$ vessel. What is the pressure of the mixture of gases?

3. A mixture of gases at a pressure of $1.01 \times 10^5 \, N \, m^{-2}$ contains 25.0% by volume of oxygen. What is the partial pressure of oxygen in the mixture?

4. Into a $5.00 \, dm^3$ vessel are introduced $2.50 \, dm^3$ of methane at a pressure of $1.01 \times 10^5 \, N \, m^{-2}$, $7.50 \, dm^3$ of ethane at a pressure of $2.525 \times 10^5 \, N \, m^{-2}$ and $0.500 \, dm^3$ of propane at a pressure of $2.02 \times 10^5 \, N \, m^{-2}$. What is the resulting pressure of the mixture?

5. A mixture of gases at a pressure $7.50 \times 10^4 \, N \, m^{-2}$ has the volume composition 40% N_2; 35% O_2; 25% CO_2.
   a) What is the partial pressure of each gas?
   b) What will the partial pressures of nitrogen and oxygen be if the carbon dioxide is removed by the introduction of some sodium hydroxide pellets?

6. A mixture of gases at $1.50 \times 10^5 \, N \, m^{-2}$ has the composition 40% NH_3; 25% H_2; 35% N_2 by volume.
   a) What is the partial pressure of each gas?
   b) What will the partial pressures of the other gases become if the ammonia is removed by the addition of some solid phosphorus(V) oxide?

HENRY'S LAW

Henry's law is concerned with the solubility of gases in liquids. It states that the mass of gas dissolved at constant temperature per unit volume of solvent is directly proportional to the partial pressure of the gas. This law may be expressed as

$$m_s = kp$$

where m_s is the mass of gas dissolved, p is its partial pressure, and k is a constant.

The *solubility* of a gas is the volume that will dissolve in unit volume of the solvent under stated conditions of temperature and pressure.

The *absorption coefficient* is the volume of gas (at s.t.p.) that will dissolve in unit volume of liquid at a stated temperature, under a pressure of 1 atmosphere.

Henry's law does not apply to gases which combine chemically with the solvent, for example, hydrogen chloride in solution in water. It applies only to solution as a physical process.

EXAMPLE   Taking the composition by volume of air as 80% nitrogen, 20% oxygen, calculate the volume composition of the air which is dissolved in water at 298 K. The absorption coefficients at 298 K are: nitrogen, 0.016; oxygen, 0.031.

METHOD   Air is composed of 80% by volume of nitrogen and 20% by volume of oxygen.

The mole fractions are therefore: nitrogen, 0.80; oxygen, 0.20.

The partial pressure of each gas is proportional to its mole fraction.

The absorption coefficient must be multiplied by the mole fraction to give the absorption of the gas at its partial pressure.

Vol. of N_2 at s.t.p. dissolving in $100 \, cm^2$ water

$$= 100 \times 0.016 \times 0.80 = 1.28 \, cm^3$$

Vol. of O_2 at s.t.p. dissolving in $100 \, cm^3$ water

$$= 100 \times 0.031 \times 0.20 = 0.62 \, cm^3$$

ANSWER   The dissolved air has the volume composition $1.28 \, cm^3 \, N_2$, $0.62 \, cm^3 \, O_2$

$$= 67\% \text{ nitrogen: } 33\% \text{ oxygen}$$

EXERCISE 24   Problems on Solubility of Gases

Use the following room-temperature coefficients of solubility:

N_2 = 0.0150   CO = 0.0380 O_2 = 0.0300
Cl_2 = 2.26 CO_2 = 1.00

1. If $1 \, dm^3$ of water is shaken with a gaseous mixture of 75% N_2, 20% O_2, 5% CO_2, calculate a) the volume of the dissolved gas, and b) its percentage by volume composition.

2. A mixture consisting of 79.0% N_2, 20.9% O_2 and 0.030% CO_2 was shaken with $1 \, dm^3$ of water. Calculate: a) the volume of gas dissolved, and b) the percentage composition by volume.

3. A mixture of CO and CO_2 which has a 50 : 50 by volume composition is shaken with $1 \, dm^3$ of water. What is the percentage by volume composition of the dissolved gas?

4. A mixture of 20% chlorine and 80% by volume of oxygen is shaken with water. What is the composition of the dissolved gas?

THE IDEAL GAS EQUATION

Gases which obey Boyle's law and Charles' law are called *ideal gases*. Combining these two laws gives the equation:

$$\frac{P \times V}{T} = \text{Constant for a given mass of gas}$$

It follows from Avogadro's law that, if a mole of gas is considered, the constant will be the same for all gases. It is called the universal gas constant, and given the symbol R, so that the equation becomes

$$PV = RT$$

This equation is called the *ideal gas equation*. For n moles of gas, the equation becomes

$$\boxed{PV = nRT}$$

The value of the constant R can be calculated. Consider 1 mole of gas at s.t.p. Its volume is 22.414 dm³. Inserting values of P, V and T in SI units into the ideal gas equation.

$P = 1.0132 \times 10^5\,\mathrm{N\,m^{-2}}$ $T = 273.15\,\mathrm{K}$
$V = 22.414 \times 10^{-3}\,\mathrm{m^3}$ $n = 1\,\mathrm{mol}$

gives $1.0132 \times 10^5 \times 22.414 \times 10^{-3} = 1 \times 273.15 \times R$

$$R = 8.314$$

The units of R are PV/nT, i.e.

$$\frac{\mathrm{N\,m^{-2}\,m^3}}{\mathrm{mol\,K}} = \mathrm{N\,m\,mol^{-1}\,K^{-1}} = \mathrm{J\,K^{-1}\,mol^{-1}}$$

Thus, $R = 8.314\,\mathrm{J\,K^{-1}\,mol^{-1}}$ (joules per kelvin per mole).

THE KINETIC THEORY OF GASES

To explain the gas laws, the kinetic theory of gases was put forward. The kinetic theory considers that the molecules of gas are in constant motion in straight lines. The pressure which the gas exerts results from the bombardment of the walls of the container by the molecules.

The kinetic energy of a molecule $= \frac{1}{2}mc^2$ (m = mass, c = velocity).

The kinetic energy of the gas $= \frac{1}{2}mN\overline{c^2}$ (N = number of molecules, $\overline{c^2}$ = average value of the square of the velocity for all the molecules; $\sqrt{\overline{c^2}}$ = root mean square velocity).

From the kinetic theory can be derived the equation

$$PV = \tfrac{1}{3}mN\overline{c^2}$$

Since the kinetic energy of the molecules is proportional to T (kelvins)

$$PV = \mathrm{Constant} \times T$$

This is the ideal gas equation. The agreement between theory and experimental results is good support for the kinetic theory.

The kinetic theory can be used to calculate the root mean square velocity of gas molecules.

EXAMPLE Calculate the root mean square velocity of hydrogen molecules at s.t.p.

METHOD 1 Use $M(H_2) = 2.02 \, \text{g mol}^{-1}$. In the equation $PV = \frac{1}{3}mN\overline{c^2}$, substitute $PV = RT$ for 1 mole of gas, and $mN = M$, the molar mass of gas in kg. Substituting $mN = 2.02 \times 10^{-3} \, \text{kg mol}^{-1}$ in $\frac{1}{3}mN\overline{c^2} = RT$, gives

$$\overline{c^2} = 3 \times 8.31 \times 273/(2.02 \times 10^{-3})$$

$$\sqrt{\overline{c^2}} = 1.84 \times 10^3 \, \text{m s}^{-1}$$

ANSWER The root mean square velocity of hydrogen molecules at s.t.p. is $1.84 \times 10^3 \, \text{m s}^{-1}$.

METHOD 2 Use the density of hydrogen ($9.00 \times 10^{-2} \, \text{kg m}^{-3}$ at s.t.p.). Since $mN/V = \rho$, the density of the gas, substituting in $P = \frac{1}{3}\rho\overline{c^2}$ gives

$$\sqrt{\overline{c^2}} = \sqrt{\frac{3 \times 1.01 \times 10^5}{9.00 \times 10^{-2}}} = 1.84 \times 10^3 \, \text{m s}^{-1}$$

ANSWER As before, the root mean square velocity is $1.84 \times 10^3 \, \text{m s}^{-1}$.

EXERCISE 25 Problems on the Kinetic Theory and the Ideal Gas Equation

Use $R = 8.314 \, \text{J K}^{-1} \text{mol}^{-1}$.

1. Krypton has a density of $3.44 \, \text{g dm}^{-3}$ at $25 \, ^\circ\text{C}$ and $1.01 \times 10^5 \, \text{N m}^{-2}$. Calculate its molar mass.

2. The density of hydrogen at $273 \, \text{K}$ and $1.01 \times 10^5 \, \text{N m}^{-2}$ is $8.96 \times 10^{-2} \, \text{g dm}^{-3}$. Calculate the root mean square velocity of the hydrogen molecules under these conditions.

3. Using the equation $PV = \frac{1}{3}mN\overline{c^2}$ calculate the kinetic energy of the molecules in one mole of an ideal gas at $0 \, ^\circ\text{C}$.

4. Calculate the root mean square velocity for argon at s.t.p. ($M_r(\text{Ar}) = 40.0$).

5. A volume of $1.00 \, \text{dm}^3$ is occupied by $1.798 \, \text{g}$ of a gas at $298 \, \text{K}$ and $101 \, \text{kPa}$. Calculate the molar mass of the gas.

6. Calculate the ratio of the root mean square velocities of oxygen and xenon molecules at $27 \, ^\circ\text{C}$. ($A_r(\text{O}) = 16.0$, $A_r(\text{Xe}) = 131$.)

7. Calculate the root mean square velocity of hydrogen iodide molecules at $27 \, ^\circ\text{C}$. ($A_r(\text{I}) = 127$.)

8. a) Calculate the ratio of the root mean square velocity of hydrogen molecules to the root mean square velocity of argon molecules at the same temperature.

   b) At what temperature will argon molecules have the same root mean square velocity as hydrogen molecules at $0 \, ^\circ\text{C}$? ($A_r(\text{Ar}) = 40.0$.)

EXERCISE 26   Questions from A-level Papers

1. a) The equation of state for an ideal (perfect) gas is $pV = nRT$.
       i) Write the van der Waals equation for a real gas.
       ii) State which two properties of real gases are taken into account in the van der Waals equation.

    b) It is estimated that the decay of 1 g of radium produces 0.043 cm³ of helium (measured at s.t.p.) per year, and counting methods show that 1 g of radium produces α-particles at a rate of 1.16×10^{18} per year. Given that the molar volume of a gas at s.t.p. is 22 400 cm³, calculate:
       i) the number of moles of helium atoms produced by 1 g of radium per year,
       ii) the value of the Avogadro constant.

    c) For the new element formed when the radioactive atom $^{226}_{88}Ra$ emits an α-particle, give:
       i) its name
       ii) its mass number
       iii) its atomic number.   (AEB78)

2. Define the term *partial pressure*. 200 cm³ of hydrogen at 100 kPa (1 atm) and 20 °C, and 150 cm³ of helium at 200 kPa (2 atm) and 20 °C were mixed in a total volume of 500 cm³. Calculate the partial pressure at 20 °C of each gas in the mixture.   (AEB76, p)

3. a) State Graham's law.

    b) Two gases X and Y effuse separately from a container under identical conditions of constant temperature and constant pressure difference. The total volumes effused (measured at s.t.p.) were recorded and the results are plotted in Fig. 5.1.

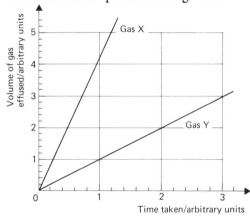

Fig. 5.1

       i) State, giving reasons, which gas has the larger relative molecular mass.
       ii) Given that gas X is deuterium, $^{2}_{1}H_2$, deduce the nature of *element* Y.

c) Without carrying out any further calculations, show qualitatively on a tracing of Fig. 5.2 the form of the graphs you would expect to obtain if, under the same conditions as in b), effusion experiments were carried out using i) hydrogen, $^1_1H_2$, ii) an equimolecular mixture of gases X and Y. Label the graphs 'H$_2$' and 'X + Y' respectively.   (C80)

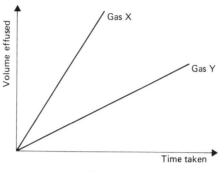

Fig. 5.2

4. a)   i) What do you understand by the term *diffusion*?
       ii) Two gases can often be separated by diffusion. Explain why this is so, stating any relevant law. Why would it be easier to separate a mixture of argon and helium by diffusion than one of nitrogen and ethane?

   b) 0.254 g of iodine are introduced into an evacuated vessel of volume 100 cm^3. The vessel is closed and heated to 1473 K. It is found that the pressure of the iodine vapour in the vessel at this temperature is $1.71 \times 10^5 \, N\,m^{-2}$.
       i) Calculate the pressure expected under these conditions if it is assumed that the iodine is wholly in the form of I_2 molecules.
       ii) What qualitative inference can be drawn from the difference between your result in i) and the observed pressure? Make any further deductions you can.   (O78)

5. When water is saturated with nitrogen at 25 °C at a nitrogen pressure of 2 atm (1 atm = $1.013 \times 10^5 \, Pa$), the solubility of the gas is 25 cm^3 of nitrogen at s.t.p. per dm^3 of the solution. Estimate the solubility of nitrogen in water at 25 °C when the nitrogen pressure is 60 atm. State and name any law you assume.   (O77)

6. Water from a certain spring gave on boiling a gaseous mixture containing 21.0% oxygen and 43.6% nitrogen, the remaining volume being carbon dioxide. Calculate the percentage by volume of the mixture of gases in which the spring water had been in contact under conditions of s.t.p. (The absorption coefficients for oxygen, nitrogen and carbon dioxide at s.t.p. are 0.04, 0.02, 1.79.)   (AEB78)

7. Gases exert a pressure on a container, they may be compressed, they interdiffuse and they diffuse at different rates through a membrane. How may the kinetic theory of gases be used to explain these properties?

Explain, *concisely*, how real gases deviate in their behaviour from that expected on simple kinetic theory.

Under comparable conditions, $200 \, cm^3$ of oxygen diffused through a membrane in 600 seconds and $60 \, cm^3$ of an unknown gas diffused through the same membrane in 300 seconds. Find:

a) the mean molecular mass of the gas,

b) the temperature at which the gas has the same root-mean-square velocity as oxygen at 273 K, both gases being at the same pressure.

(WJEC78)

8. a) State Dalton's law of partial pressures. If $200 \, cm^3$ of hydrogen, $100 \, cm^3$ of oxygen and $50 \, cm^3$ of nitrogen each at 273 K and 101.3 kPa (1 atmosphere) were introduced into an evacuated vessel of $500 \, cm^3$ capacity, calculate the pressure of the gas mixture (temperature remaining constant).

b) i) State Henry's law.

ii) If this same gas mixture was then allowed to come into contact with a water surface, what would be the composition by volume of the dissolved gases at 273 K (given the absorption coefficients for H_2, N_2 and O_2 at s.t.p. are 0.01, 0.02, 0.04 respectively).

iii) How would an increase in temperature affect the total volume of gas absorbed? How can the trend you suggest be explained?

(AEB79, p)

6 Liquids

DETERMINATION OF MOLAR MASS

The Victor Meyer method

The Victor Meyer apparatus can be used for liquids which can easily be vaporised. A weighed amount of liquid is vaporised and allowed to displace its own volume of air. The volume of air is measured at atmospheric pressure, and the temperature is recorded. The partial pressure of the air is equal to atmospheric pressure minus the saturation vapour pressure of water. Since both the volume and the mass of the vapour are recorded, its density and hence its molar mass can be found.

EXAMPLE 1 Calculate the molar mass of a liquid, 0.215 g of which displaced 92.0 cm^3 of air at 16 °C and 1.02 × 10^5 Pa. The saturation vapour pressure of water at 16 °C is 1.83 × 10^3 Pa.

METHOD The partial pressure of the measured air = $1.02 \times 10^5 - 1.83 \times 10^3$ Pa

$$= 0.999 \times 10^5 \, \text{Pa}$$

Into the equation $PV = \dfrac{m}{M} RT$, substitute this value of P, together with $V = 92.0 \times 10^{-6} \, \text{m}^3$, $T = 298 \, \text{K}$ and $R = 8.314 \, \text{J K}^{-1} \text{mol}^{-1}$.

Then, $\qquad 0.999 \times 10^5 \times 92.0 \times 10^{-6} = \dfrac{0.215}{M} \times 8.314 \times 289$

$$M = 56.8$$

ANSWER The molar mass is 57 g mol^{-1}.

The values of molar mass obtained by the Victor Meyer method are not very accurate. A knowledge of the empirical formula enables the Victor Meyer value to be corrected. For example, if the compound used above has the empirical formula CH_2O and an experimental value of 57 g mol^{-1} for the molar mass, one can see that $C_2H_4O_2$ is the molecular formula, and 60 g mol^{-1} is the correct molar mass.

The gas syringe method

This is another method of finding the molar mass of a liquid with a low boiling point. A small weighed quantity of liquid is injected into a gas syringe. The volume of vapour formed is measured, and its temperature and pressure are noted. From the values of mass and volume, the molar mass can be calculated.

EXAMPLE 2   A gas syringe contains $18.4 \, cm^3$ of air at $57 \, °C$. $0.187 \, g$ of a volatile liquid is injected into the syringe. The volume of gas in the syringe is then $54.6 \, cm^3$ at $57 \, °C$ and $1.01 \times 10^5 \, Pa$. Calculate the molar mass of the liquid.

METHOD   Using the values

$$P = 1.01 \times 10^5 \, Pa$$
$$V = 36.2 \, cm^3 = 36.2 \times 10^{-6} \, m^3$$
$$T = 273 + 57 = 330 \, K$$
$$R = 8.314 \, J \, K^{-1} mol^{-1}$$

in the equation $PV = \dfrac{m}{M} RT$ gives

$$1.01 \times 10^5 \times 36.2 \times 10^{-6} = \frac{0.187}{M} \times 8.314 \times 330$$

$$M = 140$$

ANSWER   The molar mass is $140 \, g \, mol^{-1}$.

ANOMALOUS RESULTS FROM MEASUREMENTS OF MOLAR MASS

Sometimes, an unexpectedly low result for molar mass is obtained. This happens when the molecules of the vapour on which measurements are being made dissociate, causing an increase in the actual number of particles present. If 1 mole of molecules of XY dissociate partially into X and Y, and α is the degree of dissociation, then

Species: XY $\rightleftharpoons$ X + Y
Number of moles: $(1 - \alpha)$ α α Total = $(1 + \alpha)$

$(1 - \alpha)$ moles of XY remain, and α moles of X and α moles of Y are formed.

Thus $$\frac{\text{Actual number of moles}}{\text{Expected number of moles}} = \frac{1 + \alpha}{1}$$

Since the volume occupied by a gas is proportional to the number of moles of gas,

$$\frac{\text{Actual volume of gas}}{\text{Expected volume of gas}} = \frac{1 + \alpha}{1}$$

Since we are finding molar mass from the equation, given on p. 80,

$$PV = nRT = \frac{m}{M} RT$$

where m = mass of substance, and M = its molar mass, if the volume, V, is greater than expected, M, the molar mass, is less than expected.

Thus

$$\frac{\text{Actual volume}}{\text{Expected volume}} = \frac{\text{Molar mass calculated from formula}}{\text{Measured molar mass}} = \frac{1 + \alpha}{1}$$

If the volume is kept constant, the pressure increases instead of the gas expanding and

$$\frac{\text{Calculated molar mass}}{\text{Measured molar mass}} = \frac{\text{Measured pressure}}{\text{Calculated pressure}} = \frac{1 + \alpha}{1}$$

If one molecule dissociates into n particles, the expression becomes:

$$\frac{\text{Calculated molar mass}}{\text{Measured molar mass}} = \frac{\text{Measured pressure}}{\text{Calculated pressure}} = \frac{1 + (n - 1)\alpha}{1}$$

EXAMPLE 1 The molar mass of phosphorus(V) chloride at 140 °C is 166. Calculate the degree of dissociation.

METHOD The molar mass of $PCl_5 = 31.0 + (5 \times 35.5) = 208.5 \, \text{g mol}^{-1}$
The dissociation which occurs is

$$PCl_5 \rightleftharpoons PCl_3 + Cl_2$$

Thus, $n = 2$, and

$$\frac{\text{Calculated molar mass}}{\text{Measured molar mass}} = \frac{1 + \alpha}{1}$$

$$\frac{\text{Calculated molar mass}}{\text{Measured molar mass}} = \frac{208.5}{166} = \frac{1 + \alpha}{1}$$

Therefore $\alpha = 0.26$ (26%).

ANSWER The degree of dissociation is 0.26.

EXAMPLE 2 When 1.00 g of iodine is heated at 1200°C in a 500 cm³ vessel a pressure of 1.51×10^2 kPa develops. Calculate the degree of dissociation.

METHOD Since $PV = nRT$,

$$P = \frac{1.00}{254} \times \frac{8.314 \times 1473}{500 \times 10^{-6}}$$

$$= 9.64 \times 10^4 \, \text{Pa}$$

$$\frac{\text{Observed pressure}}{\text{Calculated pressure}} = \frac{1.51 \times 10^5}{9.64 \times 10^4} = \frac{1 + (n - 1)\alpha}{1}$$

Since the dissociation is

$$I_2(g) \rightleftharpoons 2I(g)$$

$$n = 2 \quad \text{and} \quad \frac{1.51 \times 10^5}{9.60 \times 10^4} = 1 + \alpha$$

Solving this equation gives $\alpha = 0.58$ (or 58%).

ANSWER   The degree of dissociation is 0.58.

A measurement of molar mass higher than the value calculated from the formula is a sign that molecules are associated. In 1 mole of A, if 2 molecules of A form a dimer, and if the degree of dimerisation is α,

Species: $2A \rightleftharpoons A_2$

No. of moles: $(1 - \alpha)$ $\alpha/2$ Total $= (1 - \alpha/2)$

$$\frac{\text{Actual no. of moles}}{\text{Expected no. of moles}} = \frac{1 - \alpha/2}{1}$$

$$\frac{\text{Actual volume}}{\text{Calculated volume}} = 1 - \alpha/2$$

$$\frac{\text{Calculated molar mass}}{\text{Measured molar mass}} = 1 - \alpha/2$$

In general, if n molecules associate,

$$\frac{\text{Calculated molar mass}}{\text{Measured molar mass}} = 1 - \frac{(n - 1)\alpha}{n}$$

EXAMPLE 3 A value of 200 is obtained for the molar mass of aluminium chloride. Calculate the degree of dimerisation of aluminium chloride at the temperature at which the measurement was made.

METHOD   Calculated molar mass $= 133.5 \, \text{g mol}^{-1}$

$$\frac{\text{Calculated molar mass}}{\text{Measured molar mass}} = 1 - \frac{\alpha}{2}$$

$$133.5/200 = 1 - \frac{\alpha}{2}$$

$$\alpha = 0.67$$

ANSWER   The degree of association is 0.67.

EXERCISE 27   Problems on Molar Masses of Volatile Liquids

The gas constant, $R = 8.314 \, \text{J K}^{-1} \text{mol}^{-1}$.
1 atm $= 1.01 \times 10^5 \, \text{N m}^{-2} = 1.01 \times 10^5 \, \text{Pa} = 101 \, \text{kPa}$.

1. Calculate the molar mass of B, given that 0.850 g of B displaced 55.5 cm³ of air, measured in a Victor Meyer apparatus and corrected to s.t.p.

2. A compound of phosphorus and fluorine contains 24.6% by mass of phosphorus. 1.000 g of this compound has a volume of 178 cm³ at s.t.p. Deduce the molecular formula of the compound.

3. 0.110 g of a liquid produced 42.0 cm³ of vapour, measured at 147 °C and $1.01 \times 10^5 \, N\,m^{-2}$. What is the molar mass of the liquid?

4. 0.228 g of liquid was injected into a gas syringe. The volume of vapour formed was 84.0 cm³ at 17 °C and $1.01 \times 10^5 \, N\,m^{-2}$. Calculate the molar mass of the substance.

5. In a Victor Meyer apparatus, 0.125 g of X displaced 23.4 cm³ of air, measured at 77 °C and $9.97 \times 10^4 \, Pa$ (corrected for the vapour pressure of water). Calculate the molar mass of X.

6. In a Victor Meyer apparatus, 0.125 g of a volatile liquid, A, displaced 36.5 cm³ of air, measured at 20 °C and $9.98 \times 10^4 \, Pa$. The saturation vapour pressure of water at 20 °C is $2.4 \times 10^3 \, Pa$. Calculate the molar mass of A.

7. 0.452 g of a volatile solid displaced 82.0 cm³ of air, collected at 20 °C and $1.023 \times 10^5 \, N\,m^{-2}$. If the saturated vapour pressure of water at 20 °C is $2.39 \times 10^3 \, N\,m^{-2}$, calculate the molar mass of the solid.

8. Fig. 6.1 shows the results of gas syringe measurements on ethanol (○), propanone (□) and ethoxyethane (●), all at 80 °C and 1 atm. For each liquid, several measurements of the volume of vapour formed after the injection of a known mass of liquid were made.

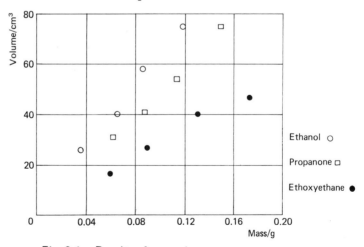

Fig. 6.1   Results of gas syringe measurements

Trace the results on to a piece of paper, and plot the best straight line through the points for each vapour. Find the slope of each line.

The reciprocal of the slope, mass/volume, is the density of the vapour, ρ. From the equation

$$PV = \frac{m}{M} RT$$

$$\left(\text{since} \ \frac{m}{V} = \rho \right)$$

$$M = \rho \frac{RT}{P}$$

a) Insert the value you have obtained for the density of ethanol vapour into the equation, and find the molar mass of ethanol. Do the same for b) propanone, and c) ethoxyethane.

EXERCISE 28   Problems on Association and Dissociation

1. 20.85 g of phosphorus(V) chloride are allowed to vaporise in a 5.00 dm³ vessel at 175 °C. A pressure of $1.04 \times 10^5 \ \mathrm{N \, m^{-2}}$ develops. Calculate the degree of dissociation of PCl_5 into PCl_3 and Cl_2.

2. 10.32 g of aluminium chloride are allowed to vaporise in a 1.00 dm³ vessel at 80 °C. A pressure of $1.70 \times 10^5 \ \mathrm{N \, m^{-2}}$ develops. What is the degree of association of $AlCl_3$ into Al_2Cl_6 molecules?

3. Nitrogen dioxide exists in an equilibrium mixture:

$$N_2O_4(g) \ \rightleftharpoons \ 2NO_2(g)$$

The relative molar mass of nitrogen dioxide at 25 °C is 80.0. What percentage of the molecules in the mixture is N_2O_4?

4. A sample of iodine of mass 25.4 g is vaporised in a 2.00 dm³ vessel at 800 K. A pressure of $4.32 \times 10^5 \ \mathrm{N \, m^{-2}}$ develops. Calculate the degree of dissociation of iodine molecules into atoms.

5. The molar mass of iron(III) chloride measured at 900 K is 246 g mol⁻¹. Calculate the degree of dimerisation of $FeCl_3$ molecules.

VAPOUR PRESSURE

In a liquid, the molecules are in constant motion. Some molecules, those with energy considerably above average, will have enough energy to escape from the liquid to the vapour state. If a liquid is introduced into a closed container, some of the liquid will evaporate. The molecules in the vapour state will exert a pressure. When equilibrium is reached between the liquid state and the vapour state, the pressure exerted by the vapour is called the *vapour pressure* of the liquid. To be correct, one should call it the *saturated vapour pressure* or the *equilibrium vapour pressure*. The magnitude of the vapour pressure depends on the identity of the liquid and on the temperature: it does not depend on the amount of liquid present.

EXAMPLE   The saturated vapour pressure of water at $65\,°C$ is $25.05\,kN\,m^{-2}$. What mass of water will be present in the vapour phase if $10.0\,cm^3$ of water are injected into a $1.000\,dm^3$ vessel?

METHOD   Use the ideal gas equation, $PV = nRT$, and substitute

$$P = 25.05 \times 10^3\,N\,m^{-2} \qquad R = 8.314\,J\,K^{-1}\,mol^{-1}$$

$$T = 338\,K \qquad V = 1.000\,dm^3 = 1.000 \times 10^{-3}\,m^3$$

giving $25.05 \times 10^{-3} \times 1.000 \times 10^{-3} = n \times 8.314 \times 338$

No. of moles of water, $n = 8.92 \times 10^{-3}$

Mass of water $= 18.0 \times 8.92 \times 10^{-3} = 0.161\,g$

ANSWER   The mass of water that evaporates is $0.161\,g$.

EXERCISE 29   Problems on Vapour Pressures of Liquids

1. $10.0\,cm^3$ of ethyl ethanoate are introduced into an evacuated $10.0\,dm^3$ vessel at $25\,°C$. What mass of ethyl ethanoate will vaporise? The saturated vapour pressure of ethyl ethanoate at $25\,°C$ is $9.55 \times 10^3\,N\,m^{-2}$.

2. At $95\,°C$, the saturated vapour pressure of bromobenzene is $1.54 \times 10^4\,N\,m^{-2}$. What mass of bromobenzene will vaporise when a small amount of liquid bromobenzene is introduced into a $2.50\,dm^3$ flask at $95\,°C$?

3. At $0\,°C$, the saturated vapour pressure of water is $6.10 \times 10^2\,N\,m^{-2}$. How many molecules of water vapour will be present in each cm^3 of air in a vessel containing ice at $0\,°C$?

4. If analysis shows that $0.0230\,g$ of water are present in $1.00\,dm^3$ of air at $25\,°C$, what is the saturated vapour pressure of water at $25\,°C$?

SOLUTIONS OF LIQUIDS IN LIQUIDS

How do you express the composition of a liquid–liquid mixture? One way is by stating the mole fraction of each constituent:

$$\text{Mole fraction of A in A--B mixture} = \frac{\text{No. of moles of A}}{\text{Total no. of moles}}$$

$$= \frac{n_A}{n_A + n_B}$$

The vapour above a mixture of the liquids A and B will contain both A and B.

Raoult's law states that the saturated vapour pressure of each component in the mixture is equal to the product of the mole fraction of that component and the saturated vapour pressure of that component when pure, at the same temperature.

If p_A = vapour pressure of A,

p_A^0 = saturated vapour pressure of pure A,

and x_A = mole fraction of A, then

$$p_A = x_A \times p_A^0 \qquad p_B = x_B \times p_B^0$$

Raoult's law is obeyed by mixtures of similar compounds. They are said to form *ideal solutions*. The vapour above a mixture of liquids does not have the same composition as the mixture. It is richer in the more volatile component. The mole fractions of A and B in the vapour phase are in the ratio of their mole fractions in the liquid phase multiplied by the ratio of the saturated vapour pressures of the two liquids. If x_A' and x_B' are the mole fractions of A and B in the vapour phase,

$$\frac{x_A'}{x_B'} = \frac{x_A}{x_B} \times \frac{p_A^0}{p_B^0}$$

EXAMPLE 1   Calculate the vapour pressure of a solution containing 50.0 g heptane and 38.0 g octane at 20 °C. The vapour pressures of the pure liquids at 20 °C are heptane 473 Pa; octane 140 Pa.

METHOD   No. of moles of heptane = 50/100 = 0.50

No. of moles of octane = 38/114 = 0.33

Mole fraction of heptane = 0.50/0.83

Mole fraction of octane = 0.33/0.83

$$p\,(\text{heptane}) = p^0(\text{heptane}) \times x(\text{heptane})$$

$$= 473 \times 0.50/0.83 = 284.0$$

$$p\,(\text{octane}) = p^0(\text{octane}) \times x(\text{octane})$$

$$= 140 \times 0.33/0.83 = 55.9$$

ANSWER   Total vapour pressure = 284.0 + 55.9 = 340 Pa.

EXAMPLE 2   Two pure liquids A and B have vapour pressures $1.50 \times 10^4\,\text{N m}^{-2}$ and $3.50 \times 10^4\,\text{N m}^{-2}$ at 20 °C. If a mixture of A and B obeys Raoult's law, calculate the mole fraction of A in a mixture of A and B which has a total vapour pressure of $2.90 \times 10^4\,\text{N m}^{-2}$ at 20 °C.

METHOD   If n_A is the mole fraction of A, $(1 - n_A)$ is the mole fraction of B. Then,

$$(n_A \times 1.50 \times 10^4) + (1 - n_A)(3.50 \times 10^4) = 2.90 \times 10^4$$

$$1.50n_A + 3.50 - 3.50n_A = 2.90$$

$$2.00n_A = 0.60$$

$$n_A = 0.30$$

ANSWER   The mole fraction of A is 0.30.

EXERCISE 30 Problems on Vapour Pressures of Solutions of Two Liquids

1. Two pure liquids A and B have vapour pressures of respectively 17 000 and 35 000 N m^{-2} at 25 °C. An equimolar mixture of A and B has a vapour pressure of 26 000 N m^{-2} at 25 °C. Calculate the vapour pressure of a mixture containing four moles of A and one mole of B at 25 °C.

2. Hexane and heptane are totally miscible and form an ideal two component system. If the vapour pressures of the pure liquids are 56 000 and 24 000 N m^{-2} at 50 °C calculate:
   a) the total vapour pressure, and
   b) the mole fraction of heptane in the vapour above an equimolar mixture of hexane and heptane.

3. The vapour pressure of water at 298 K is 3.19×10^3 Pa. What are the partial vapour pressures of water in mixtures of:
   a) 27 g water and 69 g ethanol
   b) 9.0 g of water and 92 g of ethanol
   at this temperature?

4. A and B are two miscible liquids which form an ideal solution. The vapour pressures at 20 °C are: A, 40 kPa, B, 32 kPa. Calculate the total pressure of the vapour in equilibrium with mixtures of:
   a) 3 moles of A and 1 mole of B at 20 °C
   b) 1 mole of A and 4 moles of B at 20 °C.

IMMISCIBLE LIQUIDS: SUM OF VAPOUR PRESSURES

Steam distillation

In a system of immiscible liquids, each liquid exerts its own vapour pressure independently of the other. The vapour pressure of the system is equal to the sum of the vapour pressures of the pure components. This is the basis for steam distillation. Phenylamine will distil over in steam at 98 °C, although its boiling point is 184 °C. At 98 °C, the sum of the vapour pressures of phenylamine and water is equal to atmospheric pressure. The ratio of the amounts of the two liquids in the distillate is equal to the ratio of their vapour pressures:

$$\frac{n_A}{n_W} = \frac{p_A}{p_W}$$

where n_A and n_W are the amounts of phenylamine and water in the distillate, and p_A and p_W are the vapour pressures of phenylamine and water at 98 °C.

Since $n = m/M$ (where m = mass, M = molar mass)

$$\frac{m_A}{M_A} \times \frac{M_W}{m_W} = \frac{p_A}{p_W}$$

This equation can be used to find m_A/m_W, the ratio of masses of amine and water in the distillate. Alternatively, steam distillation can be used as a method of determining molar masses. In this case, the masses of the liquid and water in the distillate must be measured and inserted into the equation to give the unknown molar mass.

EXAMPLE 1   Bromobenzene distils in steam at 95 °C. The vapour pressures of bromobenzene and water at 95 °C are $1.59 \times 10^4 \, N \, m^{-2}$ and $8.50 \times 10^4 \, N \, m^{-2}$. Calculate the percentage by mass of bromobenzene in the distillate.

METHOD   Let the percentage of bromobenzene = y.

In the equation $\dfrac{n_{C_6H_5Br}}{n_{H_2O}} = \dfrac{p_{C_6H_5Br}}{p_{H_2O}}$

$$\frac{y/157}{(100-y)/18} = \frac{1.59 \times 10^4}{8.50 \times 10^4}$$

$$y = 62.0$$

ANSWER   The distillate contains 62.0% by mass of bromobenzene.

EXAMPLE 2   An organic liquid which does not mix with water distils in steam at 96 °C under a pressure of $1.01 \times 10^5 \, N \, m^{-2}$. The vapour pressure of water at 96 °C is $8.77 \times 10^4 \, N \, m^{-2}$. The distillate contains 51.0% by mass of the organic liquid. Calculate its molar mass.

METHOD   Let M be the molar mass of the organic compound, n_A and n_W be the amounts (mol) of organic compound and water in the distillate, and p_A and p_W be the vapour pressures of the organic compound and water.

Then $n_A/n_W = p_A/p_W$

$$\frac{51.0/M}{49.0/18.0} = \frac{(1.01 \times 10^5) - (8.77 \times 10^4)}{8.77 \times 10^4}$$

$$M = 123 \, g \, mol^{-1}$$

ANSWER   The molar mass of the organic compound is $123 \, g \, mol^{-1}$.

EXERCISE 31   Problems on Steam Distillation

1. An organic liquid distils in steam. The partial pressures of the two liquids at the boiling point are X, 5.30 kPa; H_2O, 96.0 kPa. The distillate contains the liquids in a ratio of 0.480 g X to 1.00 g water. What is the molar mass of X?

2. The liquid A distils in steam. At the boiling point, the partial pressures of the two liquids are $A = 6.59 \times 10^3 \, N \, m^{-2}$; $H_2O = 9.44 \times 10^4 \, N \, m^{-2}$. If the molar mass of A is $95 \, g \, mol^{-1}$, what is the percentage by mass of A in the distillate?

3. Compound B distils in steam at 96 °C and $1.01 \times 10^5 \, N \, m^{-2}$. The vapour pressure of water at this temperature is $7.24 \times 10^4 \, N \, m^{-2}$. The distillate contains 80.0% by mass of B. Calculate the molar mass of B.

4. Phenylamine, $C_6H_5NH_2$, distils in steam at 98 °C and $1.01 \times 10^5 \, N \, m^{-2}$. If the saturation vapour pressure of water is $9.40 \times 10^4 \, N \, m^{-2}$, what is the percentage by mass of phenylamine in the distillate?

5. Naphthalene, $C_{10}H_8$, distils in steam at 98 °C and $1.01 \times 10^5 \, N \, m^{-2}$. If the vapour pressure of water is $9.50 \times 10^4 \, N \, m^{-2}$, calculate the mass of distillate that contains 10.0 g of naphthalene.

6. A substance X distils in steam at 98 °C and 97 kPa. The percentage of X by mass in the distillate is 15.8%. The saturated vapour pressure of water at 98 °C is 94.7 kPa. Calculate the molar mass of X.

DISTRIBUTION OF A SOLUTE BETWEEN TWO IMMISCIBLE SOLVENTS

Consider a solid which is appreciably soluble in both of a pair of immiscible liquids. When the solid is shaken with the two liquids, it distributes itself between the two layers. It is found that the ratio of the solute concentrations in the two layers is always the same. If C_U and C_L are the concentrations in the upper and lower layers, then

$$C_U/C_L = k$$

The constant k is called the *partition coefficient* or *distribution coefficient*, and is constant for a particular temperature.

EXAMPLE 1   The partition coefficient for iodine between water and carbon disulphide at 20 °C is 2.43×10^{-3}. A 100 cm^3 sample of a solution of iodine in water, of concentration $1.00 \times 10^{-3} \, mol \, I_2 \, dm^{-3}$ is shaken with 10.0 cm^3 of carbon disulphide. What fraction of the iodine is extracted by carbon disulphide?

METHOD Let x be the number of moles of I_2 extracted by CS_2.

No. of moles of I_2 (total) $= 100 \times 10^{-3} \times 10^{-3} = 1.00 \times 10^{-4}$

Use $C_U/C_L = k$ (C_U for water layer, C_L for CS_2 layer):

Then, $$\frac{(1.00 \times 10^{-4}) - x}{100} \Big/ \frac{x}{10.0} = 2.43 \times 10^{-3}$$

$$(1.00 \times 10^{-4}) - x = 2.43 \times 10^{-2} x$$

giving $$x = 9.76 \times 10^{-5}$$

ANSWER Fraction of iodine extracted $= (9.76 \times 10^{-5})/(1.00 \times 10^{-4})$

$$= 0.98.$$

EXAMPLE 2 The partition coefficient of X between ether and water is 25.0 at 20 °C. Calculate the mass of X extracted from a solution containing 10.0 g of X in 1.00 dm³ of water by a) 100 cm³ of ether, b) two successive portions of 50.0 cm³ of ether.

METHOD Let the mass of X extracted by 100 cm³ of ether be m_1.

Use $C_U/C_L = k$ (C_U for ether layer, C_L for water layer)

Concn of X in ether $= m_1/100 \, \text{g cm}^{-3}$

Concn of X in water $= (10.0 - m_1)/1000 \, \text{g cm}^{-3}$

$$\frac{m_1}{100} \Big/ \frac{(10.0 - m_1)}{1000} = 25.0$$

giving $m_1 = 7.14$ g.

Let $m_2 =$ mass of X extracted by the first 50.0 cm³ of ether, and $m_3 =$ mass of X extracted by the second 50.0 cm³ of ether.

Then $$\frac{m_2}{50.0} \Big/ \frac{(10.0 - m_2)}{1000} = 25.0$$

giving $m_2 = 5.55$ g.

If 5.55 g of X are extracted by ether, 4.45 g remain in the aqueous solution.

$\therefore$ $$\frac{m_3}{50} \Big/ \frac{(4.45 - m_3)}{1000} = 25.0$$

giving $m_3 = 2.47$ g.

ANSWER Total mass of X extracted by ether in two portions $= 5.55 \, \text{g} + 2.47 \, \text{g}$
$= 8.02$ g.

(*Note* that this is greater than the value of 7.14 g calculated for the mass of X extracted by using all the ether at once.)

Partition can be used to investigate an equilibrium in aqueous solution between a covalent species and an ionic species, for example, the equilibrium

$$I_2(aq) + I^-(aq) \rightleftharpoons I_3^-(aq)$$

Only the covalent I_2 molecules will dissolve in an organic solvent. If an aqueous solution of iodine in iodide ions is shaken with an organic solvent, the concentration of iodine in the solvent can be measured and divided by the partition coefficient to give the concentration of iodine molecules in the aqueous layer. The concentration of iodine combined as I_3^- ions is obtained by subtracting the free iodine from the total iodine concentration. The concentration of I^- ions is obtained by subtracting $[I_3^-]$ from the original concentration of I^- ions.

EXAMPLE 3 Iodine is dissolved in water containing $0.160\,mol\,dm^{-3}$ of potassium iodide, and the solution is shaken with tetrachloromethane. The concentration of iodine in the aqueous layer was found to be $0.080\,mol\,dm^{-3}$; that in the organic layer $0.100\,mol\,dm^{-3}$. The partition coefficient for iodine between tetrachloromethane and water is 85. Calculate the equilibrium constant for the reaction:

$$I_2(aq) + I^-(aq) \rightleftharpoons I_3^-(aq).$$

METHOD Since $[I_2]$ in $CCl_4 = 0.100\,mol\,dm^{-3}$

$[I_2]$ free in water $= 0.100/85 = 0.001\,18\,mol\,dm^{-3}$

$[I_2]$ total $= 0.080\,mol\,dm^{-3}$

$[I_2]$ combined as $I_3^- = 0.080 - 0.001\,18 = 0.0788\,mol\,dm^{-3}$

$[I^-]$ total $= 0.160\,mol\,dm^{-3}$

$[I^-]$ free $= 0.160 - 0.0788 = 0.0812\,mol\,dm^{-3}$

Putting these values into the expression

$$\frac{[I_3^-]}{[I_2][I^-]} = K$$

gives $$K = \frac{0.0788}{0.001\,18 \times 0.0812} = 822\,mol^{-1}\,dm^3$$

ANSWER The equilibrium constant is $820\,mol^{-1}\,dm^3$.

EXAMPLE 4 $100\,cm^3$ of aqueous copper(II) sulphate solution of concentration $0.100\,mol\,dm^{-3}$ were mixed with $100\,cm^3$ of ammonia solution of concentration $1.12\,mol\,dm^{-3}$. The mixture was shaken with $100\,cm^3$ of trichloromethane. The trichloromethane layer required $12.0\,cm^3$ of hydrochloric acid of concentration $0.100\,mol\,dm^{-3}$ for neutralisation.

Assume that the partitition coefficient

$$\frac{\text{Concn of free ammonia in water}}{\text{Concn of free ammonia in trichloromethane}} = 25.0$$

Deduce the formula for the complex formed by ammonia and copper(II) ions.

METHOD   Amount of NH_3 in $CHCl_3$ = $12.0 \times 10^{-3} \times 0.100$ = 1.20×10^{-3} mol
$[NH_3]$ in $CHCl_3$ = $1.20 \times 10^{-3}/0.100$ = 0.0120 mol dm^{-3}
$[NH_3]$ free in water = 25.0×0.0120 = 0.300 mol dm^{-3}
Amount of NH_3 free in water = 0.300×0.200 = 0.0600 mol
Total amount of NH_3 = $100 \times 10^{-3} \times 1.12$ = 0.112 mol
Amount of combined NH_3 in water = $0.112 - 0.0600 - 0.0120$ mol
$$= 0.0400 \text{ mol}$$
Amount of Cu^{2+} ions in water = $100 \times 10^{-3} \times 0.100$ = 0.0100 mol
Ratio (moles of Cu^{2+})/(moles of NH_3) = $0.0100/0.0400$ = $1/4$

ANSWER   The formula of the complex is $[Cu(NH_3)_4]^{2+}$.

*Distribution law for association and dissociation

If the solute has a different molecular formula in the two solvents, then the distribution law has to be modified. For example, ethanoic acid is dimerised in benzene. When ethanoic acid is partitioned between benzene and water, the equilibrium is

$$(CH_3CO_2H)_2 \text{ (benzene)} \rightleftharpoons 2CH_3CO_2H \text{ (water)}$$

and $$\frac{[CH_3CO_2H \text{ (benzene)}]}{[CH_3CO_2H \text{ (water)}]^2} = C, \quad \text{where } C \text{ is a constant}$$

$\therefore$ $\lg [CH_3CO_2H \text{ (benzene)}] = \lg C + 2 \lg [CH_3CO_2H \text{ (water)}]$

In general, if a solute is monomeric in solvent 1 and polymeric in solvent 2:

$$nA \text{(solvent 1)} \rightleftharpoons A_n \text{ (solvent 2)}$$

then $\lg [A \text{(solvent 2)}] = n \lg [A \text{(solvent 1)}] + \lg C$

A plot of $\lg [A \text{(solvent 2)}]$ against $\lg [A \text{(solvent 1)}]$ gives a straight line of gradient n and intercept $\lg C$ (see Fig. 6.2).

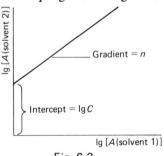

Fig. 6.2

EXERCISE 32   Problems on Partition

1. X is 12.0 times more soluble in trichloromethane than in water. What mass of X will be extracted from $1.00 \, dm^3$ of an aqueous solution containing 25.0 g by shaking with $100 \, cm^3$ of trichloromethane?

2. The partition coefficient of Y between ethoxyethane (ether) and water is 80. If $200 \, cm^3$ of an aqueous solution containing 5.00 g of Y is shaken with $50.0 \, cm^3$ of ethoxyethane, what mass of Y is extracted from the solution?

3. Z is allowed to reach an equilibrium distribution between the liquids ethoxyethane and water. The ether layer is $50.0 \, cm^3$ in volume and contains 4.00 g of Z. The aqueous layer is $250 \, cm^3$ in volume and contains 1.00 g of Z. What is the partition coefficient of Z between ethoxyethane and water?

4. $500 \, cm^3$ of an aqueous solution of concentration $0.120 \, mol \, dm^{-3}$ is shaken with $50.0 \, cm^3$ of ethoxyethane. The partition coefficient of the solute between ethoxyethane and water is 60.0. Calculate the amount (in mol) of solute which will be extracted by ethoxyethane.

5. The distribution coefficient of A between ethoxyethane and water is 90. An aqueous solution of A with a volume of $500 \, cm^3$ contains 5.00 g. What mass of A will be extracted by:
   a) $100 \, cm^3$ of ethoxyethane, and
   b) two successive portions of $50.0 \, cm^3$ of ethoxyethane?

6. An organic acid is allowed to reach an equilibrium distribution in a separating funnel containing $50.0 \, cm^3$ of ethoxyethane and $500 \, cm^3$ of water. On titration, $25.0 \, cm^3$ of the ethoxyethane layer required $22.5 \, cm^3$ of $1.00 \, mol \, dm^{-3}$ sodium hydroxide solution, and $25.0 \, cm^3$ of the aqueous layer required $9.0 \, cm^3$ of $0.100 \, mol \, dm^{-3}$ sodium hydroxide solution. Calculate the partition coefficient for the acid between ethoxyethane and water.

7. A small amount of iodine is shaken in a separating funnel containing $50.0 \, cm^3$ of tetrachloromethane and $500 \, cm^3$ of water. On titration, $25.0 \, cm^3$ of the aqueous layer require $6.7 \, cm^3$ of a $0.0550 \, mol \, dm^{-3}$ solution of sodium thiosulphate. $25.0 \, cm^3$ of the organic solvent require $27.2 \, cm^3$ of $1.15 \, mol \, dm^{-3}$ sodium thiosulphate solution. Calculate the distribution coefficient for iodine between water and tetrachloromethane.

8. A solid A is three times as soluble in solvent X as in solvent Y. A has the same relative molecular mass in both solvents. Calculate the mass of A that would be extracted from a solution of 4 g of A in $12 \, cm^3$ of Y by extracting it with: a) $12 \, cm^3$ of X, and b) three successive portions of $4 \, cm^3$ of X.

9. a) The partition coefficient of T between ethoxyethane and water at room temperature is 19. A solution of $5.0\,g$ of T in $100\,cm^3$ of water is extracted with $100\,cm^3$ of ethoxyethane at room temperature. What mass of T will be present in the ethoxyethane layer?

   b) What would be the total mass of T extracted if the ethoxyethane were used in two separate $50\,cm^3$ portions, instead of the single $100\,cm^3$ portion?

*10. When an excess of ammonia is added to an aqueous solution of copper(II) sulphate, a complex ion, $Cu(NH_3)_n{}^{2+}$ is formed.

   A $100\,cm^3$ portion of a solution of ammonia in $0.75\,mol\,dm^{-3}$ copper(II) sulphate solution is added to $100\,cm^3$ of trichloromethane. After equilibration, the two layers were titrated against standard alkali. The concentrations of ammonia in the two layers were: aqueous layer, $3.25\,mol\,dm^{-3}$; trichloromethane layer, $0.0100\,mol\,dm^{-3}$. Ammonia dissolves in trichloromethane, but the complex ion does not. For every mole of ammonia in the trichloromethane layer, there are 25 moles of ammonia not complexed with copper(II) ions in the aqueous layer.

   Calculate the number of moles of ammonia in the trichloromethane layer; the number of moles of ammonia not complexed in the aqueous layer; and the number of moles of ammonia combined with the copper(II) ions. Deduce the formula of the complex ion.

EXERCISE 33   Questions from A-level Papers

1. a) State Raoult's Law for a mixture of two volatile liquids.

   b) At $80\,^\circ C$, the vapour pressure of benzene is $10.0 \times 10^4\,N\,m^{-2}$, and that of toluene is $4.00 \times 10^4\,N\,m^{-2}$. Assuming that a mixture of these two liquids conforms to Raoult's Law, estimate the fraction of toluene in the vapour in equilibrium with a liquid mixture of benzene and toluene at $80\,^\circ C$ in which the mole fraction of toluene is 0.60.   (O78)

2. a) Define *mole fraction* of a substance A in a solution composed of three substances, A, B and C.

   b) Two pure liquids A and B have vapour pressures of respectively $1.70 \times 10^4\,N\,m^{-2}$ and $3.50 \times 10^4\,N\,m^{-2}$ at $25\,^\circ C$. Given that a mixture of A and B obeys Raoult's Law, calculate the mole fraction of A in a mixture of A and B which has a total vapour pressure of $2.78 \times 10^4\,N\,m^{-2}$ at $25\,^\circ C$.   (O78)

3. Heptane and octane form an ideal solution. Calculate the vapour pressure of a solution containing $50\,g$ heptane (C_7H_{16}) and $38\,g$ octane (C_8H_{18}) at $20\,^\circ C$. The vapour pressure of heptane at $20\,^\circ C = 473.2\,Pa$. The vapour pressure of octane at $20\,^\circ C = 139.8\,Pa$.   (AEB79, p)

4. Explain, using a vapour pressure-composition diagram and giving important practical details, how two ideally miscible liquids may be separated by fractional distillation.

   Show, concisely, using another diagram, how the fractional distillation of an ethanol–water mixture differs from the one you have described.

   Methanol–ethanol solutions are almost ideal. If the vapour pressures of methanol and ethanol at $335\,K$ are 8.1×10^4 and $4.5 \times 10^4\,N\,m^{-2}$ respectively, calculate the volume composition of the vapour over a mixture of $64\,g$ of methanol and $46\,g$ of ethanol at $335\,K$, assuming ideality. [(H) $= 1$; (C) $= 12$; (O) $= 16$.]   (WJEC77)

5. Explain the meaning of the following: (i) the mole; (ii) the ideal gas law; (iii) saturated vapour pressure; (iv) the density of a liquid.

   A $1\,dm^3$ flask at $27\,^{\circ}C$ is half filled with water. The density of water at $27\,^{\circ}C$ is $1.00\,kg\,dm^{-3}$ and its saturated vapour pressure is $3570\,N\,m^{-2}$.

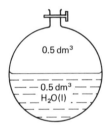

   Assuming ideal behaviour of the vapour, calculate:

   a) the number of molecules of water in the bottom half of the flask
   b) the number of molecules of water in the top half of the flask.

   [A_r(H) $= 1$; A_r(O) $= 16$; $R = 8.31\,J\,mol^{-1}K^{-1}$; $0\,^{\circ}C = 273\,K$; Avogadro constant $= 6.02 \times 10^{23}\,molecules\,mol^{-1}$.]   (WJEC80)

6. a) Give details of an experiment you would perform in order to measure the relative molar mass of a liquid with a low boiling point.

   b) A determination of the relative molar mass of nitrogen dioxide at $65\,^{\circ}C$ and pressure $101.3\,kPa$ gave a value of 58. Calculate:
      i) the degree of dissociation of dinitrogen tetroxide at this temperature;
      ii) the volumes of dinitrogen tetroxide (N_2O_4) and nitrogen dioxide (NO_2) contained in $1\,dm^3$ of vapour at this temperature and under a pressure of $101.3\,kPa$;
      iii) the partial pressure of each component gas in the mixture.

   c) Give the names and formulae of two gaseous oxides of nitrogen other than nitrogen dioxide (NO_2) and outline how each can be prepared in the laboratory. Give *one* test in each case which would identify that particular oxide.   (AEB79)

7. 0.600 g of iron(III) chloride are introduced into a vessel of internal volume 200 cm^3. The vessel is evacuated and closed, and then heated to 600 K. The iron(III) chloride completely vaporises, and the pressure in the vessel is found to be 4.60 × 10^4 Pa. Calculate the relative molecular mass of iron(III) chloride in the gaseous state at this temperature, and comment briefly on your result. (1 Pa = 1 N m^{-2}.)   (O79, p)

8. a) Describe, with the aid of a diagram, how you would prepare a sample of anhydrous aluminium(III) chloride in the laboratory.

   b) Aluminium(III) chloride vapour has a density of 2.28 × 10^{-3} g cm^{-3} at 1070 K and 1 atmosphere pressure.
      i) Calculate the apparent relative molecular mass of the sample under these conditions.
      ii) Account for your answer to (i) by discussing the structures of and bonding in aluminium(III) chloride under these conditions and calculate the relative amounts of any species present.
      iii) Describe, giving reasons, how the apparent value of the relative molecular mass will be influenced by increases in temperature and in pressure. (Gas constant R = 82.06 cm^3 atm mol^{-1}K^{-1} = 8.314 J K^{-1}mol^{-1}.)   (JMB79)

9. a) Sketch a graph representing the distributions of velocities of the molecules in a gas at two different temperatures. By reference to this graph discuss the effect of temperature on the rate of a gaseous reaction.

   b) State the principles underlying the liquefaction of gases and outline the process used for liquefying air.

   c) Write the equation which includes the laws governing the behaviour of an ideal gas and define the symbols used.

   In a mercury barometer the space above the mercury is normally considered to be 'empty'. Yet mercury has a definite vapour pressure which is 0.2 Pa (1.5 × 10^{-3} mm Hg) at 25 °C. How many atoms of mercury are there per cm^3 in this 'empty' space at 25 °C? [Standard atmospheric pressure = 101 kPa (760 mm Hg). The Avogadro constant = 6.02 × 10^{23}/mol.]   (AEB77)

10. If a mixture of two immiscible liquids is boiled, the boiling point is below that of each of the liquids.
    a) Explain why this is so.
    b) When the two liquids are water and nitrobenzene, the mixture boils at 372 K and the vapour pressures at this temperature are 97.7 kN m^{-2} (water) and 3.6 kN m^{-2} (nitrobenzene). Calculate the percentage of nitrobenzene in the vapour. (The relative molecular masses of water and nitrobenzene are 18 and 123 respectively.)
    c) What separation process used in organic chemistry utilises this principle?   (WJEC78)

11. a) State the distribution law for the partition of a solute between two immiscible liquids.

    b)   i) Give an example of a case where the law is not obeyed.

        ii) Explain why the law is not obeyed in this case.

    c) $100 \, cm^3$ of butanedioic acid of concentration $0.3 \, mol \, dm^{-3}$ were added to $100 \, cm^3$ of ethoxyethane (diethyl ether) and the mixture shaken and allowed to reach equilibrium. A $10 \, cm^3$ sample of the aqueous layer required $25.00 \, cm^3$ of sodium hydroxide of concentration $0.2 \, mol \, dm^{-3}$ for complete neutralisation. Calculate the distribution coefficient for the acid, assuming complete immiscibility of the two solvents.

    d) Name and give the operating pH range of an indicator suitable for carrying out the titration in the above experiment.   (JMB79)

12. a) What is meant by a *complex ion*? What experimental evidence exists to show the presence of these ions in solution? Give the names and structures of *five* such ions.

    b) $50 \, cm^3$ of ammonia solution of concentration $0.40 \, mol/l$ was mixed with $50 \, cm^3$ of copper(II) sulphate solution of concentration $0.05 \, mol/l$. This mixture was then shaken with $100 \, cm^3$ of trichloromethane at room temperature. $50 \, cm^3$ of the trichloromethane layer required $10 \, cm^3$ of hydrochloric acid of concentration $0.025 \, mol/l$ for neutralisation, using screened methyl orange as indicator. If the distribution (partition) coefficient of ammonia between water and trichloromethane is 20.0 at the temperature of the experiment, calculate the value of x in the ion $[Cu(NH_3)_x(H_2O)_{6-x}]^{2+}$.   (AEB78, S)

13. a) State the meaning of the term *partition coefficient* of a solute between two immiscible solvents.

    b) Outline the principles of an experiment to measure the partition coefficient for a system of your choice.

    c) The partition coefficient of a substance X between methylbenzene (toluene) and water is 4.0, X being more soluble in methylbenzene. Starting with $100 \, cm^3$ of an aqueous solution containing $6.0 \, g$ of X, compare the amounts of X extracted (i) by shaking with one portion of $100 \, cm^3$ of methylbenzene, (ii) by shaking with two successive portions of $50 \, cm^3$.

    d) Explain the principles involved in any one chromatographic technique for the separation of the components of a mixture.   (WJEC80)

*14. What is the partition coefficient of a solute between two immiscible liquids? Give two examples (other than the one below) of cases where the partition law does not appear to be obeyed. Describe one practical use of partition other than chromatography.

Some iodine is dissolved in an aqueous solution of potassium iodide of concentration $0.102 \, mol \, dm^{-3}$, and the solution is then shaken with tetrachloromethane (carbon tetrachloride) until equilibrium is attained (at $15\,^{\circ}C$). The amount of iodine at equilibrium, as determined by titration, is found to be $0.048 \, mol \, dm^{-3}$ in the aqueous layer and $0.089 \, mol \, dm^{-3}$ in the tetrachloromethane layer. The distribution coefficient of iodine between tetrachloromethane and water is 85. Calculate the equilibrium constant at $15\,^{\circ}C$ for the reaction

$$I_3^-(aq) \rightleftharpoons I_2(aq) + I^-(aq) \qquad (O \& C 78)$$

*15. Benzoic acid (175 mg) is equilibrated between methylbenzene ($50 \, cm^3$) and water ($100 \, cm^3$) at 285 K. Titration shows that the aqueous layer then contains 50 mg of the acid. Benzoic acid exists as a dimer in the methylbenzene layer. In a second experiment benzoic acid (550 mg) is dissolved in a mixture of methylbenzene ($50 \, cm^3$) and water $100 \, cm^3$) and allowed to equilibrate at 285 K. Calculate the mass of benzoic acid which will be present in the aqueous layer in the second experiment explaining carefully the physico-chemical basis of your calculation.

What intermolecular bonding is responsible for the association of the benzoic acid in the organic solvent, and why is the benzoic acid not associated in aqueous solution?

Describe in outline one chromatographic method which involves the partition of the components of a mixture between two separate phases. (O80, S)

7 Solutions of Solids in Liquids: Colligative Properties

LOWERING OF THE VAPOUR PRESSURE OF A SOLVENT BY A SOLUTE

For a solution of two liquids, Raoult's law states that the vapour pressure of each component is equal to the product of its mole fraction and the saturated vapour pressure of the pure component. In a solution of a solid in a liquid, the solid is involatile, and the vapour pressure of the solution is equal to the product of the mole fraction of the solvent and the saturated vapour pressure of the pure solvent. It follows that the lowering of the vapour pressure (vapour pressure of solvent minus vapour pressure of solution) is equal to the mole fraction of the solute. This can be expressed as

$$\frac{p^0 - p}{p^0} = \frac{n_1}{n_2 + n_2}$$

where p^0 = vapour pressure of pure solvent, p = vapour pressure of solution, n_1 = number of moles of solute, and n_2 = number of moles of solvent.

In most solutions, $n_2 \gg n_1$, and the expression becomes

$$\frac{p^0 - p}{p^0} = \frac{n_1}{n_2}$$

Substituting $n = m/M$, where m = mass, M = molar mass gives

$$\frac{p^0 - p}{p^0} = \frac{m_1}{M_1} \bigg/ \frac{m_2}{M_2} = \frac{m_1 M_2}{m_2 M_1}$$

If the masses of solute and solvent are known, and the molar mass of the solvent is known, this expression can be used to give M_1, the molar mass of the solute.

EXAMPLE 1 The vapour pressure of water at $18\,°C$ is $2.06 \times 10^3\,N\,m^{-2}$. The vapour pressure of a solution of $10.0\,g$ of a solute in $100\,g$ of water at $18\,°C$ is $1.98 \times 10^3\,N\,m^{-2}$. Calculate its molar mass.

METHOD

$$\frac{p^0 - p}{p^0} = \frac{n_1}{n_2} = \frac{m_1}{m_2} \times \frac{M_2}{M_1}$$

$$\frac{2.06 - 1.98}{2.06} = \frac{10.0}{100} \times \frac{18.0}{M_1}$$

$$M_1 = \frac{18.0 \times 2.06}{0.08 \times 10.0} = 46.35$$

ANSWER   The molar mass of the solute is 46 g mol^{-1}.

EXAMPLE 2   The vapour pressure of water at $18\,^\circ C$ is 2.053×10^3 Pa. The vapour pressure of a solution containing 5.132 g of a solute in 100 g of water is 2.021×10^3 Pa. Calculate the molar mass of the solute.

METHOD

$$\frac{p^0 - p}{p^0} = \frac{n_1}{n_2} = \frac{m_1}{m_2} \times \frac{M_2}{M_1}$$

$$\frac{(2.053 \times 10^3) - (2.021 \times 10^3)}{2.053 \times 10^3} = \frac{5.132}{100} \times \frac{18.0}{M_1}$$

$$\frac{0.032}{2.053} = \frac{5.132 \times 18.0}{100 M_1}$$

$$M_1 = 59.46$$

ANSWER   The molar mass of the solute is 59 g mol^{-1}.

The transpiration method is often used to measure the relative vapour pressure lowering. A stream of dry air is passed through absorption tubes, a first set containing solution and a second set containing solvent. The amount of water absorbed by the dry air in passing through the first set of tubes is proportional to the vapour pressure of the solution. The vapour pressure of the pure solvent is higher than that of the solution, and the air passing through the second set of tubes will therefore absorb more water. The amount is proportional to the difference in vapour pressure between the two liquids. The loss in mass of each set of bulbs is measured.

If p^0 and p are the vapour pressures of the solvent and solution,

Loss in mass in solution bulbs, w_1, is proportional to p

Loss in mass of solvent bulbs, w_2, is proportional to $p^0 - p$

$$\therefore \qquad \frac{w_2}{w_1} = \frac{p^0 - p}{p}$$

A little manipulation gives

$$\frac{w_2}{w_2 + w_1} = \frac{p^0 - p}{p^0}$$

$$\therefore \qquad \frac{w_2}{w_2 + w_1} = \frac{n_1}{n_1 + n_2} = \frac{n_1}{n_2}$$

$$\frac{\text{Loss in mass of solvent bulbs}}{\text{Loss in mass of both bulbs}} = \frac{\text{Mass of solute/Molar mass of solute}}{\text{Mass of solvent/Molar mass of solvent}}$$

EXAMPLE 3   A current of dry air is passed through a solution containing 4.00 g of phenol in 50.0 g of water, then through pure water. The loss in mass in the solution is 2.76 g; in the water the loss is 0.0420 g. Calculate the molar mass of phenol.

METHOD
$$\frac{\text{Loss in mass of solvent bulbs}}{\text{Loss in mass of both bulbs}} = \frac{\text{Mass of solute/Molar mass of solute}}{\text{Mass of solvent/Molar mass of solvent}}$$

$$\frac{0.0420}{0.0420 + 2.76} = \frac{4.00/M}{50.0/18.0}$$

$$\frac{0.0420}{2.802} = \frac{4.00 \times 18.0}{50.0 \times M}$$

$$M = \frac{4.00 \times 18.0 \times 2.802}{50.0 \times 0.0420} = 96$$

ANSWER   The molar mass of phenol is 96 g mol^{-1}.

EXERCISE 34   Problems on Vapour Pressure Lowering

1. The vapour pressure of water at $20\,^{\circ}\text{C}$ is $2.34 \times 10^3 \text{ N m}^{-2}$. Calculate the vapour pressure of the following solutions:
   a) 34.2 g of sucrose, $C_{12}H_{22}O_{11}$, in 500 g of water
   b) 45.0 g of glucose, $C_6H_{12}O_6$, in 250 g of water
   c) 15.0 g of urea, $CO(NH_2)_2$, in 200 g of water
   d) 50.0 g of urea in 50.0 g of water.

2. Which of the following, dissolved in 1 dm^3 of water at $25\,^{\circ}\text{C}$, results in the biggest lowering of the vapour pressure?
   a 1 mole $C_{12}H_{22}O_{11}$ b 0.5 mole KCl
   c 0.5 mole $CuSO_4$ d 0.5 mole $Cu(NH_3)_4SO_4$
   e 0.5 mole $KAl(SO_4)_2$

3. The vapour pressure of a solution of a non-volatile compound X (29.0 g) in 100 g of water is $1.12 \times 10^4 \text{ Pa}$ at $50\,^{\circ}\text{C}$. At the same temperature, the vapour pressure of water alone was $1.22 \times 10^4 \text{ Pa}$. Deduce the molar mass of X.

4. At $25\,^{\circ}\text{C}$, the vapour pressure of water is $3.15 \times 10^3 \text{ Pa}$. Calculate the vapour pressure of a solution of 6.00 g of urea in 100 g of water.

5. A stream of dry air was bubbled through an aqueous solution containing 24.5 g dm^{-3} of a non-volatile solute and then through pure water. The loss in mass of the solution was 2.7320 g, and the loss in mass of the water was 0.0203 g. The temperature was $20\,^{\circ}\text{C}$. Calculate the molar mass of the solute.

6. A stream of dry air was passed through an aqueous solution containing $25.170\,\text{g}\,\text{dm}^{-3}$ of a non-volatile solute. The loss in mass of the solution was $3.0322\,\text{g}$, and of the water $0.0188\,\text{g}$. Calculate the molar mass of the solute.

7. A stream of dry air was bubbled through an aqueous solution of a non-volatile solute and then through pure water. The loss in mass of the solution was $5.0304\,\text{g}$ and of the water $0.0312\,\text{g}$. Calculate the concentration of the solution in $\text{mol}\,\text{dm}^{-3}$.

ELEVATION OF THE BOILING POINT OF A SOLVENT BY A SOLUTE

The lowering of the vapour pressure of a solvent depends on the mole fraction of solute in the solution. It does not depend on the nature of the solute. Properties such as vapour pressure lowering, which depend on the concentration of dissolved particles and not on their nature, are called *colligative properties*.

Since the boiling point of a liquid is the temperature at which its vapour pressure is equal to atmospheric pressure, the lowering of the vapour pressure which results from the presence of a solute is accompanied by a rise in boiling point. The boiling points of solutions are simpler to measure than their vapour pressures. A determination of the elevation of the boiling point of a solvent produced by a solute is a useful way of finding the molar mass of the solute.

The boiling point elevation is proportional to the concentration of solute and to a constant which has a certain value for each solvent. The elevation produced by 1 mole of solute in 1 kg of water is 0.52 K. This value is called the ebullioscopic constant for water.

> 1 mole of solute in 1 kg of water gives a boiling point elevation of 0.52 K.

For other solutions, the boiling point elevation depends on the concentration of solute and the magnitude of the ebullioscopic constant for the solvent.

$$\begin{pmatrix}\text{Boiling point}\\\text{elevation}\end{pmatrix} = \begin{pmatrix}\text{Boiling point}\\\text{constant}\end{pmatrix} \times \frac{\text{Moles of solute}}{\text{Mass of solvent in kg}}$$

$$\text{Boiling point elevation} = k \times \frac{m}{M} \times \frac{1}{W}$$

where k is the ebullioscopic constant for the solvent (or boiling point constant), m and M are the mass and molar mass of the solute, and W is the mass of the solvent.

If W is expressed in kg, then

$$\text{Boiling point elevation (K)} = k \times \frac{m\,(\text{g})}{M\,(\text{g mol}^{-1})} \times \frac{1}{W\,(\text{kg})}$$

and k has the units of $K\,kg\,mol^{-1}$.

EXAMPLE 1 3.76 g of solute in 50.0 g of water give a boiling point elevation of 0.95 K. Calculate the molar mass of the solute.

METHOD 1 Since 0.95 K elevation is produced by 3.76 g of solute in 50.0 g of water, $1/20 \times 0.95$ K elevation is produced by 3.76 g of solute in 1000 g of water and 0.52 K elevation is produced by

$$\frac{0.52 \times 3.76}{1/20 \times 0.95}\,\text{g of solute} = 41.16\,\text{g of solute}$$

ANSWER The molar mass of the solute is $41\,g\,mol^{-1}$.

METHOD 2 $$\text{Boiling point elevation} = k \times \frac{m}{M} \times \frac{1}{W}$$

$$0.95 = 0.52 \times \frac{3.76}{M} \times \frac{1}{0.0500}$$

ANSWER $M = 41\,g\,mol^{-1}$ (as before)

EXAMPLE 2 Pure propanone boils at $56.38\,^{\circ}\text{C}$ at 1 atm. A solution of 2.256 g of phenol in 100 g of propanone boils at $56.79\,^{\circ}\text{C}$, and a solution of 0.635 g of a compound, X, in 50.0 g of propanone boils at $56.52\,^{\circ}\text{C}$. What is the molar mass of X?

METHOD There are two steps in this calculation: a) find the ebullioscopic constant for propanone, using the data for the compound of known molar mass; b) use the ebullioscopic constant to find the unknown molar mass.

$$\text{Boiling point elevation} = k \times \frac{m}{M} \times \frac{1}{W}$$

a) $$0.400 = k \times \frac{2.256}{94.1} \times \frac{1}{0.100}$$

$$k = 1.67\,K\,kg\,mol^{-1}$$

b) $$0.140 = 1.67 \times \frac{0.635}{M} \times \frac{1}{0.0500}$$

ANSWER $$M = 152\,g\,mol^{-1}$$

EXERCISE 35   Problems on Boiling Point Elevation

1. Calculate the boiling points of the following solutions:
   a) 0.200 mol sucrose, $C_{12}H_{22}O_{11}$, in 500 g water
   b) 0.150 mol sucrose, $C_{12}H_{22}O_{11}$, in 250 g water
   c) 0.850 mol urea, $CO(NH_2)_2$, in 750 g water
   d) 20.0 g glucose, $C_6H_{12}O_6$, in 500 g water
   e) 20.0 g urea, $CO(NH_2)_2$, in 500 g water
   f) 30.0 g propanamide, $C_2H_5CONH_2$, in 400 g water.

2. Calculate the molar masses of the solutes in the following solutions:
   a) 6.00 g A in 100 g water, boiling at 100.52 °C
   b) 518 g B in 750 g water, boiling at 101.05 °C
   c) 60.0 g C in 500 g water, boiling at 100.85 °C
   d) 33.0 g D in 250 g water, boiling at 100.76 °C
   e) 79.0 g E in 200 g water, boiling at 101.14 °C.

DEPRESSION OF THE FREEZING POINT OF A SOLVENT BY A SOLUTE

A solution has a lower freezing point than the pure solvent. The depression of the freezing point of 1 kg of solvent produced by 1 mole of solute is constant for a particular solvent. Freezing point depression is another colligative property, depending on the concentration of particles in solution and not on their nature.

> 1 mole of solute in 1 kg of water gives a freezing point depression of 1.86 K

The molar depression of the freezing point is called the cryoscopic constant or freezing point constant of the solvent.

$$\begin{pmatrix} \text{Freezing point} \\ \text{depression} \end{pmatrix} = \begin{pmatrix} \text{Freezing point} \\ \text{constant} \end{pmatrix} \times \frac{\text{Moles of solute}}{\text{Mass of solvent}}$$

$$= k \times \frac{m}{M} \times \frac{1}{W}$$

where m and M are the mass and molar mass of the solute, W is the mass (kg) of solvent and k is the cryoscopic constant (freezing point constant), which has the units $K\,kg\,mol^{-1}$.

Measurements of freezing point depression can be used to calculate molar masses of solutes in the same way as boiling point elevation.

EXAMPLE   What is the molar mass of a solute which, when 1.00 g is dissolved in 15.0 g of water, gives a freezing point of -0.370 °C? The cryoscopic constant of water is $1.86\,K\,kg\,mol^{-1}$.

METHOD 1   Since 0.370 °C depression is produced by 1.00 g solute in 15.0 g water, 0.370 × 15.0/1000 °C depression is produced by 1.00 g in 1000 g water and 1.86 °C depression is produced by

$$\frac{1.86 \times 1000 \, g}{0.370 \times 15.0} = 335 \, g$$

ANSWER   The molar mass of the solute is $335 \, g \, mol^{-1}$.

METHOD 2 $$\text{Freezing point depression} = k \times \frac{m}{M} \times \frac{1}{W}$$

$$0.370 = 1.86 \times \frac{1}{M} \times \frac{1}{0.0150}$$

$$M = 335 \, g \, mol^{-1} \quad \text{(as before)}$$

EXERCISE 36   Problems on Freezing Point Depression

1. The freezing point constant for water is $1.86 \, K \, kg \, mol^{-1}$. Calculate the molar masses of the solutes in the following solutions:
   a) 21.0 g P in 200 g water, freezing at -3.72 °C
   b) 24.0 g Q in 200 g water, freezing at -2.79 °C
   c) 12.0 g R in 250 g water, freezing at -1.49 °C
   d) 100.0 g S in 500 g water, freezing at -2.07 °C
   e) 200.0 g T in 333 g water, freezing at -3.26 °C.

2. A solution of 3.75 g of sucrose, $C_{12}H_{22}O_{11}$, in 100 g of water freezes at -0.204 °C. A solution of X containing $27.3 \, g \, dm^{-3}$ freezes at -0.282 °C. Calculate the molar mass of X.

EXERCISE 37   Problems on Freezing Point Depression and Boiling Point Elevation

Use the following data:
$$\text{Freezing point constant for water} = 1.86 \, K \, kg \, mol^{-1}$$
$$\text{Boiling point constant for water} = 0.52 \, K \, kg \, mol^{-1}$$
$$\text{Freezing point of benzene} = 5.48 \, °C$$
$$\text{Freezing point constant for benzene} = 5.12 \, K \, kg \, mol^{-1}$$
$$\text{Boiling point of benzene} = 80.1 \, °C$$
$$\text{Boiling point constant for benzene} = 2.53 \, K \, kg \, mol^{-1}$$

1. 12.8 g of naphthalene, $C_{10}H_8$, are dissolved in 100 g of benzene.
   a) What is the freezing point of the solution?
   b) What is the boiling point of the solution?
   c) If the vapour pressure of pure benzene is $1.33 \times 10^4 \, Nm^{-2}$ at 25 °C, what is the vapour pressure of this solution at 25 °C?

2. 8.55 g of sucrose, $C_{12}H_{22}O_{11}$, are dissolved in 100 g of water.
   a) What is the boiling point of this solution?
   b) What is the freezing point of this solution?
   c) What is the vapour pressure of the solution at 20 °C if the vapour pressure of pure water at this temperature is $2.34 \times 10^3 \, N \, m^{-2}$.

3. 0.386 g of A in 50.0 g of benzene depresses the freezing point by 0.270 °C. Calculate the molar mass of A.

4. Calculate the freezing points and the boiling points of the following solutions:
   a) 28.0 g of ethanamide, CH_3CONH_2, in 500 g water
   b) 9.0 g of glucose, $C_6H_{12}O_6$, in 250 g water
   c) 17.1 g of sucrose, $C_{12}H_{22}O_{11}$, in 250 g water
   d) 50.0 g of urea, $CO(NH_2)_2$, in 750 g water
   e) 100.0 g of glycerol, $C_3H_8O_3$, in 500 g water.

5. Ethylene glycol (ethane-1, 2-diol, $C_2H_6O_2$) is added to car radiators to prevent freezing. What mass of ethylene glycol must be added to a radiator which holds 8 dm³ of water in order to protect against freezing down to $-15 \, °C$?

OSMOTIC PRESSURE

A semipermeable membrane is a film of material which can be penetrated by a solvent but not by a solute. When two solutions are separated by a semipermeable membrane, solvent passes from the more dilute to the more concentrated. This phenomenon is called *osmosis*. The pressure which must be applied to a solution to prevent the solvent from diffusing in is called the osmotic pressure of the solution. There is an analogy with gas pressure. One mole of a solid, A, when vaporised, occupies a volume of 22.4 dm³ at 0 °C and $1.01 \times 10^5 \, N \, m^{-2}$. One mole of A dissolved in 22.4 dm³ of solvent at 0 °C exerts an osmotic pressure of $1.01 \times 10^5 \, N \, m^{-2}$.

> 1 mole of solute in 22.4 dm³ of solvent at 0 °C has an osmotic pressure of $1.01 \times 10^5 \, N \, m^{-2}$ (1 atmosphere).

The expression which relates osmotic pressure to concentration and temperature is similar to the Ideal Gas Equation,

$$\pi V = nRT$$

where π is the osmotic pressure, V is the volume, T is the temperature (Kelvin), n is the amount (in mol) of solute, and R is a constant which has the same value as the gas constant, $8.314 \, J \, K^{-1} \, mol^{-1}$. This equation is obeyed by ideal solutions.

The osmotic pressure of a solution depends on the concentration of solute present: it is a colligative property. Measurements of osmotic pressure can be used to give the molar masses of solutes.

EXAMPLE Calculate the molar mass of a solute, given that 35.0 g of the solute in 1.00 dm³ water have an osmotic pressure of $5.15 \times 10^5 \, N \, m^{-2}$ at 20 °C.

METHOD
$$\pi V = nRT$$

where $\pi = 5.15 \times 10^5 \, N \, m^{-2}$, $V = 1.00 \times 10^{-3} \, m^3$,
$R = 8.314 \, J \, K^{-1} \, mol^{-1}$, and $T = 293 \, K$.

$$\therefore \qquad 5.15 \times 10^5 \times 1.00 \times 10^3 = n \times 8.314 \times 293$$

$$n = 0.211$$

$$n = \frac{35}{M}$$

$$M = 165 \, g \, mol^{-1}.$$

ANSWER The solute has a molar mass of $165 \, g \, mol^{-1}$.

EXERCISE 38   Problems on Osmotic Pressure

1. Find the osmotic pressure of the following aqueous solutions at 25 °C:
   a) a sucrose solution of concentration $0.213 \, mol \, dm^{-3}$
   b) a solution containing $144 \, g \, dm^{-3}$ of glucose
   c) a solution which freezes at $-0.60 \, °C$
   d) a solution which boils at 100.30 °C
   e) a solution containing 12.0 g of urea in 200 cm³ of solution.

2. Find the molar masses of the following solutes:
   a) 1.50 g of A in 200 cm³ of aqueous solution, having an osmotic pressure of $2.66 \times 10^5 \, N \, m^{-2}$ at 20 °C
   b) 20.0 g of B in 100 cm³ of aqueous solution, having an osmotic pressure of $3.00 \times 10^6 \, N \, m^{-2}$ at 27 °C
   c) 5.00 g of C in 200 cm³ of solution, having an osmotic pressure of $2.39 \times 10^5 \, N \, m^{-2}$ at 25 °C.

3. The osmotic pressure of blood at 37 °C is $8.03 \times 10^5 \, N \, m^{-2}$. What is its freezing point?

4. A polysaccharide has the formula $(C_{12}H_{22}O_{11})_n$. A solution containing $5.00 \, g \, dm^{-3}$ of the sugar has an osmotic pressure of $7.12 \times 10^2 \, N \, m^{-2}$ at 20 °C. Find n in the formula.

5. A solution of PVC $(CH_2CHCl)_n$, in dioxan has a concentration of $4.00 \, g \, dm^{-3}$ and an osmotic pressure of $65 \, N \, m^{-2}$ at 20 °C. Calculate the value of n.

6. A solution of 2.00 g of a polymer in 1 dm³ of water has an osmotic pressure of $300\,N\,m^{-2}$ at 20 °C. Calculate the molar mass of the polymer.

7. a) Calculate the osmotic pressure of a solution of glucose containing $15.0\,g\,dm^{-3}$ at 17 °C.

   b) What concentration of urea solution (in $g\,dm^{-3}$) would be isotonic with this solution?

8. A stream of dry air was bubbled through an aqueous solution of a non-volatile solute and then through pure water. The loss in mass of the solution was 5.0304 g and of the water 0.0312 g. Calculate the osmotic pressure of the solution at 22 °C.

ANOMALOUS RESULTS FROM MOLAR MASS DETERMINATIONS FROM COLLIGATIVE PROPERTIES

Colligative properties depend on the number of particles in solution. If the number of particles increases through dissociation of the solute, the observed colligative property increases. If association occurs, the number of particles is less than expected, and the colligative property is less than expected.

Dissociation

Electrolytes give abnormally high values of colligative properties. Van't Hoff discovered this, and the ratio

$$\frac{\text{Observed colligative property}}{\text{Calculated colligative property}}$$

is called the *Van't Hoff factor*, usually represented as i.

The reason for the high value of i for electrolytes is dissociation into ions. If a solute, such as sodium chloride, NaCl, is completely dissociated, there are two moles of ions for every mole of salt, and the colligative properties are twice the values calculated in the absence of ionisation. If copper(II) nitrate, $Cu(NO_3)_2$ is completely dissociated, there are three moles of ions per mole of solute, and the colligative properties are three times the values calculated in the absence of ionisation.

If the solute is incompletely dissociated, the degree of dissociation can be calculated. If a solute AB dissociates into two ions, and α is the degree of dissociation,

$$\text{AB} \; \rightleftharpoons \; \text{A}^+ + \text{B}^-$$
$$(1-\alpha) \qquad\qquad \alpha \qquad \alpha$$

1 mole of AB will produce α moles of A^+ and α moles of B^-, leaving $(1-\alpha)$ moles of AB. The total number of particles present is $(1+\alpha)$.

Thus,

$$\frac{\text{Actual no. of particles}}{\text{Expected no. of particles}} = \frac{\text{Observed colligative property}}{\text{Calculated colligative property}} = \frac{1+\alpha}{1}$$

In general, if 1 mole of solute dissociates into n moles of ions,

$$\frac{\text{Observed colligative property}}{\text{Expected colligative property}} = \frac{1 + (n-1)\alpha}{1}$$

$$\therefore \qquad \alpha = \frac{i-1}{n-1}$$

It is now believed that strong electrolytes are always completely dissociated in solution. The Debye-Hückel and Onsager theory offers a different explanation for the low values of α obtained in measurements on concentrated solutions of strong electrolytes. The value of α calculated from colligative properties is called the *apparent degree of ionisation*.

Association

A value for a colligative property lower than the calculated value may indicate association of molecules of solute. If n molecules of solute associate to form X_n, and the degree of association is α,

$$nX \rightleftharpoons X_n$$
$$1-\alpha \qquad\quad \alpha/n$$

The total amount of particles $= 1 - \alpha + \alpha/n = 1 - \dfrac{(n-1)\alpha}{n}$ mol

and

$$\frac{\text{Observed colligative property}}{\text{Calculated colligative property}} = i = 1 - \frac{(n-1)\alpha}{n}$$

If X dimerises, $n = 2$, and $i = 1 - \alpha/2$.

EXAMPLE 1 When 2.15 g of calcium nitrate are dissolved in 100 g of water, the solution freezes at $-0.62\,°C$. Find the apparent degree of dissociation of the salt.

METHOD The molar mass of $Ca(NO_3)_2 = 164\,g\,mol^{-1}$.

For 1 mole of solute in 1000 g of water, the expected depression $= 1.86\,°C$.

Thus, 164 g of $Ca(NO_3)_2$ in 100 g water should give a depression $= 18.6\,°C$

and 2.15 g of $Ca(NO_3)_2$ in 100 g water should give $18.6 \times \dfrac{2.16}{164} = 0.24\,°C$

$$i = 0.62/0.24 = 2.58$$

Number of ions per formula unit of $Ca(NO_3)_2 = n = 3$

$$\therefore \quad \alpha = \frac{i-1}{n-1} = \frac{2.58-1}{3-1} = 0.78$$

ANSWER   The apparent degree of dissociation of the salt is 0.78 (78%).

EXAMPLE 2 A solution of 3.78 g of ethanoic acid in 75.0 g of benzene boiled 1.14 °C above the boiling point of the pure solvent. The ebullioscopic constant of benzene is $2.53 \ K \ kg \ mol^{-1}$. What do these results indicate about the molecular condition of ethanoic acid?

METHOD   The molar mass of CH_3CO_2H is $60.0 \ g \ mol^{-1}$

60 g of ethanoic acid in 1 000 g of benzene should give an elevation of 2.53 °C.

3.78 g of ethanoic acid in 75.0 g of solvent should give

$$\frac{3.78}{60.0} \times \frac{1000}{75} \times 2.53 = 2.12 \ ^\circ C$$

The observed elevation is only just over half this; therefore the molecules are associated in pairs. The degree of association can be calculated, using the expression:

$$\frac{\text{Observed colligative property}}{\text{Calculated colligative property}} = 1 - \frac{(n-1)\alpha}{n} = 1 - \frac{\alpha}{2}$$

$$\therefore \quad \frac{1.14}{2.12} = 1 - \frac{\alpha}{2} \quad \text{and} \quad \alpha = 0.92$$

ANSWER   The degree of association is 0.92 (92%).

EXERCISE 39   Problems on Electrolytes

Use the physical constants on p. 111.

1. Assume that sodium chloride is completely ionised in aqueous solution. What are the freezing point and the boiling point of a $0.220 \ mol \ dm^{-3}$ solution of sodium chloride?

2. Assume that sodium nitrate is completely ionised in aqueous solution. Calculate the concentration of a sodium nitrate solution that freezes at $-2.15 \ ^\circ C$.

3. A solution of calcium nitrate freezes at $-2.45 \ ^\circ C$. What is the concentration of the solution, assuming calcium nitrate to be completely ionised?

4. What mass of anhydrous calcium chloride must be dissolved in $1.00 \ dm^3$ of water to make a solution with the same freezing point as a solution containing 1.86 g of potassium chloride in $500 \ cm^3$ of water?

5. What is the boiling point of a solution containing 150.0 g of potassium sulphate in 750 cm^3 of aqueous solution?

6. A solution of ethanoic acid has a concentration of 0.100 mol dm^{-3} and a freezing point of $-0.190\,°C$. What is the degree of ionisation of ethanoic acid at this temperature?

7. A solution of potassium bromide has a concentration of 0.250 mol dm^{-3} and a freezing point of $-0.83\,°C$. What is the apparent degree of ionisation?

8. A solution of magnesium nitrate has a concentration of 0.500 mol dm^{-3} and a boiling point of 100.665 $°C$. What is the apparent degree of ionisation?

9. Barium hydroxide has an apparent degree of ionisation of 0.92. What is the freezing point of a solution containing 2.50 g of barium hydroxide in 1.00 dm^3?

10. An organic compound has the empirical formula C_3H_2Br. 10.0 g of the compound dissolved in 1000 g benzene made a solution with a boiling point of 80.21 $°C$. What is the molecular formula of the compound?

11. A 0.200 mol dm^{-3} aqueous solution of the acid HA freezes at $-0.656\,°C$. Calculate the degree of ionisation of the acid.

12. Nitrous acid is 4.25% ionised in a 0.200 mol dm^{-3} solution. What is the freezing point of the solution?

13. Benzoic acid has the empirical formula $C_7H_6O_2$. When 0.2095 g of benzoic acid is dissolved in 100 g of water, the freezing point of the solution is $-0.0320\,°C$. A solution of 2.057 g of benzoic acid in 100 g of benzene froze at 5.05 $°C$. Find the molecular formula for benzoic acid in these two solutions. Explain any differences observed.

14. A solution of car radiator antifreeze ($C_2H_6O_2$, 1,2-dihydroxyethane) freezes at $-10.0\,°C$. What is the concentration of the antifreeze in kg per kg of water?

15. The freezing point of a solution of a salt is $-0.100\,°C$. What is the vapour pressure of the solution at 25 $°C$ if the vapour pressure of pure water at 25 $°C$ is 3.16×10^3 Pa?

16. 1.00 g of an acid dissolved in 100 g of water depressed the freezing point by 0.250 $°C$. Dissolved in 100 g of benzene, 1.00 g of the acid depressed the freezing point by 0.350 $°C$. What can you deduce from this information?

17. A solution containing 1.00 g of X in 50.0 g of water boils at 100.135 $°C$ at 10^5 Pa. Calculate: a) the molar mass of X, and b) the osmotic pressure of the solution.

18. The freezing point of a solution is $-1.150\,°C$. Calculate its osmotic pressure at 20 $°C$.

EXERCISE 40   Questions from A-level Papers

1. a) Explain what is meant by the term 'colligative property'. Describe briefly *three* colligative properties and discuss the factors which affect their magnitudes (practical details are not required).

   b) Calculate the relative molecular masses of an organic monobasic acid from each of the following sets of data and account for the results obtained.

   0.25 g of the acid required 20.5 cm^3 of 0.1 M sodium hydroxide for complete neutralisation.

   3.61 g of the acid dissolved in 25 g of benzene caused a boiling point elevation of 1.5 °C and 3.61 g of the acid dissolved in 25 g of water caused an elevation of 0.6 °C.

   [Boiling point constant for benzene = 2.53 K mol^{-1} (kg of solvent)$^{-1}$; boiling point constant for water = 0.51 K mol^{-1} (kg of solvent)$^{-1}$.]

   (JMB79)

2. a) Assuming that the vapour pressure of water at any constant temperature is reduced on adding an involatile solute such as sugar by an amount proportional to the concentration of the solute, show with the aid of a diagram that the elevation of the boiling-point of such a solution is proportional to the concentration of the solute.

   b) Give a labelled sketch of the apparatus you would use to determine the depression of the freezing-point of a solution of sugar.

   c) The freezing-point of pure benzene is 5.533 °C. The freezing-point of a solution of 6.40 g of naphthalene ($C_{10}H_8$) in 1000 g of benzene is 5.277 °C, while that of a solution of 15.25 g of benzoic acid, $C_7H_6O_2$, in 1000 g of benzene is 5.175 °C. What conclusions can you draw from this information?   (O79)

3. Describe how you would determine the relative molecular mass of a solute by measuring the depression in freezing point of water. State clearly the precautions you would take to ensure accuracy.

   When 0.775 g of white phosphorus was dissolved in 50.0 g of benzene, the solution froze at 4.89 °C. What value do these data give for the relative molecular mass of white phosphorus? What explanation can you offer for this value?

   [Freezing point of pure benzene = 5.53 °C. Cryoscopic (freezing point) constant for benzene = 5.12 °C mol^{-1} kg, i.e. 5.12 °C mol^{-1} for 1000 g of solvent.]   (C78)

4. a) The melting points of aluminium chloride and sodium chloride are 466 K (under pressure), and 1047 K respectively. Comment on the difference between these figures.

   b) 1.0 g of anhydrous aluminium chloride dissolved in 40 g of benzene, depressed the freezing point of the solvent by 0.5 °C. The molar freezing point depression for benzene is $5.1 °kg^{-1}$. [Cl = 35.5; Al = 27.] What do you conclude from the data?

   c) Show diagrammatically the bonding in aluminium chloride, when dissolved in benzene.   (WJEC77, p)

5. In working out each of the following calculations state clearly the physico-chemical principles upon which it depends.

   a) An organic compound is four times as soluble in tetrachloromethane (carbon tetrachloride) as in water. By calculation show that two successive extractions with $50 \, cm^3$ of tetrachloromethane are together more productive than one extraction with $100 \, cm^3$ of tetrachloromethane, in recovering the compound from its solution in $200 \, cm^3$ of water. State any assumptions relating to the underlying principle clearly.

   b) The molar masses of nitrobenzene and of water are 123 and 18 respectively. During steam distillation of nitrobenzene the mixture boiled at 99.25 °C under atmospheric pressure, $1.00 \times 10^5 \, Pa$ (760 mm Hg), the partial vapour pressure of water at this temperature being $9.75 \times 10^4 \, Pa$ (741 mm Hg). Calculate the percentage by weight of nitrobenzene in the steam-distillate.

   Pa (pascal, $N \, m^{-2}$) is the SI unit of pressure.
   1 mm Hg = 13.60×9.807 Pa.

   c) A complex plant sugar, a polysaccharide, develops an osmotic pressure of $7.1 \times 10^2 \, Pa$ (5.40 mm Hg) at 27 °C when a solution of concentration $5.0 \, g \, dm^{-3}$ is investigated by the Berkeley and Hartley method. Calculate the number of simple sucrose sugar units, $C_{12}H_{22}O_{11}$, in each molecule of the polysaccharide, using information from *the following data only.*

$$C_{12}H_{22}O_{11} = 342; \quad H_2O = 18$$
$$R = 8.314 \, J \, mol^{-1} K^{-1} (= 0.0821 \, dm^3 \, atm \, mol^{-1} K^{-1})$$
$$J = kg \, m^2 s^{-2} = N \, m$$

   Pa (pascal, $N \, m^{-2}$) is the SI unit of pressure.
   Standard atmospheric pressure = 760 mm Hg.   (SUJB77)

6. Explain the term *osmotic pressure.*

Outline the principles of an osmotic pressure method for the determination of the relative molecular mass of a solute.

Calculate the osmotic pressure of an aqueous solution containing $25.0 \, g \, dm^{-3}$ of a protein of relative molecular mass 5.00×10^4 at 27 °C.

Explain carefully why this method is appropriate for solutes of high relative molecular mass.   (C80, p)

7. An organic monobasic acid contains 48.6% of carbon. 0.35 g of the vapour of the acid occupies a volume of 84.5 cm³ at 157 °C and 99.7 kPa (750 mm Hg), while 0.43 g contained in 100 cm³ of aqueous solution had an osmotic pressure of 142 kPa (1069 mm Hg) at 17 °C. [Standard pressure = 101 kPa (760 mm Hg).]

a) Calculate the apparent relative molecular mass of the acid i) in the vapour state, ii) in aqueous solution.

b) Explaining your reasoning, identify the acid.

c) Calculate the degree of ionisation of the acid in aqueous solution.

d) Draw a fully labelled diagram of the apparatus you would use to determine the ratio mass/volume at constant temperature for the acid vapour. List the measurements you would make.   (AEB76)

8 Electrochemistry

ELECTROLYSIS

Electrovalent compounds, when molten or in solution, conduct electricity. The conductors which connect the melt or solution with the applied voltage are called the *electrodes*. The positive electrode is called the *anode*; the negative electrode is the cathode. Chemical reactions occur at the electrodes, and elements are deposited as solids or evolved as gases. These reactions are called *electrolysis*.

If the compound is a salt of a metal low in the electrochemical series, the metal ions are discharged, and a layer of metal is deposited on the cathode. Silver, copper and gold are such metals. During electrolysis of their salts, the cathode processes can be represented by

$$Ag^+(aq) \; + \; e^- \longrightarrow Ag(s)$$
$$Cu^{2+}(aq) \; + \; 2e^- \longrightarrow Cu(s)$$
$$Au^{3+}(aq) \; + \; 3e^- \longrightarrow Au(s)$$

One can see from these equations that

1 mole of e^- discharge 1 mole of Ag^+ ions to give 1 mole of Ag atoms,

2 moles of e^- discharge 1 mole of Cu^{2+} ions to give 1 mole of Cu atoms,

3 moles of e^- discharge 1 mole of Au^{3+} ions to give 1 mole of Au atoms.

Thus, in general,

$$\left(\begin{array}{l}\text{No. of moles of} \\ \text{element discharged}\end{array}\right) = \frac{\text{No. of moles of electrons}}{\text{No. of charges on one ion of the element}}$$

It is found by experiment that 96 500 coulombs of charge are required to discharge 1 mole of silver (108 g); therefore 96 500 coulombs must be the charge on a mole of electrons. The ratio $96\,500\,C\,mol^{-1}$ is called the *Faraday constant*. Note that

No. of coulombs (C) = No. of amperes (A) × Time in second (s)

EXAMPLE 1 A direct current of 10.0 mA flows for 4.00 hours through three cells in series. They contain solutions of silver nitrate, copper(II) sulphate, and gold(III) nitrate. Calculate the mass of metal deposited in each.

METHOD   Coulombs $=$ Amperes $\times$ Seconds $=$ $0.0100 \times 4.00 \times 60 \times 60$
$$= 144\,C$$

Electrical charge passed $=$ $144/96\,500$ moles of electrons
No. of moles of Ag deposited $=$ $144/96\,500$ mol
Mass of Ag deposited $=$ $108 \times 144/96\,500$ $=$ $0.161\,g$
No. of moles of Cu deposited $=$ $\frac{1}{2} \times 144/96\,500$ mol
Mass of Cu deposited $=$ $63.5 \times \frac{1}{2} \times 144/96\,500$ $=$ $0.0474\,g$
No. of moles of Au deposited $=$ $\frac{1}{3} \times 144/96\,500$ mol
Mass of Au deposited $=$ $197 \times \frac{1}{3} \times 144/96\,500$ $=$ $0.0980\,g$

ANSWER   Deposited are: $0.161\,g$ silver; $0.0474\,g$ copper; $0.0980\,g$ gold.

EXAMPLE 2 A metal of relative atomic mass 27 is deposited by electrolysis. If $0.176\,g$ of the metal is deposited on the cathode when $0.15\,A$ flows for $3\frac{1}{2}$ hours, what is the charge on the cations of this metal?

METHOD   Coulombs $=$ Amperes $\times$ seconds $=$ $0.15 \times 3\frac{1}{2} \times 60 \times 60$ $=$ $1890\,C$
If $1890\,C$ deposit $0.176\,g$ of metal,

then $96\,500\,C$ deposit $\dfrac{96\,500 \times 0.176}{1890}\,g$ $=$ $8.98\,g$

1 mole of metal $=$ $27\,g$
Since $8.98\,g$ of metal are discharged by 1 mole of electrons,
          $27\,g$ are discharged by $27/8.98$ $=$ 3 moles of electrons

ANSWER   If 1 mole of metal needs 3 moles of electrons, the charge on a metal ion must be $+3$.

EVOLUTION OF GASES

Solutions of salts of metals high in the electrochemical series evolve hydrogen at the cathode on electrolysis. The cathode process is:
$$H^+(aq) \;+\; e^- \longrightarrow H(g)$$
followed by $2H(g) \longrightarrow H_2(g)$

Thus, 2 moles of electrons are needed to evolve 1 mole of hydrogen molecules; that is, $2\,g$ of hydrogen (the molar mass) or $22.4\,dm^3$ at s.t.p. (the molar volume).

At the anode, solutions of halides evolve the halogen, and other salts evolve oxygen. When chlorine is evolved, the anode process is
$$Cl^-(aq) \longrightarrow Cl(g) \;+\; e^-$$
followed by $2Cl(g) \longrightarrow Cl_2(g)$

Thus, 2 moles of electrons are required for the evolution of 1 mole of chlorine molecules, $71.0\,g$ or $22.4\,dm^3$ at s.t.p.

When oxygen is evolved, the anode process is the discharge of hydroxide ions, derived from the water in the solution:

$$OH^-(aq) \longrightarrow OH(aq) + e^-$$

followed by $4OH(aq) \longrightarrow O_2(g) + 2H_2O(l)$

Thus, 4 moles of electrons are required for the evolution of 1 mole of oxygen molecules ($22.4\ dm^3$ oxygen at s.t.p.).

EXAMPLE 1 State the names and calculate the volumes of gases formed at the cathode and anode at s.t.p. when $0.0250\ A$ of current are passed for 4.00 hours through a solution of sulphuric acid.

METHOD At the anode, oxygen is evolved, and at the cathode hydrogen.

Coulombs = Amperes $\times$ Seconds = $0.0250 \times 4.00 \times 60 \times 60 = 360\ C$

No. of moles of electrons = $360/96\,500$

No. of moles of H_2 = $\frac{1}{2} \times 360/96 \cdot 500$

Vol. of H_2 = $22.4 \times \frac{1}{2} \times 360/96\,500\ dm^3$ = $0.0417\ dm^3$ = $41.7\ cm^3$

Vol. of O_2 = $\frac{1}{2} \times 41.7$ = $20.9\ cm^3$

ANSWER $41.7\ cm^3$ hydrogen is formed at the cathode and $20.9\ cm^3$ of oxygen at the anode.

EXAMPLE 2 A current of $1.00\ A$ flowing for 1 hour 50 minutes deposits $2.15\ g$ of copper from an aqueous solution of copper(II) sulphate. If the Avogadro constant is $6.02 \times 10^{23}\ mol^{-1}$, calculate the charge on an electron.

METHOD Coulombs = Amperes $\times$ Seconds = $1.00 \times 110 \times 60 = 6600\ C$

No. of moles of Cu = $2.15/63.5$

No. of atoms of Cu = $6.02 \times 10^{23} \times 2.15/63.5$

No. of electrons = $2 \times 6.02 \times 10^{23} \times 2.15/63.5$ = 0.406×10^{23}

$\dfrac{\text{No. of coulombs}}{\text{No. of electrons}}$ = $6600/(0.406 \times 10^{23})$ = $1.63 \times 10^{-19}\ C\ electron^{-1}$

ANSWER The charge on an electron is $1.63 \times 10^{-19}\ C$.

EXERCISE 41   Problems on Electrolysis

In the following problems use Faraday constant = $96\,500\ C\,mol^{-1}$.

1. Calculate the mass of copper that would be deposited on a copper cathode from an aqueous solution of copper(II) sulphate, if the same current, passed for the same time, liberated $0.900\ g$ of silver from an aqueous solution of silver nitrate, $AgNO_3$.

2. A current of 100 mA was passed for 1.00 h through an aqueous solution of 1.00 mol dm^{-3} silver nitrate. A silver cathode was used. Calculate the increase in mass of the cathode. State how the change in mass would be affected by:
   a) passing a current of 200 mA
   b) passing a current of 100 mA for 2.00 h
   c) using a 2.00 mol dm^{-3} solution of silver nitrate.

3. An electric current of 5.00 A was passed through molten anhydrous calcium chloride, $CaCl_2$, for 20.0 minutes between graphite electrodes. Calculate the mass of each product liberated.

4. A current of electricity liberated 6.22×10^{-3} mol of silver from a silver nitrate solution. What mass in grams, of aluminium, would be liberated from a suitable aluminium compound, using the same quantity of electricity?

5. A current is passed for 45 minutes through three solutions in series, using platinum electrodes. In one, 0.203 g of silver is deposited from silver nitrate solution. In a second, hydrogen is evolved from a solution of dilute sulphuric acid, and, in a third, lead is deposited from a solution of lead nitrate. Calculate: a) the current passed, b) the volume of hydrogen collected at $0.983 \times 10^5 \, N\,m^{-2}$ and 18 °C, c) the mass of lead deposited.

6. A current of 0.750 A passes through 250 cm^3 solution of 0.250 mol dm^{-3} copper(II) sulphate solution. How long (in minutes) will it take to deposit all the copper on the cathode?

7. What current is needed to deposit 0.500 g of nickel from a nickel(II) sulphate solution in 1.00 hour?

8. A current of 1.25 A passes for 5.00 h between platinum electrodes in 500 cm^3 of copper(II) sulphate solution of concentration 2.00 mol dm^{-3}. What will be the concentration of copper(II) sulphate at the end of the time?

9. How long (in hours) will it take a current of 0.100 A to deposit 1.00 kg of silver from an unlimited source of silver ions?

10. A current of 2.05 A passes through a solution of sulphuric acid for 5.00 h. Calculate the volumes of hydrogen and oxygen produced.

11. A steady current was passed through a solution of copper(II) sulphate until 6.05 g of metallic copper were deposited on the cathode. How many coulombs of electricity had been used?

12. In the electrolysis of a solution of potassium chloride, 100 cm^3 of chlorine are produced at 20 °C and $9.9 \times 10^4 \, N\,m^{-2}$. How many seconds has a current of 0.750 A been flowing through the solution to effect this?

13. A current of 1.75 A passed for 1.00 hour through a solution of copper(II) sulphate. At the cathode, 1.245 g of copper were deposited, and some hydrogen was evolved. Calculate the volume of hydrogen (at s.t.p.) which was evolved.

ELECTROLYTIC CONDUCTIVITY

Solutions of electrolytes obey Ohm's law, from which it follows that

$$R = V/I$$

where R = Resistance, V = Potential difference, I = Current.

Resistivity is defined by the equation

$$R = \rho \times l/A$$

where l = Length and A = Cross-sectional area of the conductor.

If R is in ohms (Ω), l in metres (m), and A in square metres (m^2), then ρ has the dimensions Ω m. (If l is in cm, and A in cm^2, then ρ has the dimensions Ω cm.)

The reciprocal of resistance is conductance; the reciprocal of resistivity ρ is conductivity κ.

$$1/\rho = \kappa$$

κ has the dimensions $\Omega^{-1} m^{-1}$ or $\Omega^{-1} cm^{-1}$.

Molar conductivity Λ is defined by the equation

$$\Lambda = \kappa/c$$

---

where c = Molar concentration of solute. If κ is in $\Omega^{-1} m^{-1}$ and c is in mol m^{-3}, then $\Lambda = \Omega^{-1} m^2 mol^{-1}$, and Λ is numerically equal to the conductivity of 1 mole of the electrolyte.

The molar conductivity, Λ, of the solute increases as the concentration of the solution decreases, until further dilution has no further effect. The value of molar conductivity at this dilution is called the molar conductivity at infinite dilution, and is represented as Λ_0 or Λ_∞.

EXAMPLE 1   Calculate the molar conductivity of sodium chloride solution of concentration 0.0100 mol dm^{-3}, given that its conductivity is 0.1185 $\Omega^{-1} m^{-1}$.

METHOD $c = 0.0100$ mol dm$^{-3} = 10.0$ mol m^{-3}

$\kappa = 0.1185 \, \Omega^{-1} m^{-1}$

$\Lambda = \kappa/c = 0.1185/10.0 = 1.185 \times 10^{-2} \, \Omega^{-1} m^2 mol^{-1}$

ANSWER   The molar conductivity is $1.185 \times 10^{-2} \, \Omega^{-1} m^2 mol^{-1}$ or $118.5 \, \Omega^{-1} cm^2 mol^{-1}$.

To find the conductivity of a solution, its resistance is measured, and the equation $R = \rho \times l/A$ is inverted to give

$$1/R = \kappa \times A/l$$

The area of the electrodes, A, and the distance between them, l, can be measured and inserted in the equation to give κ. In practice, the method of finding the ratio l/A, which is called the cell constant, is to measure the resistance of a solution of known conductivity and insert κ and R into the equation to give the value of l/A. Once the cell constant is known, this cell can be used to find values of conductivity from measurements of the conductance of solutions.

EXAMPLE 2 The resistance of a conductance cell containing $0.100\,\text{mol dm}^{-3}$ KCl solution at $25\,°C$ is $47.85\,\Omega$. If the same cell contains sodium nitrate solution of concentration $0.0200\,\text{mol dm}^{-3}$, the resistance is $254\,\Omega$. The conductivity of the KCl solution is $0.0129\,\Omega^{-1}\text{cm}^{-1}$. Calculate a) the cell constant, b) the conductivity of the sodium nitrate solution and c) the molar conductivity of sodium nitrate at a concentration of $0.0200\,\text{mol dm}^{-3}$.

METHOD   a) $R = 47.85\,\Omega$, and $\kappa = 0.0129\,\Omega^{-1}\text{cm}^{-1} = 1.29\,\Omega^{-1}\text{m}^{-1}$

Cell constant $= l/A$

Since $1/R = \kappa \times A/l$

ANSWER $l/A = 47.85 \times 1.29 = 61.7\,\text{m}^{-1}$

b) $R = 254.0\,\Omega$

Since $1/R = \kappa \times A/l$

$1/254.0 = \kappa/61.7$

ANSWER $\kappa = 61.7/254.0 = 0.243\,\Omega^{-1}\text{m}^{-1}$

c) Since $\Lambda = \kappa/c$

ANSWER $\Lambda = 0.243/20.0 = 1.215 \times 10^{-2}\,\Omega^{-1}\text{m}^2\,\text{mol}^{-1}$.

EXERCISE 42   Problems on Conductivity

1. A column of electrolyte $3.00\,\text{cm}$ long and of cross-section $2.00\,\text{cm}^2$ has a resistance of $15.0\,\Omega$. Calculate: a) the resistivity and b) the conductivity of the electrolyte.

2. An electrolytic cell is $3.00\,\text{cm}$ long and has a cross-section of $1.50\,\text{cm}^2$. If the resistance is $12.0\,\Omega$, a) what is the resistivity, and b) what is the conductivity of the electrolyte?

3. The resistance of a cell containing $0.0100\,\text{mol dm}^{-3}$ KCl solution is $12.5\,\Omega$. The conductivity of KCl at this concentration is $0.1413\,\Omega^{-1}\text{m}^{-1}$. What is the cell constant?

4. An electrolyte solution of concentration $3.00 \times 10^{-3}\,\text{mol dm}^{-3}$ has a resistance of $45.0\,\Omega$ in a cell with a cell constant of $2.20\,\text{m}^{-1}$. Calculate the molar conductivity of the electrolyte.

5. A solution of an electrolyte of concentration $0.0200 \, mol \, dm^{-3}$ has a resistance of $0.357 \, \Omega$ in a cell with a cell constant of $1.50 \, m^{-1}$. Calculate the molar conductivity of the electrolyte.

MOLAR CONDUCTIVITY AND CONCENTRATION

The molar conductivity of a solute depends on its concentration. When values of molar conductivity Λ are plotted against the volume of solution containing 1 mole of solute (called the dilution), the graphs obtained for different electrolytes fall into two categories. These are shown in Fig. 8.1(a).

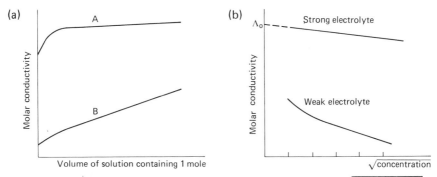

Fig. 8.1 Plots of molar conductivity against (a) dilution, (b) $\sqrt{\text{concentration}}$

Strong electrolytes are those which are completely ionised in solution. They give graphs of shape A, swiftly rising to a maximum. Weak electrolytes, which are incompletely ionised, give graphs of shape B. They do not reach a maximum at dilutions for which conductivities can be measured in practice. Arrhenius explained graphs of shape B as arising from an increase in the ionisation of the solute with dilution. The fraction of the solute which is ionised, α, gradually increases with dilution. If measurements could be made in more and more dilute solutions, α would increase up to a value of 1, and Λ would reach its limiting value of Λ_0. This is called the molar conductivity at infinite dilution since further dilution results in no further increase in Λ. In more concentrated solutions,

$$\alpha = \Lambda/\Lambda_0$$

Strong electrolytes show the steep increase in Λ with dilution illustrated in curve A. It seems to indicate that the strong electrolytes are completely ionised except at high concentrations. According to more modern views, strong electrolytes are completely ionised at all concentrations. Another explanation of the low value of Λ at high concentration is proposed. It is attributed to the attraction of oppositely charged ions slowing down the ions in a concentrated solution. The value of α given by Λ/Λ_0 is referred to as the *apparent degree of ionisation* or the *conductivity ratio*.

For strong electrolytes, the graph of Λ against $\sqrt{c}$ is a straight line, which can be extrapolated to cut the Λ axis at Λ_0. For weak electrolytes, the graph of Λ against $\sqrt{c}$ is of the shape shown in Fig. 8.1(b). You can see that it cannot be used to find Λ_0. Kohlrausch's law helps here. It enables Λ_0 to be found for weak electrolytes. Kohlrausch's law can be expressed as

$$\Lambda_0 = \Lambda_0(\text{cation}) + \Lambda_0(\text{anion})$$

For every cation and anion, values of Λ_0 can be looked up in tables and added to give Λ_0 for the salt. Alternatively, measurements on strong electrolytes can be used to give values of Λ_0 which can be combined to give Λ_0 for the required weak electrolyte.

EXAMPLE Calculate the molar conductivity at infinite dilution at 25 °C of ethanoic acid, given that the molar conductivities at infinite dilution of hydrochloric acid, sodium chloride and sodium ethanoate are 4.26×10^{-2}, 1.26×10^{-2} and $9.1 \times 10^{-3}\,\Omega^{-1}\,m^2\,mol^{-1}$ respectively.

METHOD According to Kohlrausch's law,

$$\Lambda_0(CH_3CO_2H) = \Lambda_0(CH_3CO_2^-) + \Lambda_0(H^+)$$

$$\Lambda_0(HCl) = \Lambda_0(H^+) + \Lambda_0(Cl^-)$$

$$\Lambda_0(NaCl) = \Lambda_0(Na^+) + \Lambda_0(Cl^-)$$

$$\Lambda_0(CH_3CO_2Na) = \Lambda_0(CH_3CO_2^-) + \Lambda_0(Na^+)$$

Thus,

$$\Lambda_0(CH_3CO_2H) = \Lambda_0(HCl) + \Lambda_0(CH_3CO_2Na) - \Lambda_0(NaCl)$$

ANSWER $$\Lambda_0(CH_3CO_2H) = (4.26 \times 10^{-2}) + (9.00 \times 10^{-3}) - (1.26 \times 10^{-2})$$

$$= 3.91 \times 10^{-2}\,\Omega^{-1}\,m^2\,mol^{-1}$$

CALCULATION OF SOLUBILITY FROM CONDUCTIVITY

The solubility of a sparingly soluble salt can be found by measuring the conductance of a saturated solution of the salt. The conductance depends on the concentration of the dissolved ions. From it, the conductivity can be calculated. For a sparingly soluble salt, the concentration is so low, even in a saturated solution, that the salt can be assumed to be completely ionised. The conductivity is compared with the molar conductivity at infinite dilution of the salt. This is calculated as the sum of the molar conductivities at infinite dilution of the cation and anion. Then

$$\Lambda_0 = \kappa/c$$

where Λ_0 = Molar conductivity at infinite dilution, κ = Conductivity, and c = Molar concentration of solute.

Hence c can be calculated. Since the solution is saturated, the concentration is equal to the solubility of the salt.

EXAMPLE After making allowance for the conductivity of water, the conductivity of a saturated solution of silver chloride at $25\,°C = 1.50 \times 10^{-4}\,\Omega^{-1}\,m^{-1}$. If $\Lambda_0(Ag^+) = 6.2 \times 10^{-3}$ and $\Lambda_0(Cl^-) = 7.6 \times 10^{-3}\,\Omega^{-1}\,m^2\,mol^{-1}$, what is the solubility of silver chloride in $mol\,dm^{-3}$?

METHOD The molar conductivity at infinite dilution, Λ_0, is the sum of those of the ions.

$$\Lambda_0 = \Lambda_0(Ag^+) + \Lambda_0(Cl^-)$$
$$= (6.2 \times 10^{-3}) + (7.7 \times 10^{-3}) = 1.38 \times 10^{-2}$$
$$\Lambda_0 = \kappa/c$$
$$c = \kappa/\Lambda_0$$
$$= (1.50 \times 10^{-4})/(1.38 \times 10^{-2})$$
$$= 1.09 \times 10^{-2}\,mol\,m^{-3} = 1.09 \times 10^{-5}\,mol\,dm^{-3}$$

ANSWER The solubility of silver chloride is $1.09 \times 10^{-5}\,mol\,dm^{-3}$.

IONIC EQUILIBRIA; WEAK ELECTROLYTES

Electrolytes are ionic compounds. Strong electrolytes consist entirely of ions. Weak electrolytes consist of molecules, some of which dissociate to form ions. The fraction of molecules which dissociate is called the degree of ionisation or degree of dissociation. As the concentration of a solution of a weak electrolyte decreases, the degree of ionisation increases. *Ostwald's dilution law* gives a relationship between the degree of ionisation, α, of a weak electrolyte and its concentration c. Consider a weak acid, HA, in a solution of concentration c. An equilibrium is set up between the undissociated molecules, HA, and the ions H_3O^+ and A^-. H_3O^+ is the oxonium ion, a complex ion formed by the combination of a hydrogen ion and a molecule of water. It can also be written as $H^+(aq)$. Since the degree of ionisation is α, the equilibrium concentrations of the species are: $[HA] = (1-\alpha)c$; $[H_3O^+] = [A^-] = \alpha c$ (where the square brackets stand for concentration). The equilibrium can be represented:

$$HA + H_2O \rightleftharpoons H_3O^+ + A^-$$
$$c(1-\alpha) \qquad c\alpha \qquad c\alpha$$

The ratio $\dfrac{[H_3O^+]\,[A^-]}{[HA]\,[H_2O]}$ is a constant. (See Chapter 11 on Equilibria.)

The value of $[H_2O]$ is constant as the concentration of water is not significantly altered by the ionisation of HA.

The ratio $\dfrac{[H_3O^+][A^-]}{[HA]}$ is called the dissociation constant, K_a, of the acid:

$$K_a = \frac{[H_3O^+][A^-]}{[HA]} = \frac{c\alpha \times c\alpha}{c(1-\alpha)} = \frac{c\alpha^2}{1-\alpha}$$

The Ostwald dilution law is embodied in the equation

$$K_a = \frac{c\alpha^2}{1-\alpha}$$

For some weak electrolytes, α is so small that the error involved in putting $(1-\alpha)$ equal to 1 is negligible. In this case,

$$K_a = c\alpha^2$$

$$\alpha = \sqrt{K_a/c}$$

Ostwald's dilution law can be applied to weak bases. In the case of a weak base B which is partially ionised in solution,

$$B + H_2O \rightleftharpoons BH^+ + OH^-$$

The concentration of water is not significantly altered by the dissociation: $[H_2O] = $ constant.

The base dissociation constant K_b is given by the equation

$$K_b = \frac{[BH^+][OH^-]}{[B]}$$

$$K_b = \frac{c\alpha^2}{1-\alpha}$$

If $\alpha \ll 1$, the expression approximates to $K_b = c\alpha^2$.

CALCULATION OF THE DEGREE OF DISSOCIATION AND THE DISSOCIATION CONSTANT OF A WEAK ELECTROLYTE FROM CONDUCTANCE MEASUREMENTS

From conductance measurements on a solution of known concentration, the molar conductivity Λ is found. The relationship,

$$\alpha = \Lambda/\Lambda_0$$

enables the degree of ionisation α to be found. The value of α can be substituted in the Ostwald equation to give the ionisation constant of the electrolyte.

EXAMPLE 1 **a)** Calculate the degree of dissociation of ethanoic acid of concentration 1.00×10^{-3} mol dm^{-3}. The conductivity of the solution is $4.85 \times 10^{-3} \Omega^{-1}$ m^{-1} and the values of Λ_0 for ethanoic acid is $3.91 \times 10^{-2} \Omega^{-1}$ m^2 mol^{-1}.

b) Calculate the dissociation constant of ethanoic acid at the temperature at which the measurements were made.

METHOD **a)** Conductivity, $\kappa = 4.85 \times 10^{-3} \Omega^{-1}$ m^{-1}

Concentration, $c = 1.00 \times 10^{-3}$ mol dm$^{-3} = 1.00$ mol m^{-3}

$$\Lambda = \kappa/c = (4.85 \times 10^{-3})/1.00 = 4.85 \times 10^{-3} \Omega^{-1}\,\text{m}^2\,\text{mol}^{-1}$$

$$\Lambda_0(CH_3CO_2H) = 3.91 \times 10^{-2} \Omega^{-1}\,\text{m}^2\,\text{mol}^{-1}$$

$$\alpha = \Lambda/\Lambda_0$$

$$= (4.85 \times 10^{-3})/(3.91 \times 10^{-2}) = 0.124$$

ANSWER The degree of dissociation of ethanoic acid is 0.124.

METHOD **b)** From the Ostwald equation,

$$K_a = \frac{c\alpha^2}{1-\alpha}$$

With a value of 0.124 for α, $(1-\alpha)$ cannot be put $= 1$

$$K_a = \frac{10^{-3} \times (0.124)^2}{0.876} = 1.76 \times 10^{-5}\,\text{mol dm}^{-3}$$

ANSWER The dissociation constant of ethanoic acid is 1.76×10^{-5} mol dm^{-3}.

EXAMPLE 2 Calculate the degree of dissociation of hydrogen cyanide in a solution of concentration 0.100 mol dm^{-3}. $K_a = 7.24 \times 10^{-10}$ mol dm^{-3}.

METHOD $$K_a = \frac{c\alpha^2}{(1-\alpha)}$$

$$7.24 \times 10^{-10} = \frac{0.100\alpha^2}{(1-\alpha)} \qquad \therefore\ 7.24 \times 10^{-9} = \frac{\alpha^2}{1-\alpha}$$

Inspection shows that α is of the order of 10^{-5}, and $(1-\alpha)$ can be put equal to 1.

Then $$\alpha = \sqrt{7.24 \times 10^{-9}} = 8.5 \times 10^{-5}$$

ANSWER The degree of dissociation is 8.5×10^{-5}.

EXERCISE 43   Problems on Molar Conductivity

1. The molar conductivity of a solution of a weak acid of concentration 0.0400 mol dm^{-3} at 25 °C is $8.25 \times 10^{-4} \Omega^{-1}$ m^2 mol^{-1}. If the molar conductivity at infinite dilution is $3.98 \times 10^{-2} \Omega^{-1}$ m^2 mol^{-1} at 25 °C, calculate the dissociation constant of the acid at 25 °C.

2. A solution of a weak monobasic acid has a concentration of 0.0250 mol dm^{-3}. The conductivity at 25 °C is $2.64 \times 10^{-2} \Omega^{-1} m^{-1}$. The molar conductivity at infinite dilution is $3.91 \times 10^{-2} \Omega^{-1} m^2 mol^{-1}$. Calculate: a) the degree of dissociation, b) the concentration of hydrogen ions in solution, and c) the dissociation constant of the acid.

3. A solution of a weak monobasic acid of concentration 0.0200 mol dm^{-3} has conductivity $0.184 \Omega^{-1} m^{-1}$ at 25 °C. The molar conductivity at infinite dilution is $4.00 \times 10^{-2} \Omega^{-1} m^2 mol^{-1}$. Calculate: a) the degree of dissociation, and b) the dissociation constant of the acid at 25 °C.

4. A solution of methanoic acid of concentration 0.100 mol dm^{-3} has conductivity $0.166 \Omega^{-1} m^{-1}$ at 25 °C. The molar conductivity at infinite dilution is $4.04 \times 10^{-2} \Omega^{-1} m^2 mol^{-1}$. Calculate the dissociation constant of methanoic acid at 25 °C.

5. A solution of an organic acid, RCO_2H, has a concentration of 0.0300 mol dm^{-3} and a molar conductivity of $1.00 \times 10^{-3} \Omega^{-1} m^2 mol^{-1}$ at 25 °C. The molar conductivity at infinite dilution at 25 °C is $39.0 \times 10^{-3} \Omega^{-1} m^2 mol^{-1}$. Calculate: a) the degree of dissociation, and b) the dissociation constant for the acid at 25 °C.

CALCULATION OF pH, pOH and pK

Acids react with water to give hydrogen ions. These are not simple H^+ ions; they are complex ions of formula H_3O^+, formed by combination of a molecule of water and an H^+ ion. Although they are properly called *oxonium* ions, in this text, H_3O^+ ions will be referred to as hydrogen ions. The hydrogen ion concentration of a solution can be indicated by means of a number on the pH scale.

The pH of a solution is the negative logarithm to the base 10 of the hydrogen ion concentration. pOH, pK_a and pK_b are defined below.

$$pH = -\lg [H_3O^+] \qquad pK_a = -\lg K_a$$
$$pOH = -\lg [OH^-] \qquad pK_b = -\lg K_b$$

Strong acids and bases

The pH of a solution of a strong acid or base is simply calculated. If the concentration of hydrochloric acid is 0.1 mol dm^{-3},

$$[H_3O^+] = 0.1 \text{ mol dm}^{-3} = 10^{-1} \text{ mol dm}^{-3}$$
$$\lg [H_3O^+] = -1$$
$$pH = 1$$

If the concentration of a solution of sodium hydroxide is $0.01\,\text{mol}\,\text{dm}^{-3}$,

$$[OH^-] = 0.01 = 10^{-2}\,\text{mol}\,\text{dm}^{-3}$$

The product of the hydrogen ion concentration and the hydroxide ion concentration in a solution is called the ionic product for water, K_w. At $25\,^{\circ}\text{C}$, the value of K_w is $10^{-14}\,\text{mol}^2\,\text{dm}^{-6}$.

$$[H_3O^+]\,[OH^-] = K_w = 10^{-14}\,\text{mol}^2\,\text{dm}^{-6}$$

Therefore, in this solution,

$$[H_3O^+] = 10^{-14}/10^{-2} = 10^{-12}\,\text{mol}\,\text{dm}^{-2}$$

$$pH = -\lg\,[H_3O^+] = 12$$

EXAMPLE 1   Calculate the pH of a solution of $0.020\,\text{mol}\,\text{dm}^{-3}$ hydrochloric acid.

METHOD   If the concentration of hydrochloric acid is $0.020\,\text{mol}\,\text{dm}^{-3}$,

$$[H_3O^+] = 2 \times 10^{-2}\,\text{mol}\,\text{dm}^{-3}$$

ANSWER $pH = -\lg\,[H_3O^+] = -(0.301 + \bar{2}) = 1.70$

EXAMPLE 2   Calculate the pH of a $0.010\,\text{mol}\,\text{dm}^{-3}$ solution of calcium hydroxide.

METHOD   If the concentration of $Ca(OH)_2$ is $10^{-2}\,\text{mol}\,\text{dm}^{-3}$,

$$[OH^-] = 2 \times 10^{-2}\,\text{mol}\,\text{dm}^{-3}$$

$$pOH = -\lg\,(2 \times 10^{-2}) = 1.7$$

Since $[H_3O^+]\,[OH^-] = K_w = 10^{-14}\,\text{mol}^2\,\text{dm}^{-6}$

$$\lg\,[H_3O^+] + \lg\,[OH^-] = \lg K_w = -14$$

∴ $pH + pOH = 14$

Thus, if $pOH = 1.7$,

ANSWER $pH = 14 - 1.7 = 12.3$.

(For practice in using a calculator to find logarithms, see pp. 3 and 10.)

Weak acids and bases

For calculating the pH of a weak acid or base, the concentration of the solution is not sufficient information. The degree of dissociation must be taken into account. The following examples show how the pH can be calculated from the concentration and the degree of dissociation or the dissociation constant of a weak electrolyte.

The converse is also true. If the pH can be measured experimentally, the value of the pH can be used to calculate the dissociation constant of the weak electrolyte.

EXAMPLE 1   Calculate the pH of a solution of propanoic acid of concentration $0.0100 \, mol \, dm^{-3}$, given that the degree of dissociation is 0.116.

METHOD   Let the concentration of acid be c, and the degree of dissociation be α. Then in the equilibrium

$$HA \;+\; H_2O \;\rightleftharpoons\; H_3O^+ \;+\; A^-$$

we can put the concentrations

$$[HA] \;=\; c(1-\alpha); \; [H_3O^+] \;=\; [A^-] \;=\; c\alpha.$$

Then $[H_3O^+] \;=\; 0.0100 \times 0.116 \;=\; 1.16 \times 10^{-3}$

$$pH \;=\; -lg \, [H_3O^+] \;=\; 2.9355$$

ANSWER   pH = 2.94.

EXAMPLE 2   Calculate the pH of a $0.0100 \, mol \, dm^{-3}$ solution of ethanoic acid, which has a dissociation constant of $1.76 \times 10^{-5} \, mol \, dm^{-3}$.

METHOD   When a weak acid ionises,

$$HA \;+\; H_2O \;\rightleftharpoons\; H_3O^+ \;+\; A^-$$

the dissociation constant K_a is given by the expression

$$K_a \;=\; \frac{[H_3O^+] \, [A^-]}{[HA]}$$

One A^- is formed for each H_3O^+.

∴ $[H_3O^+] \;=\; [A^-]$ and $[H_3O^+] \, [A^-] \;=\; [H_3O^+]^2$

The degree of dissociation is so small that we can put [HA] equal to the total acid concentration, $0.0100 \, mol \, dm^{-3}$. In fact, it must be slightly less than this. Say the degree of dissociation is 0.001. Then $[HA] = 0.0100 - 0.000\,01 = 0.009\,99$.   The error in assuming $[HA] = 0.0100$ is 1 part in 1000. This is smaller than the error in the practical measurements from which the values used in the calculation are obtained. For most solutions of weak acids and bases, we can make the approximation safely.

Thus, $K_a \;=\; \dfrac{[H_3O^+]^2}{10^{-2}}$

$$[H_3O^+]^2 \;=\; 1.76 \times 10^{-7}$$

$$[H_3O^+] \;=\; 4.19 \times 10^{-4} \, mol \, dm^{-3}$$

$$pH \;=\; -lg \, [H_3O^+]$$

ANSWER   pH = 3.34.

EXAMPLE 3   Calculate the dissociation constant of phenol, given that a 1.00×10^{-2} $mol \, dm^{-3}$ solution has a pH of 5.95.

METHOD   If pH $= 5.95$,

$$[H_3O^+] = \text{antilg}(-pH) = 1.13 \times 10^{-6}$$

If we write the dissociation of phenol

$$PhOH + H_2O \rightleftharpoons PhO^- + H_3O^+$$

$$K_a = \frac{[H_3O^+][PhO^-]}{[PhOH]} = \frac{[H_3O^+]^2}{[PhOH]}$$

$$K_a = \frac{(1.13 \times 10^{-6})^2}{10^{-2}} = 1.28 \times 10^{-10}\,mol\,dm^{-3}$$

ANSWER   The dissociation constant of phenol is $1.28 \times 10^{-10}\,mol\,dm^{-3}$.

EXAMPLE 4   A solution of pyridine of concentration $0.0100\,mol\,dm^{-3}$ has a pH 8.63. Calculate the dissociation constant of pyridine.

METHOD   We can write the dissociation of pyridine as

$$P + H_2O \rightleftharpoons PH^+ + OH^-$$

Thus, $$K_b = \frac{[PH^+][OH^-]}{[P]} = \frac{[OH^-]^2}{0.0100}$$

Since   pH $= 8.63$

pOH $= 14.00 - 8.63 = 5.37$

$$[OH^-] = 4.26 \times 10^{-6}$$

$$K_b = \frac{(4.26 \times 10^{-6})^2}{0.0100} = 1.82 \times 10^{-9}\,mol\,dm^{-3}$$

ANSWER   The dissociation constant of pyridine is $1.82 \times 10^{-9}\,mol\,dm^{-3}$.

EXAMPLE 5   Calculate the degree of dissociation of a $0.0100\,mol\,dm^{-3}$ solution of ethylamine, given that $K_b = 6.46 \times 10^{-4}\,mol\,dm^{-3}$; $K_w = 1.00 \times 10^{-14}\,mol^2\,dm^{-6}$.

METHOD   Since $$B + H_2O \rightleftharpoons BH^+ + OH^-$$

then

$$K_b = \frac{c\alpha^2}{(1-\alpha)} \quad (c = \text{Concentration};\ \alpha = \text{Degree of dissociation})$$

Making the approximation $(1-\alpha) = 1$ gives

$$K_b = c\alpha^2$$

$$6.46 \times 10^{-4} = 0.0100\,\alpha^2$$

$$\alpha = 0.254$$

This value of α is not small compared with 1: the approximation $(1-\alpha) = 1$ cannot be made. Since $(1-\alpha) = 0.766$,

$$6.46 \times 10^{-4} = \frac{0.0100\alpha^2}{0.766}$$

$$\alpha = 0.222$$

This second value for α makes $(1-\alpha) = 0.778$. Putting this value into the equation gives

$$6.46 \times 10^{-4} = \frac{0.0100\alpha^2}{0.778}$$

$$\alpha = 0.224$$

The value of $(1-\alpha)$ is now 0.776, and

$$6.46 \times 10^{-4} = \frac{0.0100\alpha^2}{0.776}$$

ANSWER   Degree of dissociation $\alpha = 0.224$

We have finally arrived at a value of α which satisfies the equation

$$6.46 \times 10^{-4} = \frac{0.0100\alpha^2}{(1-\alpha)}$$

This method of calculation is called the *method of successive approximations.*

It is a neater and more convenient method than solving a quadratic equation.

Conjugate pairs

Some tables list the base dissociation constants, K_b, of bases such as amines. Others list the acid dissociation constants, K_a.

The value of K_b refers to the equilibrium

$$B + H_2O \rightleftharpoons BH^+ + OH^-$$

where $$K_b = \frac{[BH^+][OH^-]}{[B]}$$

The species BH^+ is referred to as the conjugate acid of the base, B. BH^+ dissociates according to the equilibrium

$$BH^+ + H_2O \rightleftharpoons B + H_3O^+$$

One can, therefore, write the acid dissociation constant of BH^+ as

$$K_a = \frac{[B][H_3O^+]}{[BH^+]}$$

Multiplying K_a by K_b gives $[H_3O^+][OH^-]$. That is, $K_aK_b = K_w$.

This is a useful relationship between the dissociation constants of conjugate pairs.

EXAMPLE 1   Calculate the pH of a solution of ammonia ($pK_a = 9.25$) of concentration $0.0100 \, mol \, dm^{-3}$.

METHOD

$$pK_a = 9.25 \qquad pK_b = 14.00 - 9.25 = 4.75$$

$$K_b = 1.78 \times 10^{-5}$$

Since

$$K_b = \frac{[NH_4^+][OH^-]}{[NH_3]} = \frac{[OH^-]^2}{[NH_3]}$$

$$[OH^-]^2 = K_b \times c$$

$$= 1.78 \times 10^{-5} \times 0.0100$$

$$[OH^-] = 4.22 \times 10^{-4}$$

$$pOH = 3.37$$

ANSWER

$$pH = 10.6$$

EXAMPLE 2   Calculate the degree of ionisation of phenylmethylamine ($pK_a = 9.37$) in a $0.100 \, mol \, dm^{-3}$ aqueous solution. Quote your answer to two significant figures.

METHOD   Since

$$pK_a = 9.37$$

$$pK_b = 14.00 - 9.37 = 4.63$$

$$K_b = 2.34 \times 10^{-5}$$

Since

$$K_b = \frac{[C_6H_5CH_2NH_3^+][OH^-]}{[C_6H_5CH_2NH_2]}$$

$$= \frac{\alpha^2 c}{(1-\alpha)}$$

making the approximation $1 - \alpha = 1$,

$$\alpha^2 = K_b/c$$

$$= 2.34 \times 10^{-5}/0.100 = 2.34 \times 10^{-4}$$

$$\alpha = 1.53 \times 10^{-2}$$

The approximation $(1-\alpha) = 1$ is not entirely justified, but, since the answer is required to two significant figures only, this is close enough.

ANSWER   The degree of ionisation is 1.5×10^{-2}.

EXERCISE 44   Problems on pH

1. Calculate the pH of solutions with following H_3O^+ concentrations in mol dm^{-3}:

   a) 10^{-8} b) 10^{-4} c) 10^{-7}
   d) 6.8×10^{-3} e) 3.2×10^{-5} f) 0.035
   g) 0.25 h) 5.4×10^{-9} i) 7.1×10^{-7}
   j) 9.9×10^{-2}

   Now calculate the pOH of each of the solutions.

2. Calculate the pH of solutions with the following OH^- concentrations in mol dm^{-3}:

   a) 10^{-2} b) 10^{-3} c) 10^{-8}
   d) 0.055 e) 0.0010 f) 0.083
   g) 7.6×10^{-3} h) 4.9×10^{-5} i) 6.4×10^{-8}
   j) 3.7×10^{-10}

3. Calculate the H_3O^+ concentrations in solutions with the following pH values:

   a) 0.00 b) 4.30 c) 2.35
   d) 1.88 e) 4.15 f) 7.84
   g) 9.21 h) 13.7 i) 9.50
   j) 2.63

4. Calculate the pH of the solutions made by dissolving the following in distilled water and making up to 500 cm^3 of solution:

   a) $3.00\,g$ of hydrogen chloride
   b) $4.50\,g$ of chloric(VII) acid, $HClO_4$
   c) $4.00\,g$ of sodium hydroxide
   d) $1.00\,g$ of calcium hydroxide
   e) $6.30\,g$ of potassium hydroxide

5. An aqueous solution contains the acid, HX, at a concentration 0.100 mol dm^{-3}. The degree of ionisation of the acid is 0.0300. Calculate the pH of the solution.

6. If the acid HA is 1% ionised in a solution of concentration 0.0100 mol dm^{-3}, calculate: a) K_a, b) pK_a.

7. The degree of ionisation of phenol in water in a solution of concentration 1.10×10^{-2} mol dm^{-3} is 1.1×10^{-4}. Calculate the value of pK_a.

8. Calculate the dissociation constants for each of the weak acids listed below:

	Electrolyte	*Concentration/mol dm^{-3}*	*$[H_3O^+]$ /mol dm^{-3}*
*a)	HCO_2H	0.0100	1.33×10^{-3}
b)	CH_3CO_2H	0.100	1.32×10^{-3}
c)	HCN	1.00	1.99×10^{-5}
*d)	HF	0.200	5.97×10^{-3}
*e)	$HClO_2$	0.0200	1.41×10^{-2}

9. Calculate the dissociation constants of the weak acids listed below:
   a) a solution of $0.0100 \, mol \, dm^{-3}$ CH_3CO_2H has a pH of 3.38
   b) a solution of $0.200 \, mol \, dm^{-3}$ HCN has a pH of 5.05
   *c) a solution of $0.500 \, mol \, dm^{-3}$ $ClCH_2CO_2H$ has a pH of 3.16
   d) a solution of $0.0100 \, mol \, dm^{-3}$ $CH_3CH_2CO_2H$ has a pH of 3.43
   e) a solution of $0.0100 \, mol \, dm^{-3}$ HOBr has a pH of 5.35

10. Calculate the values of K_b for the following weak bases from the data:

Base	Concentration/$mol \, dm^{-3}$	$[OH^-]$/$mol \, dm^{-3}$
a) CH_3NH_2	0.0100	4.78×10^{-7}
b) NH_3	1.00	2.37×10^{-5}
c) $C_2H_5NH_2$	0.0500	9.65×10^{-7}
d) $C_3H_7NH_2$	0.0100	3.80×10^{-7}
e) $C_6H_5CH_2NH_2$	0.0100	2.06×10^{-6}

11. Calculate the degree of ionisation of each of the following in aqueous solution:
   a) $2.00 \times 10^{-3} \, mol \, dm^{-3}$ HCN $\quad$ ($pK_a = 9.40$)
   b) $0.500 \times 10^{-3} \, mol \, dm^{-3}$ HCO_2H $\quad$ ($pK_a = 3.75$)
   c) $3.00 \times 10^{-2} \, mol \, dm^{-3}$ CH_3CO_2H $\quad$ ($pK_a = 4.76$)
   d) $5.00 \times 10^{-2} \, mol \, dm^{-3}$ NH_3 $\quad$ ($pK_a = 9.25$)
   e) $1.00 \times 10^{-2} \, mol \, dm^{-3}$ $(CH_3)_3N$ $\quad$ ($pK_a = 9.80$)

   (*Note.* You are given the values of pK_a for the conjugate acids of the bases.)

12. Calculate the pH of:
   a) $0.0100 \, mol \, dm^{-3}$ hydrochloric acid,
   b) $0.0100 \, mol \, dm^{-3}$ sodium hydroxide, and
   c) the solution obtained by diluting $5.00 \, cm^3$ of hydrochloric acid of concentration $1.00 \, mol \, dm^{-3}$ with conductivity water to $1.00 \, dm^3$.

13. The dissociation constant of phenol in water at $20 \, ^{\circ}C$ is $1.21 \times 10^{-10} \, mol \, dm^{-3}$. Calculate the percentage of phenol which is ionised in a solution of concentration $0.0100 \, mol \, dm^{-3}$.

14. Given that the pK_a of the ammonium ion, NH_4^+, is 9.25 at $25 \, ^{\circ}C$, find the pH of an aqueous solution of ammonia of concentration $0.100 \, mol \, dm^{-3}$. The ionic product for water at $25 \, ^{\circ}C$ is $1.00 \times 10^{-14} \, mol^2 \, dm^{-6}$.

15. If the acid HA is 1% ionised in a solution of concentration $0.0100 \, mol \, dm^{-3}$, calculate: a) K_a, b) pK_a.

16. Calculate the pH of hydrochloric acid of concentration $1.00 \times 10^{-3} \, mol \, dm^{-3}$. Calculate how many cm^3 of a) hydrochloric acid of concentration $1.00 \, mol \, dm^{-3}$, b) sodium hydroxide of concentration $1.00 \, mol \, dm^{-3}$ are needed to change the pH of $1.00 \, dm^3$ of hydrochloric acid of concentration $1.00 \times 10^{-3} \, mol \, dm^{-3}$ by 1 unit. Ignore the small changes in volume.

BUFFER SOLUTIONS

A buffer solution is one which will resist changes in pH due to the addition of small amounts of acid and alkali. An effective buffer can be made by preparing a solution containing both a weak acid and also one of its salts with a strong base, e.g. ethanoic acid and sodium ethanoate. This will absorb hydrogen ions because they react with ethanoate ions to form molecules of ethanoic acid:

$$CH_3CO_2^- + H_3O^+ \rightleftharpoons CH_3CO_2H + H_2O$$

Hydroxide ions are absorbed by combining with ethanoic acid molecules to form ethanoate ions and water:

$$OH^- + CH_3CO_2H \rightleftharpoons CH_3CO_2^- + H_2O$$

A solution of a weak base and one of its salts formed with a strong acid, e.g. ammonia solution and ammonium chloride, will act as a buffer. If hydrogen ions are added, they combine with ammonia, and, if hydroxide ions are added, they combine with ammonium ions:

$$NH_3 + H_3O^+ \rightleftharpoons NH_4^+ + H_2O$$
$$OH^- + NH_4^+ \rightleftharpoons NH_3 + H_2O$$

The pH of a buffer solution consisting of a weak acid and its salt is calculated from the equation

$$K_a = \frac{[H_3O^+][A^-]}{[HA]}$$

$$[H_3O^+] = K_a \frac{[HA]}{[A^-]}$$

$$pH = pK_a + \lg \frac{[A^-]}{[HA]}$$

Since the salt is completely ionised and the acid only slightly ionised, one can assume that all the anions come from the salt, and put

$$[A^-] = [Salt]$$
$$[HA] = [Acid]$$

∴
$$pH = pK_a + \lg \frac{[Salt]}{[Acid]}$$

For a buffer made from a base, B, and its salt with a strong acid, BH^+X^-,

$$K_b = \frac{[BH^+][OH^-]}{[B]}$$

$$pOH = pK_b + \lg \frac{[BH^+]}{[B]}$$

and
$$pH = pK_w - pK_b + lg \frac{[B]}{[BH^+]}$$

Since the weak base is only slightly ionised, one can put

$$[B] = [Base\ added]$$
$$[BH^+] = [Salt\ added]$$

∴
$$pH = pK_w - pK_b + lg \frac{[Base]}{[Salt]}$$

EXAMPLE 1 Three solutions contain propanoic acid $(K_a = 1.34 \times 10^{-5}\ mol\ dm^{-3})$ at a concentration of $0.10\ mol\ dm^{-3}$ and sodium propanoate at concentrations of a) $0.10\ mol\ dm^{-3}$, b) $0.20\ mol\ dm^{-3}$, c) $0.50\ mol\ dm^{-3}$ respectively. Calculate the pH values of the three solutions.

METHOD
$$pH = pK_a + lg \frac{[Salt]}{[Acid]}$$

In solution a), $pH = 4.87 + lg \dfrac{0.10}{0.10} = 4.87 + lg\ 1.0$

ANSWER
$$pH(a) = 4.87$$

In solution b), $pH = 4.87 + lg \dfrac{0.20}{0.10} = 4.87 + lg\ 2.0$

ANSWER
$$pH(b) = 5.17$$

In solution c), $pH = 4.87 + lg \dfrac{0.50}{0.10} = 4.87 + lg\ 5.0$

ANSWER
$$pH(c) = 5.57$$

EXAMPLE 2 Calculate the pH of solutions of $0.25\ mol\ dm^{-3}$ methylamine $(K_b = 4.54 \times 10^{-4}\ mol\ dm^{-3})$ containing a) $0.25\ mol\ dm^{-3}$ and b) $0.50\ mol\ dm^{-3}$ methylamine hydrochloride.

METHOD
$$pH = pK_w - pK_b + lg \frac{[Base]}{[Salt]}$$

In solution a), $pH = 14.00 - 3.34 + lg \dfrac{0.25}{0.25}$

ANSWER
$$pH(a) = 10.66$$

In solution b), $pH = 14.00 - 3.34 + lg \dfrac{0.25}{0.50}$

ANSWER
$$pH(b) = 10.36$$

*Calculating the change in the pH of a buffer solution produced by the addition of acid or alkali

EXAMPLE 3 Calculate the effect of adding a) $10 \, cm^3$ of hydrochloric acid of concentration $1.0 \, mol \, dm^{-3}$ and b) $10 \, cm^3$ of sodium hydroxide of concentration $1.0 \, mol \, dm^{-3}$ to $1.0 \, dm^3$ of a buffer containing $0.10 \, mol \, dm^{-3}$ ethanoic acid and $0.10 \, mol \, dm^{-3}$ sodium ethanoate. (pK_a of ethanoic acid $= 4.75$.)

METHOD Use the equation $pH = pK_a + lg \dfrac{[Salt]}{[Acid]}$

$$pH = 4.75 + lg \frac{0.10}{0.10} = 4.75$$

a) Amount of hydrochloric acid added $= 10 \, cm^3$ of $1.0 \, mol \, dm^{-3}$ solution $= 0.010 \, mol$.

This amount of hydrogen ion combines with ethanoate ions to form undissociated molecules of ethanoic acid. The amount of salt is therefore decreased by $0.010 \, mol$, and the amount of acid is increased by $0.010 \, mol$. Thus,

$$pH = 4.75 + lg \frac{(0.10 - 0.010)}{(0.10 + 0.010)} = 4.66$$

ANSWER The pH has decreased by 0.09.

b) Amount of OH^- ion added $= 10 \times 10^{-3} \times 1.0 = 0.010 \, mol$.

The hydroxide ions react with $0.010 \, mol$ of weak acid. This results in the formation of an additional $0.010 \, mol$ of anions. Thus, [Acid] is decreased by $0.010 \, mol$, and [Salt] is increased by $0.010 \, mol$.

$$pH = 4.75 + lg \frac{(0.10 + 0.010)}{(0.10 - 0.010)} = 4.84$$

ANSWER The pH has increased by 0.09.

EXERCISE 45   Problems on Buffers

1. What fraction of a mole of sodium ethanoate must be added to $1 \, dm^3$ of ethanoic acid of concentration $0.10 \, mol \, dm^{-3}$ and $K_a = 2 \times 10^{-5}$ $mol \, dm^{-3}$ in order to produce a buffer solution of pH $= 5$?

   a 0.2   b 0.25   c 0.4   d 0.5   e 0.6

2. What amount of sodium ethanoate must be added to $1.00 \, dm^3$ of ethanoic acid of pK_a 4.73, concentration $0.100 \, mol \, dm^{-3}$, to produce a buffer of pH $= 5.73$?

3. Calculate the pH values of the following solutions.

   a) $20.0 \, cm^3$ of $1.00 \, mol \, dm^{-3}$ nitrous acid ($pK_a = 3.34$) added to $40.0 \, cm^3$ of $0.500 \, mol \, dm^{-3} \, mol \, dm^{-3}$ sodium nitrite solution.

   b) $10.0 \, cm^3$ of $1.00 \, mol \, dm^{-3}$ nitrous acid added to $20.0 \, cm^3$ of $2.00 \, mol \, dm^{-3}$ sodium nitrite solution.

4. a) What is the pH of a solution containing $0.100 \, mol \, dm^{-3}$ of ethanoic acid and $0.100 \, mol \, dm^{-3}$ of sodium ethanoate?
   ($K_a(CH_3CO_2H) = 1.86 \times 10^{-5} \, mol \, dm^{-3}$.)

   b) How many moles of sodium ethanoate must be added to $1.00 \, dm^3$ of $0.0100 \, mol \, dm^{-3}$ ethanoic acid to produce a buffer solution of pH 5.8?

*5. What is the pH of a buffer which has been made by adding $100 \, cm^3$ of chloroethanoic acid of concentration $0.500 \, mol \, dm^{-3}$ and $100 \, cm^3$ of sodium hydroxide of concentration $0.250 \, mol \, dm^{-3}$?
   ($pK_a(ClCH_2CO_2H) = 2.86$.)

*6. a) Calculate the effect of adding $10.0 \, cm^3$ of hydrochloric acid of concentration $1.00 \, mol \, dm^{-3}$ to $1.00 \, dm^3$ of the following buffers (pK_a of ethanoic acid is 4.75):

Buffer	A	B	C	D
Concentration	$mol \, dm^{-3}$	$mol \, dm^{-3}$	$mol \, dm^{-3}$	$mol \, dm^{-3}$
Ethanoic acid	0.10	0.10	0.10	0.20
Sodium ethanoate	0.10	0.050	0.020	0.20

   b) From a comparison of your values, state how the ratio of $\dfrac{[Salt]}{[Acid]}$ affects the buffering capacity of the solution and also how the concentration of the buffer affects its buffering capacity.

*7. Methanoic acid has $K_a = 1.60 \times 10^{-4} \, mol \, dm^{-3}$. Calculate the pH of:

   a) a solution of concentration $0.100 \, mol \, dm^{-3}$

   b) a solution of $3.40 \, g$ of sodium methanoate in $1.00 \, dm^3$ of methanoic acid of concentration $0.100 \, mol \, dm^{-3}$

   c) the solution obtained when $24.9 \, cm^3$ of $0.100 \, mol \, dm^{-3}$ sodium hydroxide solution has been added to $25.0 \, cm^3$ of methanoic acid of concentration $0.100 \, mol \, dm^{-3}$.

*8. a) Calculate the pH of a solution of $0.200 \, mol$ ethanoic acid ($K_a = 1.8 \times 10^{-5} \, mol \, dm^{-3}$) in $1.00 \, dm^3$ of aqueous solution.

   b) What does the pH of the solution become after the addition of $0.100 \, mol$ sodium ethanoate?

   c) If $0.010 \, mol$ hydrochloric acid is added to the ethanoic acid–sodium ethanoate solution, what change in pH results?

*SALT HYDROLYSIS

When the salt of a weak acid and a strong base, e.g. sodium ethanoate, dissolves in water, the solution is alkaline. This is because the anions react with water molecules to form undissociated molecules of ethanoic acid. This interaction is called *salt hydrolysis*. An equilibrium is set up:

$$CH_3CO_2^-(aq) + H_2O(l) \rightleftharpoons CH_3CO_2H(aq) + OH^-(aq)$$

Since CH_3CO_2H and $CH_3CO_2^-$ are a conjugate acid–base pair (see p. 136).

$$K_a(CH_3CO_2H) = K_w/K_b(CH_3CO_2^-)$$

$$K_b(CH_3CO_2^-) = \frac{[CH_3CO_2H][OH^-]}{[CH_3CO_2^-]}$$

If the degree of hydrolysis of the salt is α and its concentration is c,

$$[CH_3CO_2^-] = c(1-\alpha)$$

$$[CH_3CO_2H] = c\alpha$$

$$[OH^-] = c\alpha$$

If α is small, $(1-\alpha)$ can be put $= 1$, and

$$c\alpha^2 = K_b$$

$$\alpha = \sqrt{K_b/c} = \sqrt{K_w/cK_a}$$

The pH of the solution can now be found:

Since
$$[OH^-] = c\alpha$$

$$[H_3O^+] = K_w/c\alpha = \sqrt{K_w K_a/c}$$

$$\lg[H_3O^+] = \tfrac{1}{2}\lg K_w + \tfrac{1}{2}\lg K_a - \tfrac{1}{2}\lg c$$

$$pH = \tfrac{1}{2}pK_w + \tfrac{1}{2}pK_a + \tfrac{1}{2}\lg c$$

This equation enables one to find the pH of a solution of a salt of a weak acid from the dissociation constant of the acid and the concentration of the salt.

In the case of a salt of a weak base and a strong acid, e.g. ammonium chloride, hydrolysis of the salt will increase the concentration of hydrogen ions:

$$NH_4^+(aq) + H_2O(l) \rightleftharpoons NH_3(aq) + H_3O^+(aq)$$

A treatment similar to that above gives the relationship

$$pH = \tfrac{1}{2}pK_w - \tfrac{1}{2}pK_b - \tfrac{1}{2}\lg c$$

Thus, the pH of a solution of a salt of a weak base and a strong acid can be found from the dissociation constant of the base and the concentration of the salt.

EXAMPLE 1 Calculate the pH at 25 °C of a solution of 0.100 mol dm^{-3} solution of sodium propanoate. ($K_a = 1.34 \times 10^{-5}$ mol dm^{-3}, $K_w = 1.00 \times 10^{-14}$ mol^2 dm^{-6} at 25 °C.)

METHOD Since

$$K_a = 1.34 \times 10^{-5}, \quad pK_a = 4.87$$

$$pH = \tfrac{1}{2}pK_w + \tfrac{1}{2}pK_a + \tfrac{1}{2}\lg c$$

$$= 7.00 + 2.44 - 0.50$$

ANSWER $\quad\quad pH = 8.94$

EXAMPLE 2 Calculate the pH of a solution of methylamine hydrochloride of concentration 0.100 mol dm^{-3} at 25 °C. ($K_b = 4.54 \times 10^{-4}$ mol dm^{-3}; $K_w = 1.00 \times 10^{-14}$ mol^2 dm^{-6} at 25 °C.)

METHOD Since

$$K_b = 4.54 \times 10^{-4}, \quad pK_b = 3.34$$

$$pH = \tfrac{1}{2}pK_w - \tfrac{1}{2}pK_b - \tfrac{1}{2}\lg c$$

$$= 7.00 - 1.67 + 0.50$$

ANSWER $\quad\quad pH = 5.83$

EXERCISE 46   Problems on Salt Hydrolysis

*1. Calculate the pH of the following solutions:

   a) a 0.100 mol dm^{-3} solution of ammonium chloride
   b) a 0.0100 mol dm^{-3} solution of methylamine hydrochloride
   c) a 0.100 mol dm^{-3} solution of potassium cyanide
   d) a 0.100 mol dm^{-3} solution of sodium methanoate
   e) a 0.0100 mol dm^{-3} solution of sodium ethanoate

   (pK$_a$ values are: NH$_3$, 9.25; CH$_3$NH$_2$, 10.64; HCN, 9.40; HCO$_2$H, 3.75; CH$_3$CO$_2$H, 4.76. pK$_w$ = 14.00.)

*2. Use the following dissociation constants (mol dm^{-3}):

   HNO$_2$ $K_a = 4.6 \times 10^{-4}$
   HCN $K_a = 4.9 \times 10^{-10}$
   NH$_3$ $K_b = 1.8 \times 10^{-5}$
   HCO$_2$H $K_a = 1.8 \times 10^{-4}$

   a) Find the pH of a 0.50 mol dm^{-3} solution of sodium methanoate.
   b) What is the pH of a solution of ammonium nitrate of concentration 0.25 mol dm^{-3}?
   c) Calculate the pH of a 0.25 mol dm^{-3} solution of potassium cyanide.
   d) Calculate the pH at the end point of a titration of 0.40 mol dm^{-3} nitrous acid and 0.40 mol dm^{-3} sodium hydroxide solution.
   e) Find the pH at the end point of a titration of 0.25 mol dm^{-3} ammonia solution with 0.25 mol dm^{-3} hydrochloric acid.
   f) Calculate the concentration of a solution of ammonium chloride which has a pH of 4.75.

*3. Use the following dissociation constants (mol dm^{-3}):

Ethanoic acid	$K_a = 1.76 \times 10^{-5}$	Ethylamine	$K_b = 6.46 \times 10^{-4}$
Methanoic acid	$K_a = 1.77 \times 10^{-4}$	Pyridine	$K_b = 5.62 \times 10^{-9}$
Hydrogen cyanide	$K_a = 4.9 \times 10^{-10}$	Phenylamine	$K_b = 4.17 \times 10^{-10}$

Calculate the values of pH in the following solutions:
a) 0.100 mol dm^{-3} sodium ethanoate
b) 0.0200 mol dm^{-3} sodium methanoate
c) 0.0200 mol dm^{-3} sodium cyanide
d) 0.100 mol dm^{-3} ethylamine hydrochloride
e) 0.0200 mol dm^{-3} pyridine hydrochloride
f) 0.0500 mol dm^{-3} phenylamine hydrochloride

COMPLEX ION FORMATION

Another type of equilibrium between ions is due to complex ion formation. A complex ion is formed by the combination of an ion with an oppositely charged ion or ions or with a neutral molecule or molecules. For example, when copper(II) ions combine with ammonia to form tetraamminecopper ions, there is set up an equilibrium:

$$Cu(NH_3)_4^{2+}(aq) \rightleftharpoons Cu^{2+}(aq) + 4NH_3(aq)$$

The equilibrium constant or dissociation constant for the reaction is

$$K_d = \frac{[Cu^{2+}][NH_3]^4}{[Cu(NH_3)_4^{2+}]} = 5.0 \times 10^{-14} \, mol^4 \, dm^{-12}$$

The inverse of the dissociation constant is called the *stability constant* of the complex ion.

EXAMPLE 1   The complex $Ag(CN)_2^-$ has a dissociation constant of 1.4×10^{-20} mol^2 dm^{-6}. Find the concentration of silver ions in a 2.0×10^{-2} mol dm^{-3} solution of $KAg(CN)_2$.

METHOD $$Ag(CN)_2^- \rightleftharpoons Ag^+ + 2CN^-$$

$$K_d = \frac{[Ag^+][CN^-]^2}{[Ag(CN)_2^-]} \, mol^2 \, dm^{-6}$$

There are two CN^- ions for every Ag^+ ion

∴ $$1.4 \times 10^{-20} = \frac{[Ag^+](2[Ag^+])^2}{2.0 \times 10^{-2}}$$

ANSWER $$[Ag^+] = 4.1 \times 10^{-8} \, mol \, dm^{-3}$$

EXAMPLE 2 In a $2.0 \times 10^{-2} \, \text{mol dm}^{-3}$ solution of diammine silver nitrate, the concentration of free silver ions is $6.8 \times 10^{-4} \, \text{mol dm}^{-3}$. What is the dissociation constant of the complex ion?

METHOD

$$Ag(NH_3)_2^+(aq) \rightleftharpoons Ag^+(aq) + 2NH_3(aq)$$

$$K_d = \frac{[Ag^+][NH_3]^2}{[Ag(NH_3)_2^+]} = \frac{[Ag^+](2[Ag^+])^2}{[Ag(NH_3)_2^+]} = \frac{4[Ag^+]^3}{[Ag(NH_3)_2^+]}$$

ANSWER

$$K_d = \frac{4(6.8 \times 10^{-4})^3}{2.0 \times 10^{-2}} = 6.3 \times 10^{-8} \, \text{mol}^2 \, \text{dm}^{-6}$$

EXERCISE 47   Problems on Complex Ions

1. The dissociation constant for the hexaaquo aluminium ion is $1.0 \times 10^{-5} \, \text{mol dm}^{-3}$.

   $$Al(H_2O)_6^{3+}(aq) \rightleftharpoons Al(H_2O)_5(OH)^{2+}(aq) + H^+(aq)$$

   Calculate the pH of a $0.10 \, \text{mol dm}^{-3}$ solution of aluminium nitrate.

2. A solution was made by dissolving $0.025 \, \text{mol KAu(CN)}_2$ in $1 \, \text{dm}^3$ of solution. If the dissociation constant of the complex ion, $Au(CN)_2^-$ is $1.1 \times 10^{-47} \, \text{mol}^2 \, \text{dm}^{-6}$, calculate the concentration of free Au^+ ions in the solution.

3. Ammonia is added to a solution of $0.50 \, \text{mol dm}^{-3}$ silver nitrate solution. What is the concentration of free ammonia when 95% of the Ag^+ ions have been converted to the complex $Ag(NH_3)_2^+$ ion? The dissociation constant of the complex ion is $6.0 \times 10^{-8} \, \text{mol}^2 \, \text{dm}^{-6}$.

4. A solution is made by adding equal volumes of a solution of cadmium ions at a concentration of $0.20 \, \text{mol dm}^{-3}$ and ammonia at a concentration of $0.80 \, \text{mol dm}^{-3}$. At equilibrium, the concentration of free Cd^{2+} ions is $1.0 \times 10^{-2} \, \text{mol dm}^{-3}$. Calculate the dissociation constant for the equilibrium

   $$Cd(NH_3)_4^{2+} \rightleftharpoons Cd^{2+} + 4NH_3$$

5. The salt $Cu(NH_3)_4(NO_3)_2$ was dissolved at a concentration of $0.100 \, \text{mol dm}^{-3}$. The equilibrium concentration of ammonia was found to be $2.85 \times 10^{-3} \, \text{mol dm}^{-3}$. Find the dissociation constant for the equilibrium

   $$Cu(NH_3)_4^{2+} \rightleftharpoons Cu^{2+} + 4NH_3$$

SOLUBILITY AND SOLUBILITY PRODUCT

Many salts which we refer to as insoluble do in fact dissolve to a small extent. In a saturated solution, an equilibrium exists between the dissolved ions and the undissolved salt. For example, in a saturated solution of silver chloride in contact with undissolved silver chloride,

$$AgCl(s) \rightleftharpoons Ag^+(aq) + Cl^-(aq)$$

The product of the concentrations of silver ions and chloride ions is called the solubility product of silver chloride.

$$K_{sp} = [Ag^+][Cl^-]$$

The solubility product of a salt is the product of the concentrations of all the ions in a saturated solution of the salt.

It is different from solubility. *The solubility of a salt is expressed as either the amount of solute (in mol) or the mass of solute (in g) dissolved in 1 dm³ of solution at a stated temperature.*

Another example of a sparingly soluble salt is lead(II) chloride.

$$PbCl_2(s) \rightleftharpoons Pb^{2+}(aq) + 2Cl^-(aq)$$

$$K_{sp} = [Pb^{2+}][Cl^-]^2$$

If the solubility of $PbCl_2$ is a mol dm^{-3}, then $[Pb^{2+}] = a$ and $[Cl^-] = 2a$

$$K_{sp} = a \times (2a)^2 = 4a^3$$

Calculation of solubility product

EXAMPLE 1 The solubility of lead(II) hydroxide at 25 °C is 6.64×10^{-4} mol dm^{-3}. Calculate its solubility product.

METHOD $$Pb(OH)_2(s) \rightleftharpoons Pb^{2+}(aq) + 2OH^-(aq)$$

Since the solubility of $Pb(OH)_2$ is 6.64×10^{-4} mol dm^{-3}

$$[Pb^{2+}] = 6.64 \times 10^{-4} \text{ mol dm}^{-3}$$

$$[OH^-] = 2 \times 6.64 \times 10^{-4} \text{ mol dm}^{-3}$$

$$K_{sp} = [Pb^{2+}][OH^-]^2 = 6.64 \times 10^{-4} \times (1.33 \times 10^{-3})^2$$

ANSWER $$K_{sp} = 1.17 \times 10^{-9} \text{ mol}^3 \text{ dm}^{-9}$$

Note that the units in which the solubility product is expressed are the result of multiplying three concentrations together: (mol dm^{-3})³ = mol³ dm^{-9}.

EXAMPLE 2 Given that the solubility product of lead(II) sulphate at 25 °C is 1.60×10^{-8} mol² dm^{-6}, calculate the solubility at this temperature.

METHOD Solubility of $PbSO_4$ = Concn of $PbSO_4$ in solution.

All the dissolved $PbSO_4$ is in the form of Pb^{2+} ions and SO_4^{2-} ions.

$$\therefore \text{ Solubility of } PbSO_4 = [Pb^{2+}] = [SO_4^{2-}]$$

$$\text{Solubility product} = K_{sp} = [Pb^{2+}][SO_4^{2-}] = [PbSO_4]^2$$

$$[PbSO_4] = \sqrt{K_{sp}} = \sqrt{1.60 \times 10^{-8}} \text{ mol dm}^{-3}$$

ANSWER $$[PbSO_4] = 1.26 \times 10^{-4} \text{ mol dm}^{-3}$$

EXAMPLE 3 $1.00 \, dm^3$ of a solution of calcium chloride of concentration $0.100 \, mol$ dm^{-3} is added to $1.00 \, dm^3$ of sodium hydroxide of concentration $0.100 \, mol \, dm^{-3}$. If the solubility product of calcium hydroxide is $5.50 \times 10^{-6} \, mol^3 \, dm^{-9}$, calculate the mass of calcium hydroxide that will be precipitated.

METHOD   The maximum concentration of Ca^{2+} which can remain in solution is given by

$$[Ca^{2+}] \, [OH^-]^2 \; = \; 5.5 \times 10^{-6} \, mol^3 \, dm^{-9}$$

Let $[Ca^{2+}] \; = \; a$;   then $4a^3 \; = \; 5.5 \times 10^{-6} \, mol^3 \, dm^{-9}$

$$a \; = \; 1.11 \times 10^{-2} \, mol \, dm^{-3}$$

Amount of Ca^{2+} in $2.00 \, dm^3 \; = \; 2.22 \times 10^{-2} \, mol$
Amount of Ca^{2+} added as $CaCl_2 \; = \; 0.100 \, mol$
Amount of Ca^{2+} precipitated as $Ca(OH)_2 \; = \; 0.100 - 0.0222$
$$= \; 0.0778 \, mol$$

ANSWER   Mass of $Ca(OH)_2$ precipitated $\; = \; 0.0778 \times 74.0 \; = \; 5.76 \, g$

THE COMMON ION EFFECT

In a saturated solution of a salt, MA, in equilibrium with solid MA,

$$MA(s) \; \rightleftharpoons \; M^{2+}(aq) \; + \; A^{2-}(aq)$$
$$K_{sp} \; = \; [M^{2+}] \, [A^{2-}]$$

If a solution containing M^{2+} ions is added, $[M^{2+}]$ is increased. The solubility product, K_{sp}, remains the same, even when the ions are not present in equimolar concentrations. So that the product $[M^{2+}] [A^{2-}]$ shall not exceed K_{sp}, M^{2+} ions will be removed from solution as solid MA. Solute will be precipitated from solution. The addition of a solution containing A^{2-} ions will have the same effect. The separation of a solute from a solution on addition of an electrolyte solution which has an ion in common with the solute is an example of the *common ion effect.*

Another example of the common ion effect is the change in the concentration of ions produced by the dissociation of a weak acid in the presence of a solution of one of its ions.

EXAMPLE 1   Calculate the solubility at $25 \, ^\circ C$ of silver chloride: a) in water, and b) in $0.10 \, mol \, dm^{-3}$ hydrochloric acid. The solubility product of silver chloride at $25 \, ^\circ C$ is $1.8 \times 10^{-10} \, mol^2 \, dm^{-6}$.

METHOD a) Since $K_{sp} = [Ag^+][Cl^-]$

$$[Ag^+]^2 = 1.8 \times 10^{-10}\,mol^2\,dm^{-6}$$
$$[Ag^+] = 1.3 \times 10^{-5}\,mol\,dm^{-3}$$

Concn of AgCl $= 1.3 \times 10^{-5}\,mol\,dm^{-3}$

Molar mass of AgCl $= 143.5\,g\,mol^{-1}$

ANSWER Solubility of AgCl $= 1.3 \times 10^{-5} \times 143.5 = 1.9 \times 10^{-3}\,g\,dm^{-3}$.

b) The value of $[Cl^-]$ is the sum of that from the $0.10\,mol\,dm^{-3}$ HCl, and that from the dissolved AgCl. The latter is of the order of $10^{-5}\,mol\,dm^{-3}$, and can be neglected in comparison with $0.10\,mol\,dm^{-3}$ from the acid

$$K_{sp} = [Ag^+][Cl^-] = [Ag^+](0.10)\,mol^2\,dm^{-6}$$
$$[Ag^+] = 1.8 \times 10^{-10}/0.10 = 1.8 \times 10^{-9}\,mol\,dm^{-3}$$

Concn of AgCl $= 1.8 \times 10^{-9}\,mol\,dm^{-3}$

ANSWER Solubility of AgCl $= 1.8 \times 10^{-9} \times 143.5 = 2.6 \times 10^{-7}\,g\,dm^{-3}$.

The qualitative analysis scheme for the identification of metal cations is an application of solubility products. Many metals are precipitated as sulphides when a solution of the metal cations is treated with hydrogen sulphide. This large group of sparingly soluble sulphides can be divided into two groups. Hydrogen sulphide in acid solution brings down the least soluble sulphides in Group II of the qualitative analysis scheme. The rest are precipitated in Group IV from an alkaline solution of hydrogen sulphide. The effect of hydrogen ion concentration on the ionisation of hydrogen sulphide can be calculated.

EXAMPLE 2 In a saturated solution of hydrogen sulphide,

$$[H_3O^+]^2[S^{2-}] = 1.1 \times 10^{-23}\,mol^3\,dm^{-9}$$

Calculate the sulphide ion concentration: a) at pH 7, b) at pH 8, and c) at pH 2.

METHOD a) At pH 7,

$$[H_3O^+] = 10^{-7}$$

ANSWER $$[S^{2-}] = 1.1 \times 10^{-23}/(10^{-7})^2 = 1.1 \times 10^{-9}\,mol\,dm^{-3}$$

b) At pH 8,

ANSWER $$[S^{2-}] = 1.1 \times 10^{-23}/(10^{-8})^2 = 1.1 \times 10^{-7}\,mol\,dm^{-3}$$

c) At pH 2,

ANSWER $$[S^{2-}] = 1.1 \times 10^{-23}/(10^{-2})^2 = 1.1 \times 10^{-19}\,mol\,dm^{-3}$$

The very low sulphide ion concentration at pH 2 will precipitate only the most insoluble metal sulphides.

EXERCISE 48   Problems on Solubility Products

1. Given the following solubilities in $mol\,dm^{-3}$ of solution, calculate the solubility products of the solids listed:

   a) CaS 1.3×10^{-14} b) CoS 6.3×10^{-10}
   c) Ag_2S 1.1×10^{-17} d) Sb_2S_3 1.0×10^{-19}
   e) $Pb(OH)_2$ 5.0×10^{-6}

2. Given the following solubilities in g per dm^3 of solution, calculate the solubility products of the solids listed:

   a) PbS 1.20×10^{-11} b) AgI 2.14×10^{-6}
   c) $BaSO_4$ 2.41×10^{-3} d) CaF_2 1.47×10^{-2}
   e) AgCN 1.50×10^{-6}

   The following questions require a knowledge of the solubility products listed below:

CuS	$6.3 \times 10^{-36}\,mol^2\,dm^{-6}$	Bi_2S_3	$1.0 \times 10^{-97}\,mol^5\,dm^{-15}$
Ag_2S	$6.3 \times 10^{-51}\,mol^3\,dm^{-9}$	HgS	$1.6 \times 10^{-52}\,mol^2\,dm^{-6}$
NiS	$3.2 \times 10^{-19}\,mol^2\,dm^{-6}$	FeS	$6.3 \times 10^{-18}\,mol^2\,dm^{-6}$
$BaSO_4$	$1.1 \times 10^{-10}\,mol^2\,dm^{-6}$	$Al(OH)_3$	$6.3 \times 10^{-32}\,mol^4\,dm^{-12}$
$CaSO_4$	$2.4 \times 10^{-5}\,mol^2\,dm^{-6}$	SrF_2	$2.4 \times 10^{-9}\,mol^3\,dm^{-9}$
Ag_2SO_4	$1.7 \times 10^{-5}\,mol^3\,dm^{-9}$	$PbCl_2$	$1.6 \times 10^{-5}\,mol^3\,dm^{-9}$

3. What concentration of sulphide ion is needed to precipitate the metal as its sulphide from each of the following solutions?

   a) $CuSO_4(aq)$ $1.0 \times 10^{-2}\,mol\,dm^{-3}$
   b) $AgNO_3(aq)$ $1.0 \times 10^{-4}\,mol\,dm^{-3}$
   c) $NiSO_4(aq)$ $1.0 \times 10^{-5}\,mol\,dm^{-3}$
   d) $Bi(NO_3)_3(aq)$ $1.0 \times 10^{-4}\,mol\,dm^{-3}$
   e) $Hg(NO_3)_2(aq)$ $1.0 \times 10^{-6}\,mol\,dm^{-3}$
   f) $FeSO_4(aq)$ $1.0 \times 10^{-6}\,mol\,dm^{-3}$

4. Will a precipitate appear when the following solutions are added?

   a) $10\,cm^3\,BaCl_2$ $(0.01\,mol\,dm^{-3})$ and
      $10\,cm^3\,Na_2SO_4$ $(0.1\,mol\,dm^{-3})$

   b) $25\,cm^3\,Ca(OH)_2$ $(8 \times 10^{-3}\,mol\,dm^{-3})$ and
      $25\,cm^3\,Na_2SO_4$ $(0.01\,mol\,dm^{-3})$

   c) $50\,cm^3\,AlCl_3$ $(10^{-3}\,mol\,dm^{-3})$ and
      $50\,cm^3\,NaOH$ $(10^{-2}\,mol\,dm^{-3})$

   d) $10\,cm^3\,AgNO_3$ $(10^{-3}\,mol\,dm^{-3})$ and
      $40\,cm^3\,Na_2SO_4$ $(0.1\,mol\,dm^{-3})$

   e) $100\,cm^3\,Sr(NO_3)_2$ $(10^{-2}\,mol\,dm^{-3})$ and
      $100\,cm^3\,KF$ $(2 \times 10^{-2}\,mol\,dm^{-3})$

   f) $250\,cm^3\,Pb(NO_3)_2$ $(2 \times 10^{-2}\,mol\,dm^{-3})$ and
      $150\,cm^3\,NaCl$ $(0.01\,mol\,dm^{-3})$

   Show how you arrive at your conclusions.

5. In a saturated aqueous solution of hydrogen sulphide, the product
$$[H_3O^+]^2[S^{2-}] \; = \; 1.1 \times 10^{-23} \, mol^3 \, dm^{-9}$$
The solubility products of four sulphides are:

$$CdS, \quad 3.6 \times 10^{-29} \, mol^2 \, dm^{-6}$$
$$FeS, \quad 3.7 \times 10^{-19} \, mol^2 \, dm^{-6}$$
$$MnS, \quad 1.4 \times 10^{-15} \, mol^2 \, dm^{-6}$$
$$NiS, \quad 1.4 \times 10^{-24} \, mol^2 \, dm^{-6}$$

A solution contains each of the metal ions at a concentration of $0.10 \, mol \, dm^{-3}$ and $0.25 \, mol \, dm^{-3}$ hydrochloric acid. The solution is saturated with hydrogen sulphide. Calculate which of the sulphides will be precipitated.

6. In the estimation of chlorides by titration with a standard silver nitrate solution, using a chromate indicator, the precipitation of silver chloride is complete before the precipitation of silver chromate begins. Explain why this is so, using the solubility of silver chloride ($2.009 \times 10^{-3} \, g \, dm^{-3}$) and silver chromate ($3.207 \times 10^{-2} \, g \, dm^{-3}$) at $25\,^{\circ}C$ to calculate the solubility products and then: a) the concentration of silver ions needed to precipitate silver chloride from a neutral solution of chloride ions of concentration $0.100 \, mol \, dm^{-3}$, and b) the concentration of silver ions required to precipitate silver chromate, Ag_2CrO_4, from a neutral solution of chromate ions containing $5.00 \times 10^{-3} \, mol \, dm^{-3}$.

7. A solution contains $0.10 \, mol \, dm^{-3}$ of sodium carbonate and $0.10 \, mol \, dm^{-3}$ of sodium sulphate. To $1 \, dm^3$ of the solution is added $0.10 \, mol$ calcium chloride. The solubility products are: $CaCO_3$, 1.7×10^{-8}; $CaSO_4$, $2.3 \times 10^{-4} \, mol^2 \, dm^{-6}$. Find out which salt will be precipitated and calculate the mass of the precipitate.

8. The solubility product of mercury(II) sulphide is quoted in one reference book as $2 \times 10^{-49} \, mol^2 \, dm^{-6}$. If this value is correct, how many mercury(II) ions will be present in $1 \, dm^3$ of a saturated solution of this salt? (The Avogadro constant is $6 \times 10^{23} \, mol^{-1}$.)

9. The solubility product of $PbBr_2$ is 7.9×10^{-5}; that of PbI_2 is $1.0 \times 10^{-9} \, mol^3 \, dm^{-9}$; $1.0 \, dm^3$ of a solution containing $0.20 \, mol \, dm^{-3}$ of sodium bromide and $0.20 \, mol \, dm^{-3}$ of sodium iodide is added to $1.0 \, dm^3$ of lead(II) nitrate solution of concentration $0.10 \, mol \, dm^{-3}$. Which salt is precipitated? What is the mass of the precipitate?

10. Calculate the solubility of magnesium hydroxide: a) in water, b) in $0.10 \, mol \, dm^{-3}$ sodium hydroxide solution, c) in $0.010 \, mol \, dm^{-3}$ magnesium chloride solution, all at $25\,^{\circ}C$. The solubility product of magnesium hydroxide is $1.1 \times 10^{-11} \, mol^3 \, dm^{-9}$ at $25\,^{\circ}C$.

11. $K_{sp}(SrSO_4) = 4.0 \times 10^{-7} \, mol^2 \, dm^{-6}$. Calculate the solubility in mol dm^{-3} of $SrSO_4$ a) in water, b) in $0.10 \, mol \, dm^{-3}$ aqueous sodium sulphate.

12. $K_{sp}(MgF_2) = 7.2 \times 10^{-9} \, mol^3 \, dm^{-9}$. Calculate the solubility in mol dm^{-3} of MgF_2 a) in water, b) in a $0.20 \, mol \, dm^{-3}$ solution of NaF.

13. Calculate the solubility of calcium fluoride in $mol \, dm^{-3}$ a) in water, b) in a $0.010 \, mol \, dm^{-3}$ solution of sodium fluoride, c) in a $1.0 \, mol$ dm^{-3} solution of hydrogen fluoride. $(K_{sp}(CaF_2) = 4.0 \times 10^{-11} \, mol^3$ dm^{-9}, $K_a(HF) = 5.6 \times 10^{-4} \, mol \, dm^{-3}$.)

ELECTRODE POTENTIALS

If a strip of metal is placed in a solution of its ions, atoms of the metal may dissolve as positive ions, leaving a build-up of electrons on the metal:

$$M(s) \longrightarrow M^{2+}(aq) + 2e^-$$

The metal will become negatively charged. Alternatively, metal ions may take electrons from the strip of metal and be discharged as metal atoms:

$$M^{2+}(aq) + 2e^- \longrightarrow M(s)$$

In this case, the metal will become positively charged. The potential difference between the strip of metal and the solution depends on the nature of the metal and on the concentration of the ions involved in the equilibrium at the metal surface. Zinc acquires a more negative potential than copper, since it has a greater tendency to dissolve as ions and a smaller tendency to be deposited as metal. In order to compare electrode potentials for different metals, *standard electrode potentials* are quoted at 25 °C with an ionic concentration of 1 mol dm^{-3}. The zero on the standard electrode potential scale is the potential of a strip of platinum in contact with hydrogen gas at 1 atm pressure and hydrogen ions at a concentration of 1 mol dm^{-3}.

Metals are reducing agents. Other oxidation–reduction systems also have electrode potentials, the value of which depend on the standard electrode potential for the system and the concentrations of the ions in the equilibrium. The standard electrode potential of a redox system is the potential acquired by a piece of platinum immersed in a solution of the redox system in which the concentration of each dissolved component is 1 mol dm^{-3}. A powerful oxidising agent removes electrons and gives the platinum a high positive potential. When all the redox systems are arranged in order of their standard electrode potentials, the *electrochemical series* is obtained. Table 8.1 shows some of the redox systems in the series.

Table 8.1   Values of standard electrode potential E° at 298 K

Reaction	E°/V
$K^+(aq) + e^- \longrightarrow K(s)$	-2.92
$Ca^{2+}(aq) + 2e^- \longrightarrow Ca(s)$	-2.87
$Na^+(aq) + e^- \longrightarrow Na(s)$	-2.71
$Mg^{2+}(aq) + 2e^- \longrightarrow Mg(s)$	-2.36
$Al^{3+}(aq) + 3e^- \longrightarrow Al(s)$	-1.66
$Zn^{2+}(aq) + 2e^- \longrightarrow Zn(s)$	-0.76
$Fe^{2+}(aq) + 2e^- \longrightarrow Fe(s)$	-0.44
$Cr^{3+}(aq) + e^- \longrightarrow Cr^{2+}(aq)$	-0.41
$Ni^{2+}(aq) + 2e^- \longrightarrow Ni(s)$	-0.25
$Sn^{2+}(aq) + 2e^- \longrightarrow Sn(s)$	-0.14
$Pb^{2+}(aq) + 2e^- \longrightarrow Pb(s)$	-0.13
$2H_3O^+(aq) + 2e^- \longrightarrow H_2(g) + 2H_2O(l)$	0.00
$Sn^{4+}(aq) + 2e^- \longrightarrow Sn^{2+}(aq)$	0.15
$Cu^{2+}(aq) + 2e^- \longrightarrow Cu(s)$	0.34
$I_2(s) + 2e^- \longrightarrow 2I^-(aq)$	0.54
$Fe^{3+}(aq) + e^- \longrightarrow Fe^{2+}(aq)$	0.77
$Ag^+(aq) + e^- \longrightarrow Ag(s)$	0.80
$Br_2(l) + 2e^- \longrightarrow 2Br^-(aq)$	1.09
$MnO_2(s) + 4H^+(aq) + 2e^- \longrightarrow Mn^{2+}(aq) + 2H_2O(l)$	1.23
$Cr_2O_7^{2-}(aq) + 14H^+(aq) + 6e^- \longrightarrow 2Cr^{3+}(aq) + 7H_2O(l)$	1.33
$Cl_2(g) + 2e^- \longrightarrow 2Cl^-(aq)$	1.36
$Ce^{4+}(aq) + e^- \longrightarrow Ce^{3+}(aq) \quad (in\ H_2SO_4(aq))$	1.44
$PbO_2(s) + 4H^+(aq) + 2e^- \longrightarrow Pb^{2+}(aq) + 2H_2O(l)$	1.46
$MnO_4^-(aq) + 8H^+(aq) + 5e^- \longrightarrow Mn^{2+}(aq) + 4H_2O(l)$	1.51
$Ce^{4+}(aq) + e^- \longrightarrow Ce^{3+}(aq) \quad (in\ HNO_3(aq))$	1.61
$H_2O_2(aq) + 2H^+(aq) + 2e^- \longrightarrow 2H_2O(l)$	1.78
$F_2(g) + 2e^- \longrightarrow 2F^-(aq)$	2.85

GALVANIC CELLS

When two electrodes are combined to form a cell, their standard electrode potentials will tell you which will be the positive and which the negative electrode. An easy way to work out which of the possible reactions will happen is to use the *anticlockwise rule*. Write down the two redox systems, with the more negative standard electrode potential at the top. Then draw a circle anticlockwise. For example, when copper and silver are in contact with solutions of their ions,

$$Cu^{2+}(aq) + 2e^- \rightleftharpoons Cu(s) \quad E^{\ominus} = +0.34\,V$$
$$Ag^+(aq) + e^- \rightleftharpoons Ag(s) \quad E^{\ominus} = +0.80\,V$$

The circle tells you that the reaction which takes place is

$$Cu(s) + 2Ag^+(aq) \longrightarrow Cu^{2+}(aq) + 2Ag(s)$$

The silver electrode is positive; the copper electrode is negative.

Reaction will take place between two redox systems which differ by 0.3 V or more.

Fig. 8.2 shows two metals inserted into solutions of their ions. The two solutions are joined by a salt bridge, and the two metal electrodes are connected by an external circuit. The cell has an e.m.f. which is equal to the difference between the standard electrode potentials of the two metals, and a current flows through the external circuit.

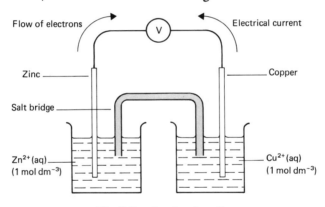

Fig. 8.2   A galvanic cell

The cell shown in Fig. 8.2 can be represented by

$$Zn(s) \mid Zn^{2+}(aq)\,(1\,mol\,dm^{-3}) \mid Cu^{2+}(aq)\,(1\,mol\,dm^{-3}) \mid Cu(s)$$

By convention, the e.m.f. of the cell is taken as

$$E = E(\text{RHS electrode}) - E(\text{LHS electrode})$$
$$= E^{\ominus}_{Cu} - E^{\ominus}_{Zn} \quad \text{where } E^{\ominus} \text{ is the standard electrode potential}$$
$$= 0.34 - (-0.76) = +1.10\,V$$

The flow of electrons is clockwise through the external circuit (from zinc to copper). Conventional electricity flows anticlockwise through the external circuit (from copper to zinc).

If the cell is written as

$$Cu(s) \mid Cu^{2+}(aq) \ (1 \ mol \ dm^{-3}) \mid Zn^{2+}(aq) \ (1 \ mol \ dm^{-3}) \mid Zn(s)$$

then the e.m.f. is given by

$$E = E_{Zn}^{\ominus} - E_{Cu}^{\ominus} = -0.76 - 0.34 = -1.10 \ V$$

In general, in a cell A | B | C | D if the reactions which occur are A ⟶ B and C ⟶ D, then the e.m.f. is positive; if B ⟶ A and D ⟶ C, then the e.m.f. is negative.

EXERCISE 49   Problems on Standard Electrode Potentials

Refer to the table of values on p.154.

1. Which of the following species are oxidised by manganese(IV) oxide?
   Br^-,   Ag, I^-, Cl^-

2. Which of the following species are reduced by Sn^{2+}?
   I_2, Ni^{2+}, Cu^{2+}, Fe^{3+}

3. Calculate the standard e.m.f.'s of the following cells at 298 K:
   a) $Ni(s) \mid Ni^{2+}(aq) \mid Sn^{2+}(aq), Sn^{4+}(aq) \mid Pt$
   b) $Pt \mid I_2(s), I^-(aq) \mid Ag^+(aq) \mid Ag(s)$
   c) $Pt \mid Cl_2(g), Cl^-(aq) \mid Br_2(l), Br^-(aq) \mid Pt$

4. Calculate the standard e.m.f. of each of the cells:
   a) $Sn(s) \mid Sn^{2+}(aq) \mid Ag^+(aq) \mid Ag(s)$
   b) $Ag(s) \mid Ag^+(aq) \mid Cu^{2+}(aq) \mid Cu(s)$
   c) $Ce^{3+}(aq) \mid Ce^{4+}(aq) \mid Fe^{3+}(aq) \mid Fe^{2+}(aq)$
   d) $Fe(s) \mid Fe^{2+}(aq) \mid Cu^{2+}(aq) \mid Cu(s)$
   e) $Zn(s) \mid Zn^{2+}(aq) \mid Pb^{2+}(aq) \mid Pb(s)$

5. Iron filings are added to a solution containg the ions Cu^{2+}, Fe^{2+}, Fe^{3+}, H_3O^+ and Zn^{2+}, all at a concentration of $1 \ mol \ dm^{-3}$. From the standard electrode potentials of the redox systems, deduce what reaction occurs, and write the equation.

6. A solution contains Fe^{2+}, Fe^{3+}. Cr^{3+} and $Cr_2O_7^{2-}$ in their standard states and dilute sulphuric acid. Deduce what happens, and write the equation for the reaction.

7. Predict the reactions between:
   a) $Fe^{3+}(aq)$ and $I^-(aq)$ b) $Ag^+(aq)$ and $Cu(s)$
   c) $Fe^{3+}(aq)$ and $Br^-(aq)$ d) $Ag(s)$ and $Fe^{3+}(aq)$
   e) $Br_2(aq)$ and $Fe^{2+}(aq)$

   From the standard electrode potentials, predict which of the halogens, Cl_2, Br_2, I_2, will oxidise i) Fe^{2+} to Fe^{3+} ii) Sn^{2+} to Sn^{4+}.

EXERCISE 50   Questions from A-level Papers

1. A current of 0.200 A is passed for 20 minutes through two voltameters in series. One voltameter has copper electrodes and contains a solution of copper(II) sulphate, the other has platinum electrodes and contains a dilute solution of sulphuric acid. Calculate to three significant figures:

   a) the mass of copper deposited on the cathode of the first voltameter,

   b) the volume of oxygen (expressed in cm^3 at s.t.p.) liberated at the anode of the second voltameter.

   (Take the Faraday constant to be $96\,500\,C\,mol^{-1}$.)   (O79)

2. a) State Faraday's laws of electrolysis. Describe an experiment to illustrate the second law.

   b) Calculate the time in minutes necessary for a current of 10 A to deposit 1 g of copper from an aqueous solution of copper(II) sulphate.

   c) Show why the ratio of the masses of copper and sodium deposited, under the appropriate conditions, by the same quantity of electricity is 1.38.

   d) Calculate the charge on an electron given that the Avogadro constant is 6.02×10^{23}/mol.   (AEB77)

3. a) Define the term *acid*.

   b) Sketch graphs to show how the molar electrical conductance of weak and strong acids depends on the square root of the concentration.

   c) Derive expressions relating α, the degree of dissociation of a *very weak acid*,
      i) to its acidity constant (K) and its molar concentration (c);
      ii) to a related expression involving molar electrical conductance (Λ). Illustrate the second expression graphically.

   d) i) Calculate the molar conductance of ethanoic acid (acetic acid) at infinite dilution given the following values at 25 °C of molar conductance at infinite dilution (zero concentration) for the electrolytes listed.

Electrolyte	*Molar conductance*/$cm^2\,mol^{-1}\,ohm^{-1}$
HCl	426.1
NaCl	126.5
CH_3COONa	91.0

   ii) If the molar conductance of a 0.001 M solution of ethanoic acid at 25 °C is $52.5\,cm^2\,mol^{-1}\,ohm^{-1}$, calculate the degree of ionization of ethanoic acid at this concentration, its acidity constant and pH.

e) Sketch the titration curve showing the changes in pH which occur on addition of 0.001 M sodium hydroxide solution to 25.0 cm³ of 0.001 M ethanoic acid and indicate on the curve where the rate of change of pH per cm³ of alkali added is at a minimum. Label axes clearly and show pH values.

f) What do you understand by the term *buffer solution*? In what proportion must solutions of 0.001 M ethanoic acid and 0.001 M sodium hydroxide be mixed in order to produce an efficient buffer solution? What is its range?   (SUJB79)

4. a) Define the terms: i) *electrolytic conductivity* (*specific conductance*), ii) *molar conductivity* of an aqueous solution of an electrolyte.

b) The electrolytic conductivity of a saturated aqueous solution of thallium(I) chloride, TlCl, at 25 °C is $2.40 \times 10^{-3} \, \text{ohm}^{-1} \, \text{cm}^{-1}$. The molar conductivities at infinite dilution of thallium(I) hydroxide, sodium hydroxide and sodium chloride are 273, 248 and $126 \, \text{ohm}^{-1} \, \text{cm}^2 \, \text{mol}^{-1}$ respectively. Estimate i) the molar conductivity of thallium(I) chloride, ii) the solubility of TlCl in water at 25 °C in mol dm^{-3}. State any law you assume in your calculation.

c) Explain concisely the *principles* involved in the purification of water by ion exchange for use in laboratory experiments.   (O77)

*5. The following data were obtained from a sample of an aqueous solution of aminoethane (ethylamine) of concentration $0.1 \, \text{mol dm}^{-3}$ at 25 °C:

i) Electrolytic conductivity $1.5 \times 10^{-1} \, \text{S m}^{-1}$ $(1 \, \text{S m}^{-1} \equiv 1 \, \Omega^{-1} \, \text{m}^{-1})$.

ii) Molar conductivity at infinite dilution $2.04 \times 10^{-2} \, \text{S m}^2 \, \text{mol}^{-1}$ $(1 \, \text{S m}^2 \, \text{mol}^{-1} \equiv 1 \, \Omega^{-1} \, \text{m}^2 \, \text{mol}^{-1})$.

iii) Addition of an equal volume of a solution of compound X of concentration $0.01 \, \text{mol dm}^{-3}$ produced a solution of pH 11.73. This pH value remained constant in spite of small additions of H_3O^+ and OH^- ions.

a) Calculate the dissociation constant and pH value for the aqueous solution of aminoethane. Comment on the pH value you obtain.

b) Suggest a named compound which could be represented by X. What is the name given to such a mixed solution?

c) Explain the action of such a solution.

d) Describe concisely how the values i) and ii) in the data above could be obtained.   (AEB80, S)

6. a) A buffer solution can be prepared by mixing solutions of a weak acid and one of its salts. Derive the relationship between the hydrogen ion concentration, the acid dissociation constant (K_a) and the concentrations of the weak acid and its salt.

   b) Calculate K_a for a weak monobasic acid given the following information. A buffer solution made up of $10 cm^3$ of acid of concentration $0.09 mol dm^{-3}$ mixed with $20 cm^3$ of the potassium salt of the acid of concentration $0.15 mol dm^{-3}$ has a pH of 5.85. (O & C80, p)

*7. a) Discuss the acid–base properties of water on both the Brønsted-Lowry and the Lewis theories.

   b) Calculate the pH of water at $60 °C$ given that $K_w = 9.61 \times 10^{-14}$ $mol^2 litre^{-2}$ at this temperature. Comment on the value you obtain.

   c) Calculate the pH at the equivalence point when $0.1 M$ sodium hydroxide is titrated with $0.1 M$ ethanoic (acetic) acid. ($K_w = 10^{-14}$ $mol^2 litre^{-2}$ (at $25 °C$), $K_a(CH_3COOH) = 1.8 \times 10^{-5} mol litre^{-1}$ (at $25 °C$).

   d) Discuss the mechanism of the colour change of an acid–base indicator. (JMB76, S)

*8. Draw a clearly labelled diagram showing the titration curve for the titration of phosphoric(V) acid with sodium hydroxide in aqueous solution. The pK_a values for phosphoric(V) acid are 2.00, 6.80 and 12.00. Explain the purpose of a buffer solution and indicate the nature of buffer action.

   A commonly used buffer solution is prepared by adding together solutions of disodium hydrogenphosphate(V) and potassium dihydrogenphosphate(V). What volumes of each of these solutions, both $0.15 mol dm^{-3}$, would have to be mixed to give $10 cm^3$ of a buffer solution having pH 7.20? To what value will the pH of the solution change if $0.1 cm^3$ of sodium hydroxide of strength $0.15 mol dm^{-3}$ is now added? (O77, S)

9. a) What is the difference between a voltaic (galvanic) process and electrolysis? Describe, with examples, the process at each electrode in both cases.

   b) Define standard electrode potential and describe how it can be measured.

   c) Half cell X consists of nickel dipping into a $1 M$ solution of Ni^{2+} and half cell Y of zinc dipping into a $1 M$ solution of Zn^{2+}. When connected to a hydrogen electrode the *numerical* values of the electrode potentials are found to be 0.23 V for X and 0.77 V for Y.
      i) Deduce the sign of each of the potentials given that the metal electrode is negative in both cases.
      ii) Will there be a reaction when zinc is dipped in a $1 M$ solution of Ni^{2+}?

iii) Will there be a reaction when nickel is dipped in a solution of Zn^{2+}?

iv) The half cells X and Y are connected together. Deduce the signs of the electrodes and write down the spontaneous cell reaction. What would be the e.m.f. of the cell?

d) In the commercial production of magnesium by electrolysis a current of approximately 50 000 A is used. What is the rate of production of magnesium?

($A_r(Mg) = 24.3$, charge of an electron $= 1.602 \times 10^{-19}$ C. Avogadro constant $= 6.022 \times 10^{23}$ mol^{-1}.)   (JMB79, S)

10. A weak monobasic acid HA has a dissociation constant, K_a, of 1×10^{-5} mol dm^{-3}. The un-ionised acid molecules are blue, the anions are yellow.

a) Calculate the pH at which a solution containing HA molecules and A$^-$ ions is green, i.e. midway between blue and yellow. Give your reasoning.

b) To a solution of HA of pH 5, sufficient of an iron(III) salt was added to make the solution 0.01 mol dm^{-3} in Fe^{3+}(aq) ions.

i) Calculate the minimum concentration of OH$^-$(aq) ions necessay to cause precipitation of iron(III) hydroxide from a solution which is 0.01 mol dm^{-3} in Fe^{3+}(aq) ions.
   (Solubility product of iron(III) hydroxide at the temperature of the experiment $= 8 \times 10^{-38}$ mol^4 dm^{-12}. Ionic product, K_w, of water at this temperature $= 1 \times 10^{-14}$ mol^2 dm^{-6}.)

ii) Hence deduce whether a precipitate of iron(III) hydroxide formed in the experiment described.   (C78)

11. a) Explain what is meant by the term *solubility product*, using calcium sulphate as your example.

b) Given that the solubility product of calcium sulphate, K_{sp}, is 2.0×10^{-5} mol^2 l^{-2} at 298 K, calculate the solubility, in mol l^{-1} at that temperature, of calcium sulphate in:

i) water,
ii) sodium sulphate solution of concentration 0.1 mol l^{-1}, and
iii) a solution containing 0.1 mol l^{-1} of calcium orthophosphate, $Ca_3(PO_4)_2$, in water containing a minimum amount of hydrochloric acid.

c) State with reasons whether or not the concept of solubility product is applicable to the following substances: sodium chloride, benzoic acid, lead(II) chloride and lead(IV) chloride.   (L79)

12. a) Write an expression for the solubility product of lead(II) chloride.

b) The solubility product of lead(II) chloride is 1.6×10^{-5} mol^3 l^{-3} at a given temperature.

i) What is the solubility in mol l^{-1} of lead(II) chloride in water at the same temperature?

ii) How many moles of chloride ion must be added to a 1.0 M solution of lead(II) nitrate at the same temperature in order just to cause a precipitate of lead(II) chloride? Assume that no change in volume occurs on adding the chloride ion.

c) When a 1.0 M solution of hydrochloric acid is added to a saturated solution of lead(II) chloride, a permanent precipitate is obtained. However, if concentrated hydrochloric acid is used, the precipitate initially formed dissolves. How do you account for this?

d) When an excess of aqueous ammonia is added to a solution containing copper(II) ions, a complex ion is formed.

i) Describe what you would see when the addition is carried out slowly and write down the equations for the reactions occurring.

ii) Write an expression for the stability constant of the complex ion.

iii) If the stability constant for the complex ion is $1 \times 10^{13} \, mol^{-4} l^4$, what will be the concentration of Cu^{2+} ions in a solution where the concentration of the complex ion is 1.0 M and that of ammonia is 3.0 M?   (L80)

*13. a) Explain what you understand by solubility product and indicate the conditions under which the principle can be applied. Discuss its application to one of the group tests in the systematic analysis of cations.

b) Calculate the solubility product of silver chloride, AgCl, from the following data. The electrolytic (specific) conductivity of a saturated solution of silver chloride in pure water at 25 °C is $3.41 \times 10^{-6} \, ohm^{-1} cm^{-1}$, and the electrolytic conductivity of the water alone is $1.60 \times 10^{-6} \, ohm^{-1} cm^{-1}$. The salts listed below have the following molar conductivities at infinite dilution at 25 °C:

$$AgNO_3 \quad 133.4 \, ohm^{-1} cm^2 mol^{-1};$$
$$KNO_3 \quad 145.0 \, ohm^{-1} cm^2 mol^{-1};$$
$$KCl \quad 149.9 \, ohm^{-1} cm^2 mol^{-1}.$$

Mention briefly any assumptions you have made in calculating your answer.   (O76, S)

14. Explain what is meant by *solubility product*.

The solubility of anhydrous calcium iodate(V) in water at 298 K is $3.07 \, g \, l^{-1}$. Calculate: a) its solubility product, b) its solubility (in $g \, l^{-1}$) in an aqueous solution containing 0.1 $mol \, l^{-1}$ of sodium iodate(V).

Explain, with full experimental details, how you would determine the solubility of calcium iodate(V) in water at 298 K, emphasising any precautions you would have to take.

$$[A_r(Ca) = 40; \quad A_r(I) = 127; \quad A_r(Q) = 16.]$$
     (WJEC78)

9 Thermochemistry

INTERNAL ENERGY AND ENTHALPY

Matter contains energy. It is the kinetic energy of molecular motion and the potential energy associated with chemical bonds. These together make up the *internal energy* of matter. Frequently during the course of a chemical reaction heat is either given out or taken in from the surroundings. The heat absorbed during a reaction is equal to the internal energy of the products minus the internal energy of the reactants plus any work done by the system on the surroundings. Since most laboratory work is done at constant pressure, any gases formed are allowed to escape into the atmosphere and work is done in expansion:

$$\begin{pmatrix} \text{Heat absorbed at} \\ \text{constant pressure} \end{pmatrix} = \begin{pmatrix} \text{Change in} \\ \text{internal energy} \end{pmatrix} + \begin{pmatrix} \text{Work done on} \\ \text{surroundings} \end{pmatrix}$$

The heat absorbed at constant pressure is given the name *change in enthalpy* and the symbol ΔH. Enthalpy is defined by the equation

$$H = U + PV$$

where H = Enthalpy, U = Internal energy, P = Pressure, and V = Volume.

Then,
$$\Delta H = \Delta U + P \Delta V$$

When expansion occurs, ΔV is positive and $\Delta H > \Delta U$.

When contraction occurs, ΔV is negative and $\Delta H < \Delta U$.

If reaction takes place at constant volume, $\Delta V = 0$, and $\Delta H = \Delta U$.

Reactions of solids and liquids do not involve large changes in volume, and ΔH is close to ΔU. Reactions in which ΔV is large are those involving gases, and the value of ΔV can be calculated from the ideal gas equation. Since

$$PV = nRT$$

$$P \Delta V = \Delta n R T$$

Δn, the increase in the number of molecules of gas, is indicated by the equation for the reaction. For example, in the reaction

$$CaCO_3(s) \longrightarrow CaO(s) + CO_2(g)$$

$\Delta n = 1$.

The enthalpy of a substance is quoted for the substance in its standard state. The *standard state* of a substance is 1 mole of the substance in a specified state (solid, liquid or gas) at 1 atmosphere pressure. The value of an enthalpy change is quoted for standard conditions: gases at 1 atmosphere, solutions at unit concentration, and substances in their normal states at a specified temperature. $\Delta H_T^\ominus$ means the standard enthalpy change at a temperature T. $\Delta H_{298}^\ominus$ is sometimes written as $\Delta H^\ominus$.

Definitions of some standard enthalpy changes follow:

Standard enthalpy of formation is the heat absorbed when 1 mole of a substance is formed from its elements *in their standard states* at constant pressure. (If the reaction is exothermic, the heat absorbed is negative, and $\Delta H_F^\ominus$ has a negative value. All elements in their standard states are assigned a value of zero for their standard enthalpies of formation.)

Standard enthalpy of combustion is the heat absorbed when 1 mole of a substance is completely burned in oxygen at constant pressure.

Standard enthalpy of hydrogenation is the heat absorbed when 1 mole of an unsaturated compound is converted to a saturated compound by reaction with gaseous hydrogen at constant pressure.

Standard enthalpy of neutralisation is the heat absorbed when an acid and a base react at constant pressure to form 1 mole of water.

Standard enthalpy of reaction is the heat absorbed in a reaction at constant pressure between the number of moles of reactants shown in the equation for the reaction. In the reaction

$$4H_2O(g) \ + \ 3Fe(s) \ \longrightarrow \ Fe_3O_4(s) \ + \ 4H_2(g)$$

the standard enthalpy change refers to the reaction between 4 moles of steam and 3 moles of iron.

Standard enthalpy of solution is the heat absorbed when 1 mole of a substance is dissolved at constant pressure in a stated amount of solvent. This may be $100\,g$ or $1\,000\,g$ of solvent or it may be an 'infinite' amount of solvent, i.e. a volume so large that on further dilution there is no further heat change.

STANDARD ENTHALPY CHANGE FOR A CHEMICAL REACTION

The standard enthalpy change for a chemical reaction can be calculated from the standard enthalpies of formation of all the products and reactants involved. For example, in the addition of hydrogen chloride to ethene,

$$CH_2 = CH_2(g) \ + \ HCl(g) \ \longrightarrow \ C_2H_5Cl(g)$$
$$(+52.3) \qquad\quad (-92.3) \qquad\qquad (-105)$$

The standard enthalpies of formation in $kJ\,mol^{-1}$ are shown under each species.

The standard enthalpy of reaction $\Delta H^{\ominus}$ is given by

$$\Delta H^{\ominus} = (-105) - (52.3 + (-92.3)) = -65 \, \text{kJ mol}^{-1}$$

The negative sign means that the products contain less energy than the reactants and the difference is the heat energy given out in the reaction: the reaction is exothermic. A positive value for $\Delta H^{\ominus}$ indicates an endothermic reaction.

The standard enthalpy of reaction depends only on the difference between the standard enthalpy of the reactants and the standard enthalpy of the products and not on the route by which the reaction occurs.

This idea is embodied in Hess's law, which states that, if a reaction can take place by more than one route, the overall change in enthalpy is the same, whichever route is followed.

STANDARD ENTHALPY OF NEUTRALISATION

EXAMPLE $250 \, \text{cm}^3$ of sodium hydroxide of concentration $0.400 \, \text{mol dm}^{-3}$ were added to $250 \, \text{cm}^3$ of hydrochloric acid of concentration $0.400 \, \text{mol dm}^{-3}$ in a calorimeter. The temperature of the two solutions and the calorimeter was $17.05 \, ^\circ\text{C}$. The mass of the calorimeter was $500 \, \text{g}$, and its specific heat capacity was $400 \, \text{J kg}^{-1}\text{K}^{-1}$. The temperature rose to $19.55 \, ^\circ\text{C}$. Assuming that the specific heat capacity[†] of all the solutions is $4200 \, \text{J kg}^{-1}\text{K}^{-1}$ calculate the standard enthalpy of neutralisation.

METHOD   Mass of solutions $= 500 \, \text{g}$

Heat capacity of solutions $= 0.500 \times 4200 = 2100 \, \text{J}$

Mass of calorimeter $= 500 \, \text{g}$

Heat capacity of calorimeter $= 0.500 \times 400 = 200 \, \text{J}$

Rise in temperature $= 2.50 \, ^\circ\text{C}$

Heat evolved $= (2100 + 200) \times 2.50 = 5750 \, \text{J}$

Amount of water formed $= 250 \times 10^{-3} \times 0.400 = 0.100 \, \text{mol}$

Heat evolved per mole $= 5750/0.100 = 57500 \, \text{J}$

ANSWER   The standard enthalpy of neutralisation $= 57.5 \, \text{kJ mol}^{-1}$.

[†]The *heat capacity* of a mass of substance is the quantity of heat needed to raise its temperature by 1 K or 1 $^\circ$C.

The *specific heat capacity* of a substance is the quantity of heat required to raise the temperature of 1 kg of the substance by 1 K or 1 $^\circ$C.

Heat capacity $=$ Mass $\times$ Specific heat capacity.

EXERCISE 51   Problems on Standard Enthalpy of Neutralisation

1. 50.0 cm³ of sodium hydroxide solution of concentration 0.400 mol dm⁻³ required 20.0 cm³ of sulphuric acid of concentration 0.500 mol dm⁻³ for neutralisation. A temperature rise of 3.6 °C was observed if both solutions and the calorimeter were initially at the same temperature. Calculate the standard enthalpy of neutralisation of sodium hydroxide with sulphuric acid. The heat capacity of the calorimeter is 39.0 J K⁻¹. (The specific heat capacity of all the solutions is 4.2 J K⁻¹ g⁻¹.)

2. 100 cm³ of potassium hydroxide solution of concentration 1.00 mol dm⁻³ and 100 cm³ of hydrochloric acid of concentration 1.00 mol dm⁻³ were mixed in a calorimeter. All three were at the same temperature. The heat capacity of the calorimeter was 95 J K⁻¹, and the rise in temperature was 6.25 K. Calculate the standard enthalpy of neutralisation. (Specific heat capacity of water = 4.2 J K⁻¹ g⁻¹.)

3. 100 cm³ of 1.00 mol dm⁻³ sodium hydroxide solution and 100 cm³ of 1.00 mol dm⁻³ ethanoic acid were mixed in a calorimeter. All three were at the same temperature. The heat capacity of the calorimeter was 90 J K⁻¹, and the rise in temperature was 5.3 K. Calculate the standard enthalpy of neutralisation.

4. A calorimeter has a mass of 200 g and a specific heat capacity of 0.42 J g⁻¹. Into it are put 50 cm³ of 1.25 mol dm⁻³ hydrochloric acid and 50 cm³ of 1.25 mol dm⁻³ potassium hydroxide solution at the same temperature. The temperature of the calorimeter and contents rises by 7.0 °C. Calculate the standard enthalpy of neutralisation.

5. Fig. 9.1 shows the results of a thermometric titration to find a value of the standard enthalpy of neutralisation. 50.0 cm³ of a solution of

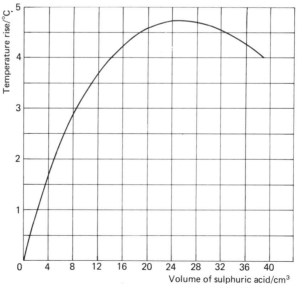

Fig. 9.1

sodium hydroxide of concentration $0.500 \, mol \, dm^{-3}$ were titrated against a $0.568 \, mol \, dm^{-3}$ solution of sulphuric acid. Calculate the standard enthalpy of neutralisation. Assume that the specific heat capacity of the solutions is $4.2 \, J \, K^{-1} g^{-1}$, and assume that no heat passes to the container.

STANDARD ENTHALPY OF COMBUSTION

EXAMPLE   The standard enthalpy of combustion of liquid ethanol is $-1370 \, kJ \, mol^{-1}$. Calculate the standard heat of combustion at constant volume.

METHOD $$C_2H_5OH(l) \; + \; 3O_2(g) \longrightarrow 2CO_2(g) \; + \; 3H_2O(l)$$

The increase in the number of gaseous molecules $= \Delta n = 2 - 3 = -1$

$$\Delta H^{\ominus} = \Delta U^{\ominus} + \Delta n R T$$

Putting $\Delta H^{\ominus} = -1370$ and $R = 8.31 \times 10^{-3} \, kJ \, mol^{-1} K^{-1}$

$$-1370 = \Delta U^{\ominus} + (-1 \times 8.31 \times 10^{-3} \times 298)$$

$$\Delta U^{\ominus} = -1367.5 \, kJ \, mol^{-1}$$

ANSWER   The standard heat of combustion at constant volume is $-1368 \, kJ \, mol^{-1}$.

EXERCISE 52   Problems on Standard Enthalpy of Combustion

1. In the combustion of $1.00 \, g$ of benzoic acid in a bomb calorimeter the temperature rises by $3.39 \, K$. The equation is
   $$C_6H_5CO_2H(s) \; + \; 7\tfrac{1}{2}O_2(g) \longrightarrow 7CO_2(g) \; + \; 3H_2O(l)$$
   a) If the heat capacity of the calorimeter and contents is $7800 \, J \, K^{-1}$, what is the molar heat of combustion of benzoic acid?
   b) What is the value of the standard enthalpy of combustion?

2. a) Calculate the value of $RT\Delta n$ in the combustion of pentane at $298 \, K$. The equation is
   $$C_5H_{12}(g) \; + \; 8O_2(g) \longrightarrow 5CO_2(g) \; + \; 6H_2O(l)$$
   b) If the standard enthalpy of combustion is $-3530 \, kJ \, mol^{-1}$, what would be the value of the molar heat of combustion obtained in a bomb calorimeter (at constant volume)?

3. a) Calculate the value of $\Delta n R T$ for the combustion of butan-1,4-dioic acid at $298 \, K$. The equation is
   $$C_4H_6O_4(s) \; + \; 3\tfrac{1}{2}O_2(g) \longrightarrow 4CO_2(g) \; + \; 3H_2O(l)$$
   b) Hence find the difference between $\Delta H^{\ominus}$ and the heat absorbed in the combustion of 1 mole of butan-1,4-dioic acid at constant volume in a bomb calorimeter.

4. Calculate the standard enthalpies of combustion of hexane and propane from the following data obtained from measurements in a bomb calorimeter:

   a) The combustion of 1.720 g of hexane released 84.06 kJ of heat.

   b) The combustion of 1.100 g of propane raised the temperature of the calorimeter and contents by 6.4 K. The heat capacity of the calorimeter and contents is $8.575 \, \text{kJ K}^{-1}$.

FINDING THE STANDARD ENTHALPY OF A COMPOUND INDIRECTLY

Sometimes, the standard enthalpy of formation of a compound can be measured directly by allowing known amounts of elements to combine and measuring the amount of heat evolved. Other reactions are difficult to study, and the standard enthalpy of reaction must be found indirectly.

To find the standard enthalpy of formation of ethyne from practical measurements is impossible as attempts to make ethyne from carbon and hydrogen,

$$2C(s) \;+\; H_2(g) \longrightarrow C_2H_2(g)$$

will result in the formation of a mixture of hydrocarbons. The standard enthalpy of combustion of ethyne can, however, be measured experimentally, and from it can be calculated the standard enthalpy of formation. The standard enthalpies of combustion of carbon and hydrogen are also required.

EXAMPLE 1 Find the standard enthalpy of formation of ethyne, given the standard enthalpies of combustion (in kJ mol^{-1}): $C_2H_2(g) = -1300$; $C(s) = -394$; $H_2(g) = -286$.

METHOD 1 The method of calculation is based on the three equations for the combustion of ethyne, carbon and hydrogen:

$$C(s) + O_2(g) \longrightarrow CO_2(g); \qquad \Delta H_1^{\ominus} = -394 \, \text{kJ mol}^{-1} \quad [1]$$

$$H_2(g) + \tfrac{1}{2}O_2(g) \longrightarrow H_2O(l) \qquad \Delta H_2^{\ominus} = -286 \, \text{kJ mol}^{-1} \quad [2]$$

$$C_2H_2(g) + 2\tfrac{1}{2}O_2(g) \longrightarrow 2CO_2(g) + H_2O(l) \qquad \Delta H_3^{\ominus} = -1300 \, \text{kJ mol}^{-1} \quad [3]$$

Looking at equation [1] one can see that the standard enthalpy of combustion of carbon is the same as the standard enthalpy of formation of carbon dioxide.

Likewise, equation [2] shows that the standard enthalpy of combustion of hydrogen is the same as the standard enthalpy of formation of water.

The standard enthalpy content of a substance is equal to the standard enthalpy of formation of the substance from its elements in their standard states.

Putting the standard enthalpy content of each substance into equation [3] gives

$$C_2H_2(g) + 2\tfrac{1}{2}O_2(g) \longrightarrow 2CO_2(g) + H_2O(l); \ \Delta H_3^\ominus = -1300 \text{ kJ mol}^{-1}$$
$$\Delta H_F^\ominus(C_2H_2) \quad 0 \qquad\qquad\quad 2(-394) \ (-286)$$

Since

$$\begin{pmatrix}\text{Standard} \\ \text{enthalpy change}\end{pmatrix} = \begin{pmatrix}\text{Standard} \\ \text{enthalpy content} \\ \text{of products}\end{pmatrix} - \begin{pmatrix}\text{Standard} \\ \text{enthalpy content} \\ \text{of reactants}\end{pmatrix}$$

$$\Delta H_3^\ominus = -1300 = 2(-394) + (-286) - \Delta H_F^\ominus(C_2H_2)$$

$$\Delta H_F^\ominus(C_2H_2) = 226 \text{ kJ mol}^{-1}$$

ANSWER The standard enthalpy of formation of ethyne is 226 kJ mol^{-1}. Since $\Delta H_F^\ominus$ is positive, ethyne is referred to as an endothermic compound.

METHOD 2 Another method of tackling the problem is to construct an enthalpy diagram:

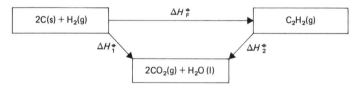

According to Hess's law, the change in standard enthalpy when carbon and hydrogen burn to form carbon dioxide and water is the same as the sum of the standard enthalpy changes when carbon and hydrogen combine to form ethyne and then ethyne burns to form carbon dioxide and water. Thus, in the above diagram,

$$\Delta H_1^\ominus = \Delta H_F^\ominus + \Delta H_2^\ominus$$

Putting

$$\Delta H_1^\ominus = 2(\Delta H^\ominus \text{ for combustion of C}) + (\Delta H^\ominus \text{ for combustion of } H_2)$$

gives

$$\Delta H_1^\ominus = 2(-394) + (-286) = -1074$$

$$\Delta H_F^\ominus = \Delta H_1^\ominus - \Delta H_2^\ominus = -1074 - (-1300)$$

ANSWER $\Delta H_F^\ominus = +226 \text{ kJ mol}^{-1}$ \quad (as before)

EXAMPLE 2 Calculate the standard enthalpy of formation of propan-1-ol, given the standard enthalpies of combustion, in kJ mol^{-1}: $C_3H_7OH(l), -2010$; $C(s), -394$; $H_2(g), -286$.

METHOD 1 Again, as the equation for combustion is the basis for the calculation, it must be carefully balanced:

$$C_3H_7OH(l) + 4\tfrac{1}{2}O_2(g) \longrightarrow 3CO_2(g) + 4H_2O(l); \quad \Delta H^\ominus = -2010 \text{ kJ mol}^{-1}$$

Putting the standard enthalpies of formation of $CO_2(g)$ and $H_2O(l)$ into the equation, as in Example 1, gives

$$C_3H_7OH(l)+4\tfrac{1}{2}O_2(g) \longrightarrow 3CO_2(g)+4H_2O(l); \quad \Delta H^\ominus = -2010\,kJ\,mol^{-1}$$
$$\Delta H_F^\ominus(C_3H_7OH) \quad 0 \qquad\qquad 3(-394)\ 4(-286)$$

Since

$$\begin{pmatrix}\text{Standard} \\ \text{enthalpy change} \\ \text{for reaction}\end{pmatrix} = \begin{pmatrix}\text{Standard} \\ \text{enthalpy content} \\ \text{of products}\end{pmatrix} - \begin{pmatrix}\text{Standard} \\ \text{enthalpy content} \\ \text{of reactants}\end{pmatrix}$$

$$-2010 = 3(-394) + 4(-286) - \Delta H_F^\ominus(C_3H_7OH(l))$$

$$\Delta H_F^\ominus(C_3H_7OH(l)) = -316\,kJ\,mol^{-1}$$

ANSWER   The standard enthalpy of formation of liquid propan-1-ol is $-316\,kJ\,mol^{-1}$.

METHOD 2   The enthalpy diagram for the formation of propanol is

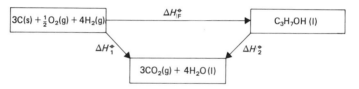

$$\Delta H_1^\ominus = 3(\Delta H^\ominus \text{ for combustion of } C) + 4(\Delta H^\ominus \text{ for combustion of } H_2)$$
$$= 3(-394) + 4(-286) = -2326$$

According to Hess's law,

$$\Delta H_1^\ominus = \Delta H_F^\ominus + \Delta H_2^\ominus$$
$$\Delta H_F^\ominus = \Delta H_1^\ominus - \Delta H_2^\ominus$$
$$\Delta H_F^\ominus = -2326 - (-2010)$$

ANSWER
$$\Delta H_F^\ominus = -316\,kJ\,mol^{-1} \quad \text{(as before)}$$

You will have noticed in both Examples 1 and 2 that

$$\begin{pmatrix}\text{Standard enthalpy} \\ \text{of reaction}\end{pmatrix} = \begin{pmatrix}\text{Sum of standard} \\ \text{enthalpies of} \\ \text{combustion of} \\ \text{reactants}\end{pmatrix} - \begin{pmatrix}\text{Sum of standard} \\ \text{enthalpies of} \\ \text{combustion of} \\ \text{products}\end{pmatrix}$$

STANDARD ENTHALPY OF REACTION FROM STANDARD ENTHALPIES OF FORMATION

The standard enthalpies of formation of the reactants and products can be used to give the standard enthalpy of a reaction.

EXAMPLE 1 Calculate the standard enthalpy of the reaction

$$CH_2 = CH_2(g) \ + \ H_2(g) \longrightarrow CH_3CH_3(g)$$

given that the standard enthalpies of formation are: ethene, $+52$, ethane, -85 kJ mol^{-1}.

METHOD Put the standard enthalpy content of each species into the equation (units kJ mol^{-1}):

$$CH_2 = CH_2(g) \ + \ H_2(g) \longrightarrow CH_3CH_3(g)$$
$$+52 \qquad\qquad\quad 0 \qquad\qquad\qquad -85$$

$$\begin{pmatrix} \text{Standard enthalpy} \\ \text{of reaction} \end{pmatrix} = \begin{pmatrix} \text{Standard enthalpy} \\ \text{of product} \end{pmatrix} - \begin{pmatrix} \text{Standard enthalpy} \\ \text{of reactants} \end{pmatrix}$$

$$= -85 - (52 + 0) = -137$$

ANSWER Standard enthalpy $= -137 \text{ kJ mol}^{-1}$

The method of calculation is simply:

$$\begin{pmatrix} \text{Standard enthalpy} \\ \text{of reaction} \end{pmatrix} = \begin{pmatrix} \text{Sum of standard} \\ \text{enthalpies of} \\ \text{formation of} \\ \text{products} \end{pmatrix} - \begin{pmatrix} \text{Sum of standard} \\ \text{enthalpies of} \\ \text{formation of} \\ \text{reactants} \end{pmatrix}$$

EXAMPLE 2 Calculate the standard enthalpy change in the reaction

$$SO_2(g) \ + \ 2H_2S(g) \longrightarrow 3S(s) \ + \ 2H_2O(l)$$

The standard enthalpy of combustion of sulphur is -297 kJ mol^{-1}, and the standard enthalpies of formation of hydrogen sulphide and water are $-20.2 \text{ kJ mol}^{-1}$ and -286 kJ mol^{-1}.

METHOD This problem is tackled by putting the standard enthalpies of formation of each species into the equation (units kJ mol^{-1}):

$$SO_2(g) \ + \ 2H_2S(g) \longrightarrow 3S(s) \ + \ 2H_2O(l)$$
$$(-297) \ + \ 2(-20.2) \qquad\quad 3 \times 0 \qquad 2(-286)$$

$$\begin{pmatrix} \text{Standard enthalpy} \\ \text{of reaction} \end{pmatrix} = \begin{pmatrix} \text{Standard enthalpy} \\ \text{of products} \end{pmatrix} - \begin{pmatrix} \text{Standard enthalpy} \\ \text{of reactants} \end{pmatrix}$$

$$= -572 + 297 + 40.4$$

ANSWER Standard enthalpy change $= -235 \text{ kJ (mol of the equation)}^{-1}$

STANDARD BOND DISSOCIATION ENTHALPIES

The standard bond dissociation enthalpy is the energy that must be absorbed to separate the two atoms in a bond. When hydrogen chloride dissociates,

$$HCl(g) \longrightarrow H(g) + Cl(g); \quad \Delta H^{\ominus} = 429.7 \, kJ \, mol^{-1}$$

The standard bond dissociation enthalpy of the H—Cl bond in HCl is $429.7 \, kJ \, mol^{-1}$.

AVERAGE STANDARD BOND ENTHALPIES

When you want to assign a value to the standard enthalpy of dissociation of the C—H bond in methane, the problem is different. The energy required to break the first C—H bond in methane is not the same as that required to remove a hydrogen atom from a methyl radical. In the dissociation,

$$CH_4(g) \longrightarrow C(g) + 4H(g); \quad \Delta H^{\ominus} = +1662 \, kJ \, mol^{-1}$$

Dividing the standard enthalpy change between the four bonds gives an average value for the C—H bond of $416 \, kJ \, mol^{-1}$. This value is called the average standard bond enthalpy for the C—H bond.

Tables of average standard bond enthalpies make the assumption that the standard enthalpy of a bond is independent of the molecule in which it exists. This is only roughly true. Since standard bond enthalpies vary from one compound to another, the use of average standard bond enthalpies gives only approximate values for standard enthalpies of reaction calculated from them. Experimental methods are used to obtain standard enthalpies of reaction whenever possible. Calculations based on average standard bond enthalpies are used only for reactions which cannot be studied experimentally — for example, the reactions of a substance which has not been isolated in a pure state.

Average standard bond enthalpy is often called the *bond energy term*. One can say that the bond energy term for the C—H bond is $416 \, kJ \, mol^{-1}$. The sum of all the bond energy terms for a compound is the standard enthalpy change absorbed in atomising that compound *in the gaseous state*. The standard enthalpy of formation of a compound includes the sum of the bond energy terms and also the standard enthalpy of atomisation of the carbon atoms and the standard enthalpy of atomisation of the hydrogen atoms.

EXAMPLE   Calculate the standard enthalpy of formation of methane. C—H bond energy term $= 416 \, kJ \, mol^{-1}$; standard enthalpies of atomisation are $C(s) = 716 \, kJ \, mol^{-1}$; $\frac{1}{2}H_2(g) = 217.5 \, kJ \, (mol \, H \, atoms)^{-1}$.

METHOD 1 The sum of the bond energy terms in methane $= 1662 \, \text{kJ mol}^{-1}$. Putting this information into the form of an equation, and writing the standard enthalpy content of each species underneath its formula, we get

$$C(g) + 4H(g) \longrightarrow CH_4(g); \quad \Delta H^{\ominus} = -1662 \, \text{kJ mol}^{-1}$$
$$(716) \quad 4(217.5) \qquad \quad \Delta H_F^{\ominus}$$

The values 716 and 217.5 are the standard enthalpies of formation of gaseous carbon and hydrogen atoms from the elements in their standard states.

Since

$$\begin{pmatrix} \text{Standard enthalpy} \\ \text{change} \end{pmatrix} = \begin{pmatrix} \text{Sum of standard} \\ \text{enthalpies of} \\ \text{products} \end{pmatrix} - \begin{pmatrix} \text{Sum of standard} \\ \text{enthalpies of} \\ \text{reactants} \end{pmatrix}$$

$$-1662 = \Delta H_F^{\ominus} - 716 - 4(217.5)$$

ANSWER
$$\Delta H_F^{\ominus} = -78 \, \text{kJ mol}^{-1}$$

METHOD 2 The information can also be represented in the form of an enthalpy diagram:

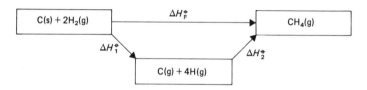

$$\Delta H_1^{\ominus} = \Delta H^{\ominus} \text{ of atomisation of } C + 4\Delta H^{\ominus} \text{ of atomisation of } H$$
$$\Delta H_2^{\ominus} = -(\text{Sum of bond energy terms for } CH_4)$$

According to Hess's law,

$$\Delta H_F^{\ominus} = \Delta H_1^{\ominus} + \Delta H_2^{\ominus}$$
$$= 716 + 4(217.5) - (4 \times 416)$$

ANSWER
$$\Delta H_F^{\ominus} = -78 \, \text{kJ mol}^{-1} \quad \text{(as before)}$$

STANDARD ENTHALPY OF REACTION FROM AVERAGE STANDARD BOND ENTHALPIES

Mean standard bond enthalpies can be used to give an approximate estimate of the standard enthalpy change which occurs in a reaction. During a reaction, energy is supplied to break the bonds in the reactants, and energy is given out when the bonds in the products form. The difference between the sum of the standard bond enthalpies of the products and the standard bond enthalpies of the reactants is the standard enthalpy of the reaction. The value obtained is less reliable than an experimental measurement.

EXAMPLE 1 Calculate the standard enthalpy of the reaction

$$CH_2 = CH_2(g) + H_2(g) \longrightarrow CH_3CH_3(g)$$

Mean standard bond enthalpies are (in $kJ\,mol^{-1}$): C—H, 416; C=C, 612; C—C, 348; H—H, 436.

METHOD Bonds broken are:

one C=C bond, of standard enthalpy = $612\,kJ\,mol^{-1}$
one H—H bond, of standard enthalpy = 436
Total enthalpy absorbed = $1048\,kJ\,mol^{-1}$
Bonds created are:
one C—C bond, of standard enthalpy = $348\,kJ\,mol^{-1}$
two C—H bonds, of standard enthalpy = 832
Total enthalpy released = $-1180\,kJ\,mol^{-1}$

ANSWER Standard enthalpy of reaction = $-1180 + 1048 = -132\,kJ\,mol^{-1}$

EXAMPLE 2 Benzene has a standard enthalpy of formation of $83\,kJ\,mol^{-1}$. Calculate the standard enthalpy of formation from the following data:

Mean standard bond enthalpies are: (C—C) = 348; (C=C) = 615; (C—H) = $412\,kJ\,mol^{-1}$.

$\Delta H^{\ominus}$ for vaporisation of carbon = $715\,kJ\,mol^{-1}$

$\Delta H^{\ominus}$ for atomisation of hydrogen (per mole of H atoms) = $217.5\,kJ\,mol^{-1}$

Compare the experimental value and the theoretical value for $\Delta H_F^{\ominus}$.

METHOD Enthalpy is absorbed in atomising carbon and hydrogen.

Standard enthalpy absorbed = $(6 \times 715) + (6 \times 217.5)$
= $5595\,kJ\,mol^{-1}$

Enthalpy is released when bonds are formed.

Standard enthalpy released = 6(C—H) = $-2472\,kJ\,mol^{-1}$
+ 3(C—C) = $-1044\,kJ\,mol^{-1}$
+ 3(C=C) = $-1845\,kJ\,mol^{-1}$
Total = $-5361\,kJ\,mol^{-1}$

ANSWER $\Delta H_F^{\ominus} = +5595 - 5361 = 234\,kJ\,mol^{-1}$.

The calculated value for the standard enthalpy of formation is higher than the experimental value: the benzene molecule is more stable than it is calculated to be. The difference is the value of the energy of electron delocalisation or 'resonance', $151\,kJ\,mol^{-1}$.

THE BORN-HABER CYCLE

The Born–Haber cycle is a technique for applying Hess's law to the standard enthalpy changes which occur when an ionic compound is formed. Consider the reaction between sodium and chlorine to form sodium chloride. The steps which are involved in this reaction are:

a) Vaporisation of sodium

$$Na(s) \longrightarrow Na(g); \quad \Delta H_s^{\ominus} = \text{standard enthalpy of sublimation}$$

b) Ionisation of sodium

$$Na(g) \longrightarrow Na^+(g) + e^-; \quad \Delta H_I^{\ominus} = \text{ionisation energy of sodium}$$

c) Dissociation of chlorine molecules

$$\tfrac{1}{2}Cl_2(g) \longrightarrow Cl(g); \quad \Delta H_D^{\ominus} = \tfrac{1}{2} \text{ standard bond dissociation enthalpy of chlorine}$$

d) Ionisation of chlorine atoms

$$Cl(g) + e^- \longrightarrow Cl^-(g); \quad \Delta H_E^{\ominus} = \text{electron affinity of chlorine}$$

e) Reaction between ions

$$Na^+(g) + Cl^-(g) \longrightarrow NaCl(s); \quad \Delta H_L^{\ominus} = \text{standard lattice enthalpy}$$

Definitions of the standard enthalpies used above are:

The *standard enthalpy of sublimation* is the heat absorbed when one mole of sodium atoms are vaporised.

The *ionisation energy* of sodium is the energy required to remove a mole of electrons from a mole of sodium atoms in the gas phase.

The *standard enthalpy of bond dissociation* of chlorine is the enthalpy required to dissociate one mole of chlorine molecules into atoms.

The *electron affinity* of chlorine is the energy absorbed when a mole of chlorine atoms form chloride ions. It has a negative value, showing that this reaction is exothermic.

The *standard lattice enthalpy* is the energy absorbed when one mole of gaseous sodium ions and one mole of gaseous chloride ions form one mole of crystalline sodium chloride. It has a negative value.

The steps in the Born–Haber cycle are represented as going upwards if they absorb energy and downwards if they give out energy (see Fig. 9.2).

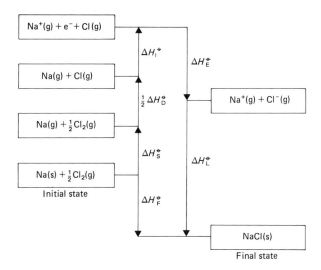

Fig. 9.2

According to Hess's law, the standard enthalpy of formation of sodium chloride is equal to the sum of the enthalpy changes in the various steps:

$$\Delta H_F^\ominus = \Delta H_S^\ominus + \tfrac{1}{2}\Delta H_D^\ominus + \Delta H_I^\ominus + \Delta H_E^\ominus + \Delta H_L^\ominus$$

$$= +109 + 121 + 494 - 380 - 755 = -411\,\text{kJ mol}^{-1}$$

In practice, it is easier to measure standard enthalpies of formation than to measure some of the other steps. The electron affinity is the hardest term to measure experimentally, and the Born–Haber cycle is often used to calculate electron affinities.

EXERCISE 53   Problems on Standard Enthalpy of Reaction and Average Standard Bond Enthalpies

1. The following are standard enthalpies of combustion at 298 K, in kJ mol^{-1}:

C(graphite)	−394	$C_2H_6(g)$	−1561	$C_4H_{10}(l)$	−3510
$H_2(g)$	−286	$CH_2{=}CH_2(g)$	−1393	$CH{\equiv}CH(g)$	−1299
$CH_3CO_2H(l)$	−876	$C_2H_5OH(l)$	−1400	$CH_3OH(l)$	−715
$C_4H_6(g)$	−2542	$CH_3OCH_3(g)$	−1455	$C_2H_5OH(g)$	−1444
$CH_4(g)$	−891	$C_3H_8(g)$	−2220		
$CH_3CO_2C_2H_5(l)$	−2246	$C_6H_{12}(l)$	−3924		

a) Calculate the standard enthalpy change for the reaction:

$$2C(\text{graphite}) + 2H_2(g) + O_2(g) \longrightarrow CH_3CO_2H(l)$$

b) Calculate the standard enthalpy change of formation of buta-1, 3-diene, $C_4H_6(g)$.

c) Calculate the standard enthalpy of formation of methane, $CH_4(g)$ and of ethene, $CH_2{=}CH_2(g)$.

d) Calculate the standard enthalpy change in the hydrogenation of ethene(g) to ethane(g).

e) Calculate the standard enthalpy change for the theoretical reaction:

$$CH_3OCH_3(g) \longrightarrow C_2H_5OH(g)$$

f) Calculate the standard enthalpy of formation of propane(g) and of butane(l).

g) Calculate the standard enthalpy of formation of methanol(l), ethanol(l), ethylethanoate(l) and cyclohexane(l).

2. Calculate the standard enthalpy change of the reaction

Anhydrous copper(II) sulphate + Water $\longrightarrow$ Copper(II) sulphate-5-water

Use the values for the standard enthalpy of solution:

a) anhydrous copper(II) sulphate, $-66.5\,kJ\,mol^{-1}$

b) copper(II) sulphate-5-water, $11.7\,kJ\,mol^{-1}$.

3. Calculate the standard enthalpies of formation of: a) sulphur dioxide, b) carbon dioxide, and c) steam. On burning in excess oxygen under standard conditions (1 atm, 298 K): 1.00 g of sulphur evolves 9.28 kJ; 1.00 g of carbon evolves 32.8 kJ; and $1.00\,dm^3$ (at 1 atm, 298 K) of hydrogen evolves 12.76 kJ of heat.

4. Calculate the standard enthalpy change in the reaction

$$PbO(s) + CO(g) \longrightarrow Pb(s) + CO_2(g)$$

The standard enthalpies of formation of lead(II) oxide, carbon monoxide and carbon dioxide are -219, -111, and $-394\,kJ\,mol^{-1}$, respectively.

5. Calculate the standard enthalpy change for the reaction

$$Fe_2O_3(s) + 2Al(s) \longrightarrow Al_2O_3(s) + 2Fe(s)$$

The standard enthalpies of formation of iron(III) oxide and aluminium oxide are -822 and $-1669\,kJ\,mol^{-1}$. State whether the reaction is exothermic or endothermic.

6. The standard enthalpy of combustion of rhombic sulphur is $-296.9\,kJ\,mol^{-1}$ and the standard enthalpy of combustion of monoclinic sulphur is $-297.2\,kJ\,mol^{-1}$. Calculate the standard enthalpy of conversion of monoclinic sulphur to rhombic sulphur.

7. The standard enthalpies of formation of $CO_2(g)$ and $H_2O(g)$ are -394 and $-242\,kJ\,mol^{-1}$. The standard enthalpy of combustion of ethane is $-1560\,kJ\,mol^{-1}$. The standard enthalpy of reduction of ethene to ethane by gaseous hydrogen is $-138\,kJ\,mol^{-1}$. Calculate the standard enthalpy of formation of ethene.

8. Given the standard enthalpy change of formation of MgO $= -602$ kJ mol^{-1} and of Al$_2$O$_3$ $= -1700$ kJ mol^{-1}, calculate the standard enthalpy change for the reaction

$$Al_2O_3 + 3Mg \longrightarrow 2Al + 3MgO$$

Does your answer tell you whether magnesium will reduce aluminium oxide?

9. The following are standard enthalpies of formation, $\Delta H_F^{\ominus}$, in kJ mol^{-1} at 298 K:

CH$_4$(g); -76; CO$_2$(g), -394; H$_2$O(l), -286; H$_2$O(g), -242; NH$_3$(g), -46.2; HNO$_3$(l), -176; C$_2$H$_5$OH(l), -278; C$_8$H$_{18}$(l), -210.

a) Calculate the standard enthalpy change at 298 K for the reaction

$$CH_4(g) + 2O_2(g) \longrightarrow CO_2(g) + 2H_2O(l)$$

b) Calculate the standard enthalpy change for the reaction

$$\tfrac{1}{2}N_2(g) + \tfrac{3}{2}H_2O(g) \longrightarrow NH_3(g) + \tfrac{3}{4}O_2(g)$$

c) Calculate the standard enthalpy change for the reaction

$$\tfrac{1}{2}N_2(g) + \tfrac{1}{2}H_2O(g) + \tfrac{5}{4}O_2(g) \longrightarrow HNO_3(l)$$

d) Calculate the enthalpy change which occurs when each of the following is burned completely under standard conditions: i) 1.00 kg hydrogen, ii) 1.00 kg ethanol(l), iii) 1.00 kg octane(l).

10. What is meant by the terms *standard bond dissociation enthalpy* and *bond energy term*?

The standard bond dissociation enthalpies for the first, second, third and fourth C—H bonds in methane are 423, 480, 425 and 335 kJ mol^{-1} respectively. Calculate the C—H bond energy term for methane.

11. Consult the average standard bond enthalpies and standard enthalpies of atomisation (in kJ mol^{-1}) listed below:

C—C	348	C=O	743	C(graphite)	718
C=C	612	H—Cl	432	$\tfrac{1}{2}$H$_2$(g)	218
C≡C	837	C—Cl	338	$\tfrac{1}{2}$O$_2$(g)	248
C—H	412	C—Br	276	$\tfrac{1}{2}$Br$_2$(g)	96.5
C—O	360	H—Br	366	$\tfrac{1}{2}$Cl$_2$(g)	121
H—O	463				

a) Calculate the standard enthalpy of formation of ethane and of ethene.

b) Find the standard enthalpy change for the reaction,

$$CH_2{=}CH{-}CH_3(g) + Br_2(g) \longrightarrow CH_2BrCHBrCH_3(g)$$

c) Find the standard enthalpy of formation of methoxymethane, CH$_3$OCH$_3$(g).

d) Calculate the standard enthalpy of formation of gaseous ethyl ethanoate, CH$_3$CO$_2$C$_2$H$_5$(g).

e) Calculate the standard enthalpy of formation of benzene, assuming its structure is

Explain the difference between the value you have calculated and the value of $83 \, kJ \, mol^{-1}$ obtained from measurements of the standard enthalpy of combustion.

f) Find the standard enthalpy of formation of gaseous buta-1,3-diene, $CH_2 = CH - CH = CH_2(g)$. How does this value compare with the value you obtained in Question 1(b) from the standard enthalpy of combustion? How do you explain the difference?

g) Estimate the standard enthalpy changes for the reactions:
   i) $Cl \cdot \; + \; CH_4 \longrightarrow CH_3Cl \; + \; H \cdot$
   ii) $Cl \cdot \; + \; CH_4 \longrightarrow CH_3 \cdot \; + \; HCl$
   Which of the two reactions will occur more readily?

12. Use the data below to draw an energy diagram for the formation of potassium chloride. Calculate the electron affinity of chlorine.

Standard enthalpy of sublimation of potassium	$= \quad 90 \, kJ \, mol^{-1}$
Standard enthalpy of ionisation of potassium	$= \quad 420 \, kJ \, mol^{-1}$
Standard enthalpy of dissociation of chlorine	$= \quad 244 \, kJ \, mol^{-1}$
Standard lattice enthalpy of potassium chloride	$= -706 \, kJ \, mol^{-1}$
Standard enthalpy of formation of potassium chloride	$= -436 \, kJ \, mol^{-1}$

13. Using the following data, which is a set of standard enthalpy changes, calculate the standard enthalpy of formation of potassium chloride, $KCl(s)$:

$\Delta H^{\ominus}/kJ \, mol^{-1}$

$KOH(aq) \; + \; HCl(aq) \longrightarrow KCl(aq) \; + \; H_2O(l)$	-57.3
$H_2(g) \; + \; \tfrac{1}{2}O_2(g) \longrightarrow H_2O(l)$	-286
$\tfrac{1}{2}H_2(g) \; + \; \tfrac{1}{2}Cl_2(g) \; + \; aq \longrightarrow HCl(aq)$	-164
$K(s) \; + \; \tfrac{1}{2}O_2(g) \; + \; \tfrac{1}{2}H_2(g) \; + \; aq \longrightarrow KOH(aq)$	-487
$KCl(s) \; + \; aq \longrightarrow KCl(aq)$	$+18$

14. Use the data below to calculate the electron affinity of chlorine:

Standard enthalpy of formation of rubidium chloride $-431\,kJ\,mol^{-1}$
Lattice energy of rubidium chloride $-675\,kJ\,mol^{-1}$
First ionisation energy of rubidium $+408\,kJ\,mol^{-1}$
Standard enthalpy of atomisation of rubidium $+86\,kJ\,mol^{-1}$
Bond dissociation enthalpy of molecular chlorine $+242\,kJ\,mol^{-1}$

15. From a Born–Haber cycle calculation, it can be estimated that the standard enthalpy of formation of magnesium(I) chloride, MgCl, would be $-130\,kJ\,mol^{-1}$. The standard enthalpy of formation of magnesium(II) chloride $MgCl_2$, is $-640\,kJ\,mol^{-1}$.

a) Why do you think that $MgCl_2$ is formed, and not MgCl, when magnesium reacts with chlorine?

b) Calculate the standard enthalpy change in the theoretical reaction

$$2MgCl(s) \longrightarrow Mg(s) + MgCl_2(s)$$

16. Calculate the lattice energy of sodium chloride from the following data:

	$\Delta H^{\ominus}/kJ\,mol^{-1}$
$Na(s) \longrightarrow Na(g)$	$+109$
$Na(g) \longrightarrow Na^+(g) + e^-$	$+494$
$Cl_2(g) \longrightarrow 2Cl(g)$	$+242$
$Cl(g) + e^- \longrightarrow Cl^-(g)$	-360
$Na(s) + \frac{1}{2}Cl_2(s) \longrightarrow NaCl(s)$	-411

17. a) Data for the Born–Haber cycle for the formation of calcium chloride are

$Ca(s) \longrightarrow Ca(g)$	$\Delta H^{\ominus} = +190\,kJ\,mol^{-1}$
$Ca(g) \longrightarrow Ca^{2+}(g) + 2e^-$	$\Delta H^{\ominus} = +1730\,kJ\,mol^{-1}$
$\frac{1}{2}Cl_2(g) \longrightarrow Cl(g)$	$\Delta H^{\ominus} = +121\,kJ\,mol^{-1}$
$Ca^{2+}(g) + 2Cl^-(g) \longrightarrow CaCl_2(s)$	$\Delta H^{\ominus} = -2184\,kJ\,mol^{-1}$
$Ca(s) + Cl_2(g) \longrightarrow CaCl_2(s)$	$\Delta H^{\ominus} = -795\,kJ\,mol^{-1}$

Calculate the electron affinity of chlorine.

b) For the reactions

$Ca(g) \longrightarrow Ca^+(g) + e^-$	$\Delta H^{\ominus} = +590\,kJ\,mol^{-1}$
$Ca^+(g) + Cl^-(g) \longrightarrow CaCl(s)$	$\Delta H^{\ominus} = -760\,kJ\,mol^{-1}$

use these standard enthalpy changes and those given in a) to calculate the standard enthalpy of formation of CaCl(s). Why do you think $CaCl_2$ is formed in preference to CaCl?

18. When an ionic compound dissolves, an amount of energy equal to the lattice energy must be supplied to separate the ions. When the ions dissolve they are hydrated by water molecules, and energy is released. If the enthalpy of hydration is greater than the lattice enthalpy, there is a net release of energy and a decrease in the enthalpy content of the system, and this favours solution.

The values below (in $kJ \, mol^{-1}$) relate to the solubility of lithium chloride, sodium chloride and sodium fluoride:

	LiCl	NaCl	NaF
Standard lattice enthalpy	−843	−775	−968
Sum of standard hydration enthalpies of separate ions	−883	−778	−965

What can you predict from these values for the standard enthalpy changes about the relative solubilities of: a) LiCl and NaCl, b) NaF and NaCl? Explain your answer.

19. Given the standard enthalpy changes for the reactions

$$H_2(g) \longrightarrow 2H(g); \qquad \Delta H^{\ominus} = 436 \, kJ \, mol^{-1}$$
$$Br_2(g) \longrightarrow 2Br(g); \qquad \Delta H^{\ominus} = 193 \, kJ \, mol^{-1}$$
$$H_2(g) + Br_2(g) \longrightarrow 2HBr(g); \qquad \Delta H^{\ominus} = -104 \, kJ \, mol^{-1}$$

calculate the standard enthalpy change for the reaction

$$H(g) + Br(g) \longrightarrow HBr(g)$$

20. The following values for standard enthalpy change relate to the hydrogenation of cyclohexene and benzene. Comment on the values of $\Delta H^{\ominus}$.

$$C_6H_{10}(l) + H_2(g) \longrightarrow C_6H_{12}(l) \qquad \Delta H^{\ominus} = -120 \, kJ \, mol^{-1}$$
$$C_6H_6(l) + H_2(g) \longrightarrow C_6H_8(l) \qquad \Delta H^{\ominus} = +31 \, kJ \, mol^{-1}$$
$$C_6H_6(l) + 3H_2(g) \longrightarrow C_6H_{12}(l) \qquad \Delta H_c^{\ominus} = -208 \, kJ \, mol^{-1}$$

FREE ENERGY AND ENTROPY

Some reactions which happen spontaneously are endothermic. The difference in enthalpy between the products and the reactants cannot be the only factor which decides whether a chemical reaction takes place. There must be an additional factor involved. It is often observed that reactions which occur spontaneously increase the randomness or disorder of the system. For example, when an ionic solid dissolves, it passes from the regular arrangement of a crystalline lattice to a random solution of ions. This is termed an increase in *entropy* of the system. The two factors combine to give the change in the *free energy* of the system:

Free energy G = Enthalpy H − Temperature/K × Entropy S

$$G = H - TS$$

It follows that $\Delta G = \Delta H - T\Delta S$

For a physical or a chemical change to occur, ΔG for that change must be negative. The change is therefore assisted by a decrease in enthalpy (ΔH negative) and by an increase in entropy (ΔS positive).

If the change takes place under standard conditions, i.e. with each reactant and product at unit concentration (or pressure), then the free energy change is equal to the standard free energy change, $\Delta G^{\ominus}$. When reaction takes place under non-standard conditions, ΔG, the free energy change differs from $\Delta G^{\ominus}$ as ΔG depends on the concentrations (or pressures) of the reactants and products. It is easy to obtain $\Delta G^{\ominus}$ from tables of standard enthalpies and standard entropies, but one really wants to know the value of ΔG for the real conditions, and this is not easy to compute. However, if $\Delta G^{\ominus}$ has a sufficiently large positive or negative value, $\Delta G^{\ominus}$ may determine the feasibility of reaction over a large range of concentrations (or pressures).

CALCULATION OF CHANGE IN STANDARD ENTROPY

One method of calculating the standard entropy change of a process is to use the expression

$$\begin{pmatrix} \text{Standard} \\ \text{entropy change} \end{pmatrix} = \begin{pmatrix} \text{Sum of standard} \\ \text{entropies of products} \end{pmatrix} - \begin{pmatrix} \text{Sum of standard} \\ \text{entropies of reactants} \end{pmatrix}$$

EXAMPLE 1 Calculate the standard entropy change for the reaction of chlorine and ethene, given the values (in $J\,K^{-1}\,mol^{-1}$):

$S^{\ominus}(Cl_2(g)) = 223$; $S^{\ominus}(CH_2{=}CH_2(g)) = 219$; $S^{\ominus}(CH_2ClCH_2Cl(l)) = 208$.

METHOD The equation for the reaction is

$$CH_2{=}CH_2(g) \;+\; Cl_2(g) \longrightarrow CH_2ClCH_2Cl(l)$$
$$S^{\ominus}(\text{product}) \;=\; 208\,J\,K^{-1}\,mol^{-1}$$
$$S^{\ominus}(\text{reactants}) \;=\; 219 + 223 \;=\; 442\,J\,K^{-1}\,mol^{-1}$$
$$\Delta S^{\ominus} \;=\; 208 - 442 \;=\; -234\,J\,K^{-1}\,mol^{-1}$$

ANSWER The standard entropy change for the reaction is $-234\,J\,K^{-1}\,mol^{-1}$. The negative sign means a decrease in disorder. Since two moles of gas have formed one mole of liquid, this is what one would expect.

The other method of calculating the change in standard entropy for a process is to derive it from the changes in standard enthalpy and standard free energy. The equation relating these quantities is

$$\Delta G^{\ominus} \;=\; \Delta H^{\ominus} - T\Delta S^{\ominus}$$

If a system is at equilibrium, $\Delta G^{\ominus} = 0$, and the standard entropy change is simply the standard enthalpy change divided by the temperature:

$$\Delta S^{\ominus} = \Delta H^{\ominus}/T$$

One process during which equilibrium obtains is the vaporisation of a liquid at its boiling point, since this process takes place under reversible conditions at a constant temperature. Another process for which $\Delta G^{\ominus} = 0$ is the melting of a solid at its melting point. For chemical reactions, $\Delta G^{\ominus} \neq 0$, and the change in standard entropy must be found from the equation

$$\Delta S^{\ominus} = \frac{\Delta H^{\ominus} - \Delta G^{\ominus}}{T}$$

*EXAMPLE 2 Calculate the standard entropy of melting (or fusion) of ice. The standard enthalpy of melting of ice is $6.00\,kJ\,mol^{-1}$.

METHOD $\Delta S^{\ominus} = \Delta H_m^{\ominus}/T = 6.00/273$

ANSWER $\Delta S^{\ominus} = 2.20 \times 10^{-2}\,kJ\,mol^{-1}\,K^{-1}$

*EXAMPLE 3 Calculate the standard entropy of vaporisation of water. The standard enthalpy of vaporisation is $41.0\,kJ\,mol^{-1}$.

METHOD $\Delta S^{\ominus} = \Delta H_v^{\ominus}/T = 41.0/373$

ANSWER $\Delta S^{\ominus} = 0.110\,kJ\,mol^{-1}\,K^{-1}$

Note. Standard enthalpy of melting was formerly called molar latent heat of fusion. Standard enthalpy of vaporisation (or evaporation) was formerly called molar latent heat of vaporisation.

CALCULATION OF CHANGE IN STANDARD FREE ENERGY

The change in standard enthalpy, the change in standard entropy and the temperature must be known and inserted into the equation

$$\Delta G^{\ominus} = \Delta H^{\ominus} - T\Delta S^{\ominus}$$

EXAMPLE Calculate the change in standard free energy and determine whether the reaction

$$Fe_2O_3(s) + 3H_2(g) \longrightarrow 2Fe(s) + 3H_2O(g)$$

will take place at a) 20 °C, **b)** 500 °C. Use the values (in $kJ\,mol^{-1}$):

	Fe_2O_3	H_2	Fe	H_2O
Standard enthalpy:	−822	0	0	−242
Standard entropy:	0.090	0.131	0.027	0.189

METHOD $\Delta G^{\ominus} = \Delta H^{\ominus} - T\Delta S^{\ominus}$

$\Delta H^{\ominus} = (0 + 3(-242)) - (-822 + 0) = +96 \, \text{kJ mol}^{-1}$

$\Delta S^{\ominus} = (2 \times 0.027) + (3 \times 0.189) - 0.090 - (3 \times 0.131)$

$= 0.054 + 0.567 - 0.090 - 0.393 = +0.138 \, \text{kJ mol}^{-1}$

a) At 20 °C,

$\Delta G^{\ominus} = \Delta H^{\ominus} - T\Delta S^{\ominus}$

$= +96 - (293 \times 0.138) = 96 - 52.74 = +43.26 \, \text{kJ mol}^{-1}$

ANSWER $\Delta G^{\ominus}$ is 42.3 kJ mol^{-1} which is positive, and the reaction will therefore not occur at 20 °C.

b) At 500 °C,

$\Delta G^{\ominus} = +96 - (773 \times 0.138) = -10.7 \, \text{kJ mol}^{-1}$

ANSWER $\Delta G^{\ominus}$ is -10.7 kJ mol^{-1} which is negative, the reaction will occur at 500 °C.

(*Note* the assumption that $\Delta H^{\ominus}$ does not vary with temperature.)

EXERCISE 54   Problems on Standard Entropy Change and Standard Free Energy Change

1. Find the standard entropy of vaporisation of lead at its boiling point, 2017 K, given that its standard enthalpy of vaporisation is 177 kJ mol^{-1}.

2. Find the standard entropy of melting of lead at its melting point, 600 K, given that its standard enthalpy of melting is 5.10 kJ mol^{-1}.

3. Calculate the standard entropies of vaporisation of the hydrogen halides at their respective boiling points from the data given.

Hydrogen halide	Boiling point/K	$\Delta H^{\ominus}_{\text{vaporisation}}$/kJ mol^{-1}
HF	293	7.5
HCl	188	16.2
HBr	206	17.6
HI	238	19.8

Comment on the values you obtain.

4. Calculate the standard entropy of sublimation of arsenic at its sublimation temperature of 886 K, given that $\Delta H^{\ominus}_{\text{sublimation}} = 32.4 \, \text{kJ mol}^{-1}$.

5. Find the standard entropy of vaporisation of ethanol at its boiling point of 352 K, given $\Delta H^{\ominus}_{\text{vaporisation}} = 43.5 \, \text{kJ mol}^{-1}$.

6. Refer to the following values of standard entropy (J mol^{-1}K^{-1}) at 298 K:

$H_2(g)$	131	$H_2O(l)$	70	$NH_4Cl(s)$	94.6
$Cl_2(g)$	223	$H_2O(g)$	189	$N_2O_4(g)$	304
$N_2(g)$	192	$HCl(g)$	187	$C_2H_4(g)$	220
$O_2(g)$	205	$NH_3(g)$	193	$C_2H_6(g)$	230
$Na(s)$	51	$NO_2(g)$	240	$HNO_3(l)$	156
		$NaCl(s)$	72.4		

Calculate the standard entropy changes for the following reactions:

a) $H_2(g) + Cl_2(g) \longrightarrow 2HCl(g)$

b) $N_2(g) + 3H_2(g) \longrightarrow 2NH_3(g)$

c) $H_2(g) + \frac{1}{2}O_2(g) \longrightarrow H_2O(l)$

d) $H_2(g) + C_2H_4(g) \longrightarrow C_2H_6(g)$

e) $N_2O_4(g) \longrightarrow 2NO_2(g)$

f) $Na(s) + \frac{1}{2}Cl_2(g) \longrightarrow NaCl(s)$

g) $NH_4Cl(s) \longrightarrow NH_3(g) + HCl(g)$

h) $4HNO_3(l) \longrightarrow 4NO_2(g) + O_2(g) + 2H_2O(l)$

7. Predict whether the following reactions will have a positive or negative value of $\Delta S^\ominus$:

a) $NH_4NO_3(s) \longrightarrow N_2O(g) + 2H_2O(g)$

b) $2H_2O_2(aq) \longrightarrow 2H_2O(l) + O_2(g)$

c) $PH_3(g) + HI(g) \longrightarrow PH_4I(s)$

d) $3O_2(g) \longrightarrow 2O_3(g)$

e) $CO_2(g) + C(s) \longrightarrow 2CO(g)$

f) $Ni(s) + 4CO(g) \longrightarrow Ni(CO)_4(g)$

8. Refer to this list of values:

Substance	$\Delta H_F^\ominus$/kJ mol^{-1}	$\Delta G_F^\ominus$/kJ mol^{-1}
$C_6H_6(g)$	82.9	130
$C_6H_6(l)$	49.0	125
$I_2(s)$	0	0
$I_2(g)$	62.6	19.4
$Hg(l)$	0	0
$Hg(g)$	60.8	31.8

a) Find the standard entropy change of vaporisation of benzene at 298 K.

b) Find the standard entropy of vaporisation of iodine at 298 K.

c) Calculate the standard entropy of vaporisation of mercury at 298 K.

9. Use the following values of standard entropy content and standard enthalpy of formation to calculate standard free energy changes:

Substance	$\Delta H_F^{\ominus}/kJ\ mol^{-1}$	$S^{\ominus}/J\ K^{-1}\ mol^{-1}$
HgO(s) (red)	−90.7	72.0
HgO(s) (yellow)	−90.2	73.0
HgS(s) (red)	−58.2	77.8
HgS(s) (black)	−54.0	83.3

a) Calculate the value of $\Delta G^{\ominus}$ for the change

$$HgO(s)\ (red) \longrightarrow HgO(s)\ (yellow)$$

at 25 °C and at 100 °C. At what temperature will the change take place?

b) Calculate the value of $\Delta G^{\ominus}$ for the change

$$HgS(s)\ (red) \longrightarrow HgS(s)\ (black)$$

at 25 °C. At what temperature will the change occur?

10. Cis-but-2-ene has $\Delta H_F^{\ominus} = -5.7\ kJ\ mol^{-1}$ and $S^{\ominus} = 301\ J\ K^{-1}\ mol^{-1}$; trans-but-2-ene has $\Delta H_F^{\ominus} = -10.1\ kJ\ mol^{-1}$ and $S^{\ominus} = 296\ J\ K^{-1}\ mol^{-1}$. Calculate

a) $\Delta G^{\ominus}$ for the transition cis-but-2-ene $\longrightarrow$ trans-but-2-ene and

b) for the transition trans-but-2-ene $\longrightarrow$ cis-but-2-ene

Which is the more stable isomer?

EXERCISE 55   Questions from A-level Papers

1. a) State Hess's law and explain what is meant by the *enthalpy change* of a reaction.

b) Calculate the *enthalpy change* of the reaction

$$CO(g)\ +\ 2H_2(g) \longrightarrow CH_3OH(l)$$

at 298 K, from the following data at 298 K:

$$\text{kJ mol}^{-1}$$

$$CO(g)\ +\ \tfrac{1}{2}O_2(g) \longrightarrow CO_2(g);\ \Delta H\ =\ -283.0$$
$$H_2(g)\ +\ \tfrac{1}{2}O_2(g) \longrightarrow H_2O(l);\ \Delta H\ =\ -285.8$$
$$CH_3OH(l)\ +\ 1\tfrac{1}{2}O_2(g) \longrightarrow CO_2(g)\ +\ 2H_2O(l);\ \Delta H\ =\ -715.0$$

c) When one mole of either hydrochloric acid or nitric acid is neutralised by one mole of either sodium hydroxide or potassium hydroxide in dilute aqueous solution, the enthalpy changes are almost identical. How is this explained?   (O76)

2. a) Describe, with the aid of a diagram, the lattice structure of crystalline sodium chloride and show how it accounts for the characteristic physical properties of the substance.

b) Define: i) *enthalpy change of formation*, ii) *lattice energy*. What factors determine the magnitude of a *lattice energy*?

Draw a Born–Haber cycle for the formation of caesium chloride and use it to calculate a value for the lattice energy of this compound.

Data

Enthalpy change of atomisation of caesium:

$$Cs(s) \longrightarrow Cs(g); \quad \Delta H = +79 \text{ kJ mol}^{-1}$$

Enthalpy change of atomisation of chlorine:

$$\tfrac{1}{2}Cl_2(g) \longrightarrow Cl(g); \quad \Delta H = +121 \text{ kJ mol}^{-1}$$

First ionisation energy of caesium:

$$Cs(g) \longrightarrow Cs^+(g) + e^-; \quad \Delta H = +376 \text{ kJ mol}^{-1}$$

Electron affinity of chlorine:

$$Cl(g) + e^- \longrightarrow Cl^-(g); \quad \Delta H = -364 \text{ kJ mol}^{-1}$$

Enthalpy change of formation of caesium chloride:

$$Cs(s) + \tfrac{1}{2}Cl_2(g) \longrightarrow Cs^+Cl^-(s); \quad \Delta H = -433 \text{ kJ mol}^{-1}.$$
(C80)

3. a) How and under what conditions do benzene and cyclohexene react with chlorine? Give equations and essential conditions and name the products for the reactions involved.

   b) i) The heat of hydrogenation of cyclohexene is -120 kJ mol^{-1}. Assuming the structural formula of naphthalene to be

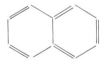

what would you predict for its heat of hydrogenation? Write an equation for the reaction involved.

ii) Use the bond energies below to calculate another value for the heat of hydrogenation of naphthalene, assuming that it has the structural formula shown above.

$E(C\!-\!C)$ general $= 346 \text{ kJ mol}^{-1}$;
$E(C\!=\!C)$ general $= 610 \text{ kJ mol}^{-1}$;
$E(C\!-\!H)$ general $= 413 \text{ kJ mol}^{-1}$;
$E(H\!-\!H) = 436 \text{ kJ mol}^{-1}$.

iii) What is the difference between your answers to i) and ii)? How may this difference be explained?

iv) The heat of hydrogenation of naphthalene, obtained by experiment, is much less than either of the theoretical values calculated in i) and ii). How are the differences between the experimental and theoretical values explained?   (SUJB80)

4. Use the following data to determine the enthalpy change for the reaction:

$$\tfrac{1}{2}I_2(s) \longrightarrow I^-(aq)$$

All steps in your calculation must be clearly shown.

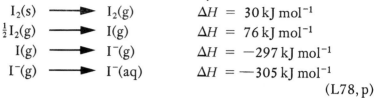

$I_2(s) \longrightarrow I_2(g)$	$\Delta H = 30\,kJ\,mol^{-1}$	
$\tfrac{1}{2}I_2(g) \longrightarrow I(g)$	$\Delta H = 76\,kJ\,mol^{-1}$	
$I(g) \longrightarrow I^-(g)$	$\Delta H = -297\,kJ\,mol^{-1}$	
$I^-(g) \longrightarrow I^-(aq)$	$\Delta H = -305\,kJ\,mol^{-1}$	

(L78, p)

5. When 10 g of metallic sodium reacts with excess water the energy evolved is 78.6 kJ. Similarly, 20 g of sodium oxide (Na_2O) reacts with excess water evolving 85.3 kJ. If the enthalpy change of formation of water is -284 kJ/mol, what is the enthalpy change of formation of sodium oxide? (AEB78, p)

6. a) Taking CH_4 as an example distinguish clearly between the concepts, bond dissociation energy and bond energy term. The following table lists bond energy terms ($kJ\,mol^{-1}$).

C—H	H—Cl	C—Cl	Cl—Cl	H—H
416	432	326	244	436

b) Estimate the enthalpy change (ΔH) for each of the following reactions:

i) $Cl\cdot + CH_4 = CH_3Cl + H\cdot$
ii) $Cl\cdot + CH_4 = HCl + \cdot CH_3$

c) Explain which energy terms in the table above will not allow you to calculate a precise value for the enthalpy change of reaction.

d) Explain, on the basis of your calculations in b) above, which of the reactions will be present in the chain mechanism for the photo-chlorination of methane.

e) Give the chain mechanism for the formation of CH_2Cl_2 by the photochlorination of methane.

f) If 8.31 dm^3 of methane gas at $10^5\,N\,m^{-2}$ pressure and 300 K produced 17 g of dichloromethane, calculate the mole percentage of methane which has been converted to dichloromethane.
($R = 8.31\,J\,mol^{-1}\,K^{-1}$; $A_r(C) = 12$; $A_r(H) = 1$; $A_r(Cl) = 35.5$.)

g) One of the isomers of pentane (C_5H_{12}) on photochlorination produced only one monochloro derivative. Give the structural formula of the isomer. (WJEC77)

7. Some bond energy terms are listed below:

Bond	Bond energy term (kJ mol^{-1})
H—H	435
C—H	415
C—Br	284
C—C	356
C=C	598
Br—Br	193

a) What do you understand by *bond energy term*?

b) Using the given data, calculate the enthalpies of formation, from gaseous atoms, of: i) gaseous propene (propylene), ii) gaseous 1,2-dibromopropane.

c) Calculate the enthalpy change, ΔH, for the following reaction:

$$CH_2{=}CH{-}CH_3(g) \; + \; Br_2(g) \longrightarrow CH_2BrCHBrCH_3(g)$$

d) The reaction in c) goes almost to completion. If it is considered as an equilibrium, what will be the effect on the equilibrium position of an increase in temperature? Give your reasoning.   (WJEC77)

8. a) Define standard enthalpy change (heat) of formation, illustrating your answer by reference to $MgCO_3(s)$.

b) When 0.20 g of magnesium ribbon was dissolved in an excess of hydrochloric acid in a vacuum flask, the temperature rose by 8.6 °C. Separate experiments showed that the vacuum flask and its contents required 500 J to raise the temperature by 1 °C.

   i) Write an equation for this reaction.
   ii) Calculate the heat released in the experiment.
   iii) Hence calculate the enthalpy (heat) change per mole of magnesium.

c) In a similar experiment, solid magnesium carbonate reacted with an excess of hydrochloric acid and the enthalpy change was found to be −90 kJ per mole of magnesium carbonate. The enthalpy change of formation of $H_2O(l)$ is −285 kJ mol^{-1} and of $CO_2(g)$ is −393 kJ mol^{-1}. Use these data and your result from b) iii) to calculate the enthalpy change of formation of $MgCO_3(s)$.

d) Would you expect the enthalpy change of formation of calcium carbonate, $CaCO_3(s)$, to be numerically greater or less than that of $MgCO_3(s)$? Explain your answer.   (C79)

*9. Explain the terms *enthalpy change* and *entropy change*, illustrating your answer with suitable examples.

Show how, under constant conditions of temperature, enthalpy change and entropy change influence the direction of chemical change. For liquid methanol in equilibrium with methanol vapour at 338 K, the boiling point of methanol,

$$CH_3OH(l) \rightleftharpoons CH_3OH(g) \qquad \Delta H = +35.3 \text{ kJ mol}^{-1}.$$

Calculate the entropy change when methanol is evaporated. (L78, S)

10. a) Giving one example in each case, explain briefly what is meant by:
   i) a Lowry–Brønsted acid, ii) a Lewis base.

b) Explain why all Lowry–Brønsted acids are also Lewis acids.

c) State two characteristics of the cation M^{n+} which influence the acidity of the species $[M(H_2O)_6]^{n+}$.

d) Fig. 9.3 represents the change in temperature which occurs when 1.00 M hydrochloric acid is added in portions to 25.00 cm^3 of sodium hydroxide solution in a polystyrene cup (the solutions are initially at the same temperature).

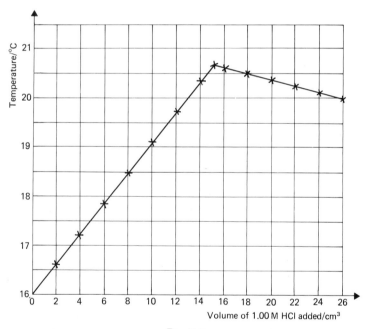

Fig. 9.3

Use the graph to:
   i) determine the molarity of the sodium hydroxide solution,
   ii) calculate a value for the enthalpy of neutralisation of sodium hydroxide.

(Assume that all the aqueous solutions have a specific heat capacity of 4.2 kJ K^{-1} litre^{-1}.) (JMB79)

11. a) What do you understand by the term *entropy*?

   b) For each of the following reactions, indicate whether the entropy is likely to *increase*, *decrease* or *stay the same*, explaining your reasoning:

   i) $H_2(g)$ + $C_2H_4(g)$ $\longrightarrow$ $C_2H_6(g)$
   ii) $N_2(g)$ + $3H_2(g)$ $\longrightarrow$ $2NH_3(g)$
   iii) $2NaNO_3(s)$ $\longrightarrow$ $2NaNO_2(s)$ + $O_2(g)$

   c) What conditions must be satisfied for a change, either physical or chemical, to be spontaneous?

   d) The following is a list of standard entropies in $J\,K^{-1}\,mol^{-1}$:

   H_2, 131; C_2H_4, 220; C_2H_6, 230; N_2, 192; NH_3, 193

   Calculate the standard entropy changes of the reactions:

   i) $H_2(g)$ + $C_2H_4(g)$ $\longrightarrow$ $C_2H_6(g)$
   ii) $N_2(g)$ + $3H_2(g)$ $\longrightarrow$ $2NH_3(g)$

   e) For the reaction

   $$H_2(g)\ +\ Cl_2(g)\ \longrightarrow\ 2HCl(g)$$
   $$\Delta S\ =\ 20\,J\,K^{-1}\quad \text{and}\quad \Delta H\ =\ -185\,kJ$$

   i) Determine whether or not this reaction is thermodynamically feasible at 300 K and constant pressure.
   ii) What other factor is important in determining whether or not the reaction will actually take place?   (L80)

12. a) State Hess's law of constant heat summation (law of conservation of energy).

   b) Define precisely the terms:
   i) *enthalpy of formation (heat of formation)*;
   ii) *enthalpy of neutralisation (heat of neutralisation)*.

   c) Using the following data, collected at 25 °C and standard atmospheric pressure, in which the negative sign indicates heat evolved, calculate the enthalpy of formation of potassium chloride, $KCl(s)$:

   $\Delta H\ (298\ K)/kJ\ mol^{-1}$

   i) $KOH(aq) + HCl(aq)$ $=$ $KCl(aq) + H_2O(l)$ -57.3
   ii) $H_2(g)$ $+ \frac{1}{2}O_2(g)$ $=$ $H_2O(l)$ -285.9
   iii) $\frac{1}{2}H_2(g)$ $+ \frac{1}{2}Cl_2(g)$ $+ aq = HCl(aq)$ -164.2
   iv) $K(s)$ $+ \frac{1}{2}O_2(g)$ $+ \frac{1}{2}H_2(g) + aq = KOH(aq)$ -487.0
   v) $KCl(s)$ $+ aq$ $=$ $KCl(aq)$ $+18.4$

   d) The enthalpy of neutralisation of ethanoic acid (acetic acid) is $-55.8\,kJ\,mol^{-1}$ while that of hydrochloric acid is $-57.3\,kJ\,mol^{-1}$, both reactions being with potassium hydroxide solution. Explain the difference in these two values and make what deductions you can.

e) Describe briefly an experiment by which the enthalpy change of a chemical reaction of *your own choice* may be determined, and outline how the various measurements would be used to calculate the final value.

f) When solid potassium chloride is dissolved in water, heat is absorbed.

$$KCl(s) + aq = KCl(aq); \quad \Delta H = +18.4\,kJ\,mol^{-1}$$

Use the ideas formulated in the kinetic-molecular theory of matter to discuss what happens during the dissolution of a crystalline salt in water, and explain why heat is absorbed in the above reaction.

(SUJB80)

10 Reaction Kinetics

Reaction Kinetics is the study of the factors which affect the rates of chemical reactions.

REACTION RATE

The rate of a chemical reaction is the rate of change of concentration. Consider a reaction of the type A ⟶ B, where one molecule of the reactant forms one molecule of the product. Fig. 10.1 shows how the concentration of product, x, increases as the time, t, which has passed since the start of the reaction increases. The initial concentration of reactant (the concentration at the start of the reaction) is a, and at any time after the start of the reaction, the concentration of reactant is $(a - x)$.

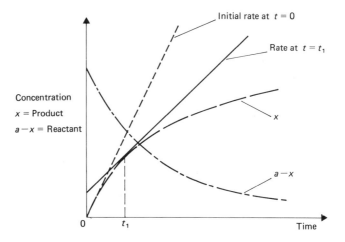

Fig. 10.1   Variation of concentrations of reactant and product with time

You can see that the rate of reaction is decreasing as the reaction proceeds and the reactant is being used up. One can only state the rate of reaction between certain times.

One can calculate the average rate of reaction over a certain interval of time in this way.

To 1 dm³ of solution containing 0.300 mol methyl ethanoate is added a small amount of mineral acid. This catalyses the hydrolysis reaction

$$CH_3CO_2CH_3(aq) \ + \ H_2O(l) \ \rightleftharpoons \ CH_3CO_2H(aq) \ + \ CH_3OH(aq)$$

After 100 seconds, the concentration has decreased to $0.292 \text{ mol dm}^{-3}$. This means that $0.008 \text{ mol dm}^{-3}$ of methyl ethanoate has reacted, and $0.008 \text{ mol dm}^{-3}$ of methanol and ethanoic acid have been formed.

$$\left(\begin{array}{l}\text{Average rate of reaction} \\ \text{over this time interval}\end{array}\right) = \frac{\text{Change in concentration}}{\text{Time}}$$

$$= \frac{0.008}{100} = 8 \times 10^{-5} \text{ mol dm}^{-3} \text{s}^{-1}$$

Since the rate of reaction varies with time, it is usual to quote the initial rate of the reaction. This is the rate at the start of the reaction when an infinitesimally small amount of the reactant has been used up. In Fig. 10.1, the gradient of the tangent to the curve at $t = 0$ gives the initial rate of the reaction.

THE EFFECT OF CONCENTRATION ON RATE OF REACTION

Consider a reaction between A and B:

$$\text{A} + \text{B} \longrightarrow \text{Products}$$

The rate of reaction depends on the concentrations of A and B, but one cannot simply say that the rate of reaction is proportional to the concentration of A and proportional to the concentration of B. The relationship is

$$\text{Reaction rate} \propto [\text{A}]^m [\text{B}]^n$$

where m and n are usually integers, often 0, 1 or 2, and are characteristic of the reaction. One says that the reaction is of order m with respect to A and of order n with respect to B. The order of reaction is $(m + n)$. One cannot tell the order simply by looking at the chemical equation for the reaction. For example, the reaction between bromate(V) ions and bromide ions and acid to give bromine

$$\text{BrO}_3^-(\text{aq}) + 5\text{Br}^-(\text{aq}) + 6\text{H}^+(\text{aq}) \longrightarrow 3\text{Br}_2(\text{aq}) + 3\text{H}_2\text{O}(\text{l})$$

has a rate

$$\frac{-\text{d}[\text{BrO}_3^-]}{\text{d}t} \propto [\text{BrO}_3^-] [\text{Br}^-] [\text{H}^+]^2$$

It is first order with respect to bromate(V), first order with respect to bromide, second order with respect to hydrogen ion and fourth order overall. The negative sign means that $[\text{BrO}_3^-]$ decreases with time.

If $\quad$ Reaction rate $\propto [\text{A}]^m [\text{B}]^n \quad$ it follows that

$\quad$ Reaction rate $= k[\text{A}]^m [\text{B}]^n$

The proportionality constant k is called the *rate constant* for the reaction or the *velocity constant* for the reaction.

ORDER OF REACTION

As a reaction proceeds, the concentrations of the reactants decrease, and the rate of reaction decreases, as shown in Fig. 10.1. The shape of the curve depends on the order of the reaction. It is obtained by integrating the rate equation.

FIRST-ORDER REACTIONS

If the reaction

$$A \longrightarrow \text{Products}$$

is a first-order reaction, the rate equation will be

$$\frac{-d[A]}{dt} = k[A]$$

If a = initial concentration of A, and x = decrease in concentration of A in time t, so that $(a - x)$ = concentration of A at time t, then

$$\frac{-d(a - x)}{dt} = k(a - x)$$

Integrating this equation gives

$$kt = \ln[a/(a-x)] = 2.303 \lg[a/(a-x)]$$

The first-order rate constant can be obtained by plotting:
a) $\ln[a/(a-x)]$ against t to give a straight line with gradient k,
b) $\lg[a/(a-x)]$ against t to give a straight line with gradient $k/2.303$,
c) $\ln(a-x)$ against t to give a straight line with gradient $-k$,
d) $\lg(a-x)$ against t to give a straight line with gradient $-k/2.303$.

The units of k, the first-order rate constant, are s^{-1}.

HALF-LIFE

Let $t_{1/2}$ be the time taken for half the amount of A to react. $t_{1/2}$ is called the *half-life* of the reaction.

After $t_{1/2}$ seconds, $x = a/2$ and $\dfrac{a}{a-x} = 2$

$\therefore$ $$kt_{1/2} = \ln 2 = 2.303 \lg 2$$

$$t_{1/2} = 0.693/k$$

The half-life of a first-order reaction is independent of the initial concentration of the reactant. Radioactive decay is an example of first-order kinetics.

PSEUDO-FIRST-ORDER REACTIONS

The acid-catalysed hydrolysis of an ester, e.g. ethyl ethanoate,

$$CH_3CO_2C_2H_5(aq) \ + \ H_2O(l) \ \longrightarrow \ CH_3CO_2H(aq) \ + \ C_2H_5OH(aq)$$

is first order with respect to ester and first order with respect to water. If water is present in excess, so that the amount of water used in the reaction is small, the concentration of water is practically constant, and, since the acid catalyst is not used up, the rate depends only on the concentration of ester:

$$\frac{-d[CH_3CO_2C_2H_5]}{dt} = k'[CH_3CO_2C_2H_5]$$

k' is constant for a certain concentration of acid, and the reaction obeys a first-order rate equation.

SECOND-ORDER REACTIONS, WITH REACTANTS OF EQUAL CONCENTRATIONS

In the simplest case, the reaction

$$A \ + \ B \ \longrightarrow \ Products$$

has a rate equation $\dfrac{-d[A]}{dt} = \dfrac{-d[B]}{dt} = k[A][B]$

Let $a = $ initial concentration of A and also of B

$x = $ decrease of concentration of A and of B in time t

$a - x = $ concentration of A and of B at time t.

The rate equation therefore becomes

$$\frac{-d(a-x)}{dt} = k(a-x)^2$$

Integration of this equation gives

$$kt = x/a(a-x)$$

A plot of $x/a(a-x)$ against t gives a straight line of gradient k.

Since $kt = \dfrac{x}{a(a-x)} = \dfrac{1}{a-x} - \dfrac{1}{a}$

a plot of $1/(a-x)$ against t gives a straight line of gradient k.

ZERO-ORDER REACTIONS

In a zero-order reaction, the rate is independent of the concentration of the reactant. In the reaction between propanone and iodine

$$CH_3COCH_3(aq) \; + \; I_2(aq) \longrightarrow CH_3COCH_2I(aq) \; + \; HI(aq)$$

the reaction rate does not change if the concentration of iodine is changed. The rate of reaction is independent of the iodine concentration, and the reaction is said to be zero order with respect to iodine.

THE EFFECT OF TEMPERATURE ON REACTION RATES

An increase in temperature increases the rate of a reaction by increasing the rate constant. A plot of the logarithm of the rate constant, k, against $1/T$ is a straight line, with a negative gradient (see Fig. 10.2).

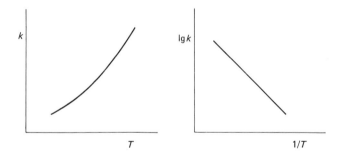

Fig. 10.2   Dependence of rate constant on temperature

The variation of rate constant with temperature obeys the Arrhenius equation

$$k \; = \; A\,e^{-E/RT}$$

A and E are constant for a given reaction; R is the gas constant. In order to react, two molecules must collide with a minimum amount of energy E, which is called the *activation energy*. The fraction of molecules possessing energy E is given by $e^{-E/RT}$. The constant A, called the *pre-exponential factor*, represents the maximum rate which the reaction can reach when all the molecules have energy equal to or greater than E.

The Arrhenius equation can be written as

$$\lg k \; = \; \lg A - \frac{E}{2.303RT}$$

A plot of $\ln k$ against $1/T$ is a straight line of gradient $-E/R$.

A plot of $\lg k$ against $1/T$ is linear with a gradient of $-E/2.303R$ and intercept $\lg A$. The values of A and E can thus be found from a plot of $\lg k$ against $1/T$. If only two values of k have been found, then

$$\lg \frac{k_2}{k_1} = -\frac{E}{2.303R} \left(\frac{1}{T_2} - \frac{1}{T_1} \right)$$

$$= \frac{E(T_2 - T_1)}{2.303RT_1T_2}$$

The activation energy can be obtained from the ratio of the two rates.

EXAMPLE 1   The rate constant of a first-order reaction is $2.0 \times 10^{-6} \, s^{-1}$. The initial concentration of the reactant is $0.10 \, mol \, dm^{-3}$. What is the value of the initial rate in $mol \, dm^{-3} s^{-1}$?

METHOD   The rate equation has the form

$$dx/dt = k[A]$$

Putting the values of $[A]$ and k into this equation gives

ANSWER $dx/dt = 2.0 \times 10^{-6} \times 0.10 = 2.0 \times 10^{-7} \, mol \, dm^{-3} s^{-1}$.

EXAMPLE 2   A first-order reaction is 40% complete at the end of 20 min. a) What is the value of the rate constant? b) In how many minutes will the reaction be 80% complete?

METHOD   a)   For a first-order reaction,

$$k = \frac{2.303}{t} \lg \frac{a}{a-x}$$

a = initial concentration = 100%
x = decrease in concentration = 40% at time t = 20 min = 1200 s
$a - x$ = remaining concentration = 60% at time t = 1200 s

$$k = \frac{2.303}{1200} \lg \frac{100\%}{60\%}$$

ANSWER $$k = 4.3 \times 10^{-4} \, s^{-1}$$

b)   For 80% reaction, $(a-x) = 20\%$,   and

$$4.25 \times 10^{-4} = \frac{2.303}{t} \lg \frac{100\%}{20\%}$$

ANSWER $$t = 3790 \, s = 63 \, minutes$$

EXAMPLE 3   The half-life for the radioactive decay of thorium-234 is 24 days. a) Calculate the rate constant for the decay. b) What time will elapse before 3/4 of the thorium has decayed?

METHOD   a) Radioactive decay follows the first-order law:

$$kt_{1/2} = 2.303 \lg 2$$

ANSWER
$$k = \frac{2.303 \lg 2}{24 \times 24 \times 60 \times 60} = 3.34 \times 10^{-7} \, s^{-1}$$

b) When 3/4 of the thorium has decayed, the initial amount of thorium, a, has become $a/4$.

$$kt = 2.303 \lg \frac{a}{a-x}$$

$$3.34 \times 10^{-7} \times t = 2.303 \lg \frac{a}{a/4}$$

ANSWER
$$t = 4.15 \times 10^6 \, s = 48 \, days$$

EXAMPLE 4   Carbon-14 is radioactive. The half-life is 5600 years. Calculate the age of a piece of wood which gives 10 counts per minute per gram of carbon, compared with 15 c.p.m. per gram of carbon from a sample of new wood.

METHOD   a) First, find the first-order rate constant for the decay.
b) Use the rate constant to find the age of the sample.

a)
$$2.303 \lg \frac{a}{a-x} = kt$$

∴
$$2.303 \lg \frac{N_0}{N} = kt$$

where N_0 = Count rate for new wood and

N = Count rate after time t

At the half-life, $2.303 \lg 2 = kt_{1/2}$

Since $t_{1/2} = 5600$ years, $k = 1.24 \times 10^{-4} \, year^{-1}$

b) Since $N_0/N = 15/10$,
$$2.303 \lg 1.5 = 1.24 \times 10^{-4} t$$

ANSWER   Age of wood, $t = 3270$ years.

EXAMPLE 5   The initial rate of a second-order reaction is $8.0 \times 10^{-3} \, mol \, dm^{-3} \, s^{-1}$. The initial concentrations of the two reactants, A and B, are 0.20 mol dm^{-3}. What is the rate constant in $dm^3 \, mol^{-1} \, s^{-1}$?

METHOD   Since
$$-\frac{d[A]}{dt} = k[A][B]$$

putting

$$\frac{d[A]}{dt} = 8.0 \times 10^{-3} \, \text{mol dm}^{-3} \, \text{s}^{-1} \text{ gives } [A] = [B] = 0.20 \, \text{mol dm}^{-3}$$

$$8.0 \times 10^{-3} = k \times 0.20 \times 0.20$$

ANSWER   Rate constant, $k = 0.20 \, \text{dm}^3 \, \text{mol}^{-1} \, \text{s}^{-1}$

EXAMPLE 6   In the alkaline hydrolysis of an ester, both ester and alkali had the initial concentration of $0.0500 \, \text{mol dm}^{-3}$. The following results were obtained.

Time/s	100	200	300	400	600
% of ester hydrolysed	29.5	44.2	55.5	62.2	70.3

Find the order of reaction, and calculate the rate constant.

METHOD   If the reaction is first order, a plot of $\lg(a-x)$ against t will be linear with a gradient of $-k/2.303$. If the reaction is second order, a plot of $1/(a-x)$ against t will be linear with a gradient of k. The table below shows values of $\lg(a-x)$ and $1/(a-x)$ calculated from the percentage hydrolysis at given times. If the original concentration of ester $= a$, then $x = \%$ hydrolysis $\times a$.

Time/s	0	100	200	300	400	600
Hydrolysis/%	0	29.5	44.2	55.5	62.2	70.3
x	0	0.0148	0.0221	0.0276	0.0311	0.0351
$a - x$	0.0500	0.0352	0.0279	0.0224	0.0189	0.0149
$\lg(a-x)$	-1.30	-1.45	-1.55	-1.65	-1.72	-1.83
$1/(a-x)$	20.0	28.4	35.8	44.6	52.9	67.1

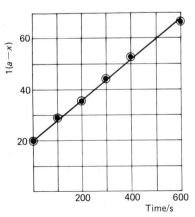
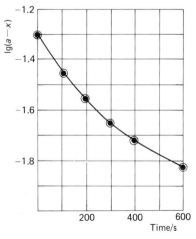

Fig. 10.3   Plots of $\lg(a-x)$ and $1/(a-x)$ against t for Example 6

The plots in Fig. 10.3 show that the second-order plot is linear.
The gradient is $40.0/500 = 8.00 \times 10^{-2}\,dm^3\,mol^{-1}\,s^{-1}$.
This is the value of k.
The reaction is second order, with a rate constant of $8.00 \times 10^{-2}\,dm^3\,mol^{-1}\,s^{-1}$.

EXAMPLE 7   The reaction

$$2N_2O_5(g) \longrightarrow 2N_2O_4(g) + O_2(g)$$

was studied at a number of temperatures, and the following values for the rate constant were obtained:

Temperature/K	293	308	318	338
Rate constant/s^{-1}	1.76×10^{-5}	1.35×10^{-4}	4.98×10^{-4}	4.87×10^{-3}

Calculate the activation energy and the pre-exponential factor for the reaction.

METHOD   Since

$$\lg k = \lg A - \frac{E}{2.303RT}$$

a plot of $\lg k$ against $1/T$ gives a straight line with a gradient of $-E/2.303R$ and an intercept on the $\lg k$ axis of $\lg A$.

The table below gives the values which must be plotted:

T/K	293	308	318	338
k/s^{-1}	1.76×10^{-5}	1.35×10^{-4}	4.98×10^{-4}	4.87×10^{-3}
$1/T\big/1/K$	3.41×10^{-3}	3.25×10^{-3}	3.14×10^{-3}	2.96×10^{-3}
$\lg k$	-4.755	-3.870	-3.303	-2.312

Fig. 10.4 shows the plot of $\lg k$ against $1/T$. The gradient is $-5\,550$ K.

$$\therefore \qquad -5\,500 = -\frac{E}{2.303 \times 8.31}$$

$$E = 106\,000\ J\,mol^{-1} = 106\,kJ\,mol^{-1}$$

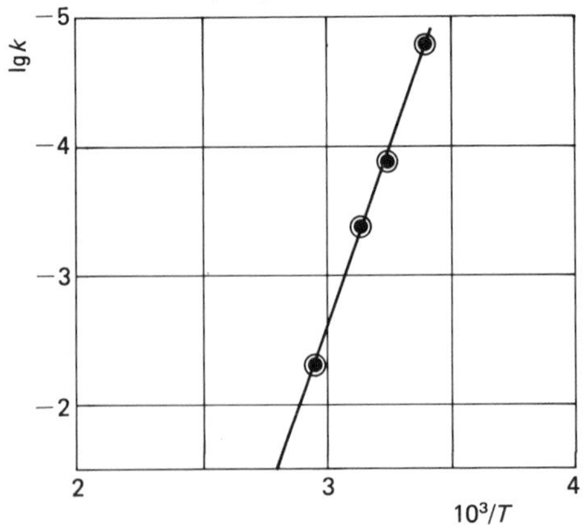

Fig. 10.4   Plot of $\lg k$ against $10^3/T$ for Example 6

A can be found from the intercept or by substituting values of $\lg k$ and $1/T$ in the equation. At $1/T = 3.28 \times 10^{-3}$, $\lg k = -4.00$.

$$\therefore \qquad \lg A = -4.00 + \frac{106 \times 10^3 \times 3.28 \times 10^{-3}}{2.303 \times 8.314} = 14.16$$

$$A = 1.45 \times 10^{14}\,\text{s}^{-1}$$

ANSWER The activation energy is $106\,\text{kJ mol}^{-1}$; the pre-exponential factor is $1.45 \times 10^{14}\,\text{s}^{-1}$.

EXAMPLE 8 Calculate the activation energy for the second-order reaction

$$2N_2O(g) \longrightarrow 2N_2(g) + O_2(g)$$

given the rate constants $1.10 \times 10^{-3}\,\text{dm}^3\,\text{mol}^{-1}\,\text{s}^{-1}$ at $828\,\text{K}$ and $1.67\,\text{dm}^3\,\text{mol}^{-1}\,\text{s}^{-1}$ at $1053\,\text{K}$.

METHOD The equation $$\lg\frac{k_2}{k_1} = \frac{E(T_2 - T_1)}{2.303RT_1T_2}$$

gives $$\lg\frac{1.67}{1.10 \times 10^{-3}} = \frac{E \times 215}{2.303 \times 8.314 \times 1053 \times 838}$$

ANSWER Activation energy, $E = 250\,\text{kJ mol}^{-1}$.

SUMMARY

The effect of concentration on rate

The type of linear plot tells the order of the reaction.

The slope of the linear plot gives the rate constant for the reaction.

Order	Rate equation	Linear plots	Gradient	Units of k
0	$k = x/t$	x against t	k	$\text{mol dm}^{-3}\text{s}^{-1}$
1	$k = \dfrac{2.303}{t}\lg\dfrac{a}{a-x}$	$\lg(a-x)$ against t	$-k/2.303$	s^{-1}
		$\ln(a-x)$ against t	$-k$	s^{-1}
2	$k = \dfrac{x}{ta(a-x)}$	$1/(a-x)$ against t	k	$\text{dm}^3\,\text{mol}^{-1}\,\text{s}^{-1}$

The effect of temperature on rate constant

Equation	Linear plot	Gradient	Intercept
$k = A\,\text{e}^{-E/RT}$			
$\lg k = \lg A - \dfrac{E}{2.303RT}$	$\lg k$ against $1/T$	$-\dfrac{E}{2.303R}$	$\lg A$
	$\ln k$ against $1/T$	$-\dfrac{E}{R}$	$\ln A$

EXERCISE 56   Problems on Finding the Order of Reaction

1. X and Y react together. For a three-fold increase in the concentration of X, there is a nine-fold increase in the rate of reaction. What is the order of reaction with respect to X?

2. A and B react to form C. In one run, the concentration of A is doubled, while B is kept constant, and the initial rate is doubled. In a second run, the concentration of B is doubled while that of A is kept constant, and the initial rate is quadrupled. What can you deduce about the order of the reaction?

3. Fig. 10.5 shows that the rate of reaction is:

   a proportional to $[I_2]$
   b proportional to $[I_2]^2$
   c proportional to $1/[I_2]$
   d independent of $[I_2]$

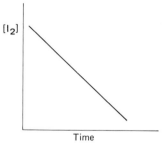

Fig. 10.5

4. X decomposes to form Y + Z. The following results were obtained in a study of the reaction:

Initial [X] /mol dm^{-3}	2.0×10^{-3}	4.0×10^{-3}	5.0×10^{-3}
Initial rate/mol dm^{-3} s^{-1}	1.3×10^{-6}	1.3×10^{-6}	1.3×10^{-6}

   What is the rate expression? What is the order of the reaction?

5. The reaction A + B $\longrightarrow$ C is first order with respect to A and to B. When the initial concentrations are $[A] = 1.5 \times 10^{-2}$ mol dm^{-3} and $[B] = 2.5 \times 10^{-3}$ mol dm^{-3}, the initial rate of reaction is found to be 3.75×10^{-4} mol dm^{-3} s^{-1}. Calculate the rate constant for the reaction.

6. In the reaction

$$A + B \longrightarrow P + Q$$

   the following results were obtained for the initial rates of reaction for different initial concentrations:

[A] /mol dm^{-3}	[B] /mol dm^{-3}	Initial rate/mol dm^{-3} s^{-1}
1.0	1.0	2.0×10^{-3}
2.0	1.0	4.0×10^{-3}
4.0	2.0	16×10^{-3}

   Deduce the rate equation and calculate the rate constant.

7. The rate of a reaction depends on the concentrations of the reactants. In the reaction between X and Y, the following results were obtained for runs at the same temperature.

Initial concentration of X/mol dm^{-3}	Initial concentration of Y/mol dm^{-3}	Initial rate/ mol dm^{-3} h^{-1}
2×10^{-3}	3×10^{-3}	3.0×10^{-3}
2×10^{-3}	6×10^{-3}	1.2×10^{-2}
4×10^{-3}	6×10^{-3}	2.4×10^{-2}

Deduce the order of the reaction with respect to: a) X, b) Y. Calculate the rate constant for the reaction.

8.

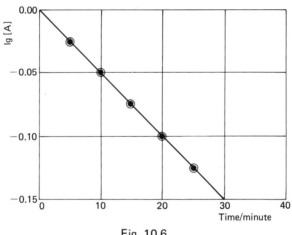

Fig. 10.6

Fig. 10.6 shows a plot of lg concentration of A (in mol dm^{-3}) against time (in minutes) for the reaction A ⟶ B. What is the order of reaction? Calculate the rate constant for the reaction.

9. The following results were obtained for the decomposition of nitrogen(V) oxide

$$2N_2O_5(g) \longrightarrow 4NO_2(g) + O_2(g)$$

Concentration of N_2O_5/mol dm^{-3}	Initial rate/mol dm^{-3} s^{-1}
1.6×10^{-3}	0.12
2.4×10^{-3}	0.18
3.2×10^{-3}	0.24

What is the rate expression for the reaction? What is the order of reaction? What is the initial rate of reaction when the concentration of N_2O_5 is:
a) 2.0×10^{-3} mol dm^{-3} b) 2.4×10^{-2} mol dm^{-3}?

EXERCISE 57   Problems on First-order Reactions

1. A isomerises to form B. The reaction is first order. If 75% of A is converted to B in 2.5 hours, what is the value of the rate constant for the isomerisation?

2. The reaction

$$2N_2O_5 \longrightarrow 4NO_2 + O_2$$

is first order. If the initial concentration of N_2O_5 is 1.25 mol dm^{-3} and the initial rate is $1.38 \times 10^{-5} \text{ mol dm}^{-3} \text{s}^{-1}$, what is the rate constant for the decomposition?

3.

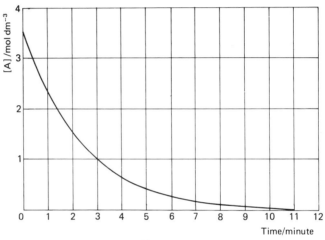

Fig. 10.7

The curve shown in Fig. 10.7 represents the decomposition of A at a certain temperature.

Calculate the gradients of the curve at: a) 1 minute, b) 3 minutes, c) 5 minutes, and d) 11 minutes. Plot a graph of the gradient against the concentration of A at 1, 3, 5 and 11 minutes. Calculate the rate constant for the reaction from the slope of the graph. What is the order of the reaction?

4. Fig. 10.8 shows the results of a study of the reaction

$$A + B \longrightarrow C$$

The experimental conditions were:

| | Initial concentrations/mol dm^{-3} | |
	[A]	[B]
Curve 1	0.10	0.10
Curve 2	0.10	0.20
Curve 3	0.10	0.30

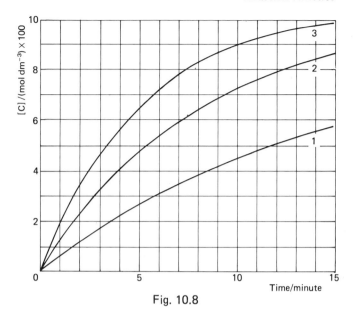

Fig. 10.8

a) Find the initial rates of curves 1, 2 and 3.
b) What is the order of reaction with respect to B?
c) Inspect curve 3. What is the time required for completion of $\frac{1}{2}$ reaction, and of $\frac{3}{4}$ reaction? What is the order with respect to A?
d) Write an overall rate equation for the reaction.
e) Find the rate constant for the reaction.

EXERCISE 58   Problems on Second-order Reactions

1. The following results were obtained from a study of the reaction between P and Q.

Concentrations/mol dm^{-3}		Initial rate/mol dm^{-3} s^{-1}
[P]	[Q]	
2.00×10^{-3}	2.00×10^{-3}	2.00×10^{-4}
1.80×10^{-3}	1.80×10^{-3}	1.62×10^{-4}
1.40×10^{-3}	1.40×10^{-3}	9.80×10^{-5}
1.10×10^{-3}	1.10×10^{-3}	6.05×10^{-5}
0.80×10^{-3}	0.80×10^{-3}	3.20×10^{-5}

Prove that the reaction is second order. Calculate the rate constant.

2. A second-order reaction occurs between X and Y, which are present in solution at concentrations of 0.01 mol dm^{-3}. The results in the table give the percentage reaction at various intervals of time after the start of the reaction. Calculate the second-order rate constant.

Time/s	40	80	120	160	200
Percentage reaction	37	53	63	69	74

3. The following results were obtained for a reaction between A and B.

| Concentrations/mol dm^{-3} | | Initial rate/mol dm^{-3} s^{-1} |
[A]	[B]	
0.5	1.0	2
0.5	2.0	8
0.5	3.0	18
1.0	3.0	36
2.0	3.0	72

What is the order of reaction with respect to A and with respect to B? What is the rate equation for the reaction? Calculate the rate constant. State the units in which it is expressed.

EXERCISE 59   Problems on Radioactive Decay

1. Plot a graph, using the following figures, to show the radioactive decay of krypton. From the graph, find the half-life.

Time/minute	0	20	40	60	80	100	120
Activity/count per second	100	92	85	78	72	66	61

2. A sample of gold was irradiated in a nuclear reactor. It gave the following results when its radioactivity was measured at various intervals. Plot the results, and deduce the half-life of the radioactive isotope of gold formed.

Time/hour	0	1	5	10	25	50	75	100
Radioactivity/count per minute	300	296	285	270	228	175	133	103

3. A sample of bromine was irradiated in a nuclear reactor. The following results were obtained when the radioactivity was measured after various time intervals. Plot the results, and deduce what you can about the decay of radioactive bromine.

Time/hour	0	0.1	0.2	0.5	1	2	5	10	25	50	75	100
Radio-activity/count per minute	500	442	399	320	268	242	225	204	154	95	55	35

4. A radioactive source, after storing for 42 days, is found to have 1/8th of its original activity. What is the half-life of the radioactive isotope present in the source?

5. Actinium B has a half-life of 36.0 min. What fraction of the original quantity of actinium remains after: a) 180.0 min, b) 1080 min?

6. The half-life of carbon-14 is 5580 years. A 10 g sample of carbon prepared from newly cut timber gave a count rate of 2.04 s^{-1}. A 10 g sample of carbon from an ancient relic gave a count rate of 1.84 s^{-1}. Calculate the age of the relic.

7. A dose of 1.00×10^{-4} g of astatine-211 is given to a patient for treatment of cancer of the thyroid gland. How much of this radio-active isotope $(t_{1/2} = 7.21$ h) will remain in the body 24 hours later?

8. In 3 hours, the activity of a radioactive compound falls from 1112 c.p.m. to 678 c.p.m. What is the half-life of the isotope?

9. In 30 minutes, the activity of a radioactive source falls from 1173 c.p.m. to 724 c.p.m. What is the half-life of the radioisotope?

10. Tritium has a half-life of 12.3 years. When tritiated water is used in tracer experiments, what percentage of the original activity will remain after: a) 5 years, b) 50 years?

11. The half-life of carbon-14 is 5600 years. A piece of wood from an ancient ship gives a count of 10 counts per minute, while carbon obtained from new wood gives 15 counts per minute. What is the age of the ship?

EXERCISE 60   Problems on Rates of Reaction

1. A first-order reaction is 50% complete at the end of 30 minutes. What is the value of the rate constant? In how many minutes is reaction 80% complete?

2. A second-order reaction is 60% complete in 1 hour. What is the value of the rate constant?

3. The half-life for the disintegration of bismuth-214 is 19.7 minutes. Calculate the rate constant for the decay in s^{-1}.

4. A second-order reaction occurs between reactants originally present at a concentration of 0.1 mol dm^{-3}. The reaction is 10% complete in 30 min. Calculate: a) the rate constant, b) the half-life, and c) the time for 10% completion of the reaction, if the original concentrations of the reactants are 0.01 mol dm^{-3}.

5. The half-life for the radioactive disintegration of bismuth-210 is 5.0 days. Calculate: a) the rate constant in s^{-1}, b) the time needed for 0.016 mg of bismuth-210 to decay to 0.001 mg.

6. Hydrogen and iodine combine to form hydrogen iodide. The reaction is first order with respect to hydrogen and first order with respect to iodine. The rate constant is 2.78×10^{-4} mol dm^{-3} s^{-1}. If the concentrations are $[H_2] = 0.85 \times 10^{-2}$ mol dm^{-3}, and $[I_2] = 1.25 \times 10^{-2}$ mol dm^{-3}, what is the initial rate of reaction?

7. The reaction $2NO(g) + Cl_2(g) \longrightarrow 2NOCl(g)$ is third order. The rate constant is 1.7×10^{-5} dm^6 mol^{-2} s^{-1}. If the concentrations of the reactants are each 0.20 mol dm^{-3}, what is the initial rate of reaction?

8. The decomposition of ethanal, $CH_3CHO \longrightarrow CH_4 + CO$, can be studied by measuring the increase in pressure of the mixture of gases at various times. Find: a) the order of the reaction, and b) the rate constant.

Time/s	0	50	100	150	250	350	500
Pressure/$10^{-4}\,N\,m^{-2}$	4.82	5.74	6.38	6.82	7.44	7.84	8.24

9. A and B react in the gas phase. In experiment 1, a glass vessel was used. In experiment 2, the glass was coated with another material. The results of the two experiments are shown below. Deduce the rate equations for the two experiments. Can you explain how they come to differ?

	$[A]/mol\,dm^{-3}$	$[B]/mol\,dm^{-3}$	Initial rate/$mol\,dm^{-3}\,s^{-1}$
Experiment 1	0.20	0.12	2×10^{-3}
	0.40	0.12	8×10^{-3}
	0.20	0.24	4×10^{-3}
Experiment 2	0.20	0.12	2×10^{-3}
	0.40	0.24	8×10^{-3}
	0.80	0.24	32×10^{-3}

10. Hydrogen peroxide decomposes in aqueous solution:
$$2H_2O_2(aq) \longrightarrow 2H_2O(l) + O_2(g)$$
The following results show how the rate of decomposition varies with the initial hydrogen peroxide concentration:

Rate/$mol\,dm^{-3}\,s^{-1}$	$[H_2O_2]/mol\,dm^{-3}$
3.64×10^{-5}	0.05
7.41×10^{-5}	0.10
1.51×10^{-4}	0.20
2.21×10^{-4}	0.30

Plot the rate of decomposition against the concentration. Deduce: a) the order of the reaction, and b) the rate constant under the conditions of the experiment.

*11. A gaseous compound C dimerises to form D:
$$2C(g) \longrightarrow D(g)$$
There is a drop in pressure as two moles of C form one mole of D. From the values of pressure and time given below, find: a) the order of reaction, and b) the rate constant in terms of $N\,m^{-2}\,s^{-1}$.

Time/s	0	5	10	15	20	25	30
Pressure/kPa	100	89	82	77	73	70.5	68

*EXERCISE 61   Problems on Activation Energy

1. The reaction

$$2A(g) \ + \ B(g) \ \longrightarrow \ C(g)$$

was studied at a number of temperatures, and the following results were obtained:

Temperature/$^{\circ}$C	12	60	112	203	292
Rate constant/$dm^6\,mol^{-2}\,s^{-1}$	2.34	13.2	52.5	316	1000

Calculate the activation energy.

2. The following results were obtained in the study of the effect of temperature on the rate of a chemical reaction.

Temperature/K	2000	1136	909	676	540
Rate constant/$dm^3\,mol^{-1}\,s^{-1}$	1.00	0.316	0.158	0.050	0.0158

From a plot of $\lg k$ against $1/T$, find the activation energy for the reaction.

3. The activation energy for a reaction is $60\,kJ\,mol^{-1}$. How much faster will the reaction be at $30\,^{\circ}$C than at $20\,^{\circ}$C?

4. If a reaction is 2.70 times faster at $40\,^{\circ}$C than at $30\,^{\circ}$C, what is its activation energy?

5. Use the following results to calculate the energy of activation of the reaction:

Temperature/K	800	625	472	377	333
Rate constant/$dm^3\,mol^{-1}\,s^{-1}$	0.562	0.316	0.132	0.0562	0.0316

EXERCISE 62   Questions from A-level Papers

1. The reaction between potassium iodide and potassium peroxodisulphate(VI), $(K_2S_2O_8)$, in aqueous solution proceeds according to the overall equation

$$2KI \ + \ K_2S_2O_8 \ = \ 2K_2SO_4 \ + \ I_2$$

The rate of this reaction is found by experiment to be directly proportional to the concentration of the potassium iodide, and directly proportional to that of the potassium peroxodisulphate(VI).

a) Write the above equation in ionic form.

b) What is the overall order of the reaction?

c) Show the meaning of the term velocity constant by writing an equation for the rate of this reaction, such that the concentration of the peroxodisulphate(VI) ions decreases with time.

d) With an initial concentration of potassium iodide of 1.0×10^{-2} mol dm^{-3} and of potassium peroxodisulphate(VI) of 5.0×10^{-4} mol dm^{-3}, it is found that at 298 K the initial rate of disappearance of the peroxodisulphate(VI) ions is 1.02×10^{-8} mol $dm^{-3} s^{-1}$. What is the velocity constant of the reaction at this temperature?   (O80)

2. The decomposition of nitrogen pentoxide (N_2O_5) dissolved in tetra-chloromethane (carbon tetrachloride) at 45 °C is first order. Using the concentrations of nitrogen pentoxide and the times given, estimate by a graphical method the rate constant for the decomposition, stating the units in which it is expressed.

Time/s	0	250	500	750	1000	1500	2000	2500
Concentration/mol dm^{-3}	2.33	1.95	1.68	1.42	1.25	0.95	0.70	0.50

(SUJB78, p)

*3. a) Explain what is meant by the terms: i) velocity constant (or rate constant, ii) activation energy.

b) Nitrogen dioxide decomposes in the gaseous phase into nitrogen oxide and oxygen. Suggest an experimental method for determining the rate of this reaction.

c) The following results were obtained for the velocity constant k of this reaction:

k/dm^3 mol^{-1}s^{-1}	Temperature/K
3.16	650
28.2	730
158	800
1120	900
5010	1000

Use this information to find the activation energy for this reaction. ($R = 8.3$ J K^{-1} mol^{-1}; $\log_e x = 2.303 \log_{10} x$.)   (O78, S)

4. The data shown in Fig. 10.9 were obtained in an investigation of the decomposition of dinitrogen pentoxide at a certain temperature and pressure.

a) Copy the following table, and complete it. You must clearly indicate how you calculate the gradients.

Time/min	20	60	100	∞ (infinity)
$N_2O_5(g)$/mol dm^{-3}				
Gradient of graph/mol dm^{-3} min^{-1}				

b) What information is given by the gradient of the graph at a given time?

c) Using the appropriate values from the table in a), plot a graph to determine a value for the rate constant k for the decomposition of dinitrogen pentoxide. Clearly label each axis. Measure the gradient of the graph, and calculate the value of k. State the units in which k is expressed.

d) State: i) the order of the reaction, ii) the time(s) at which your calculated value of k applies. Explain your reasoning.   (C78)

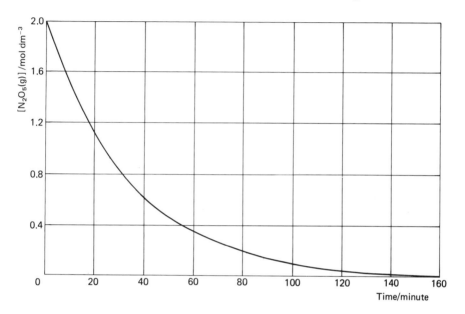

Fig. 10.9

5. Write down a general equation for the rate of a reaction and define, or explain, the terms used.

An investigation of the decomposition of nitrogen pentoxide dissolved in tetrachloromethane (carbon tetrachloride) at $318\,K$

$$N_2O_5 \longrightarrow N_2O_4 + \tfrac{1}{2}O_2$$
$$N_2O_4 \rightleftharpoons 2NO_2$$

gave the following information:

Time/s	0	250	500	750	1000
N_2O_5 concentration/mol dm^{-3}	2.33	1.95	1.68	1.42	1.25
Time/s	1500	2000	2500	3000	
N_2O_5 concentration/mol dm^{-3}	0.95	0.70	0.50	0.35	

(*Use graph paper.*)

a) Plot a graph of concentration of nitrogen pentoxide remaining against time.
   (*It is recommended that the time axis be extended to 4000s.*)

b) Measure the gradients of the curve at the following times, state the units in which these are measured and explain briefly what you have done:

Time/s   500   1000   1500   2000   2500

What information do you now possess?

c) Deduce the 1st order rate constant for the decomposition under these conditions by plotting a graph of the measured gradients against other appropriate data. State any assumption you are making. Give the units in which the rate constant is measured.

d) From the first graph, measure the time intervals for any concentration of nitrogen pentoxide to be successively halved (e.g. 2.0, 1.0, 0.5 mol dm^{-3} . . .) and average your result. Comment on what you have found.   (SUJB80)

6. What do you understand by the term *order* when applied to a chemical reaction?

Given data on the variation of concentration with time, give *two* methods which may be used to determine the order of that reaction.

Propanone (acetone) reacts with iodine in the presence of acid according to the equation

$$CH_3COCH_3 + I_2 \longrightarrow CH_3COCH_2I + H^+ + I^-$$

In an experiment, 5 cm^3 of propanone, 10 cm^3 of sulphuric acid of concentration 1.0 mol dm^{-3} and 10 cm^3 of a solution of iodine of concentration 1.0×10^{-3} mol dm^{-3} were mixed, made up to 100 cm^3 with distilled water and placed in a thermostat. Every five minutes samples were removed and titrated against a solution of sodium thiosulphate. The experiment was repeated but using 20 cm^3 of sulphuric acid in the reaction mixture. The following results were obtained.

Time/minute		5	10	15	20	25
Titre/cm^3	for 10 cm^3 acid	18.5	17.0	15.5	14.0	12.5
	for 20 cm^3 acid	17.0	14.0	11.0	8.0	5.0

From plots of titre against time, determine the order of reaction with respect to iodine and with respect to [H$^+$].

How would you determine the order with respect to propanone?
                                                     (O & C80)

7. Explain the meaning of *each* of the following terms: *law of mass action*; *order of reaction*; *rate constant*; *rate expression*.

Briefly describe how you could follow the course of a chemical reaction of your choice.

For the hydrolysis of methyl ethanoate (acetate) in aqueous hydrochloric acid solutions, the following initial rates of reaction were measured at 30 °C.

Initial rate/ mol dm^{-3} min^{-1}	Initial concentrations/ mol dm^{-3}	
	Methyl ethanoate	HCl
0.001	1.0	0.5
0.002	1.0	1.0
0.001	0.5	1.0

Deduce the order of the reaction with respect to both methyl ethanoate and hydrogen ions. Explain the role of hydrogen ions in the reaction.

In a further experiment one mole amounts of both methanol and ethanoic (acetic) acid were mixed giving a total volume of 100 cm^3. When the reaction was complete 0.35 moles of each remained. Make appropriate deductions from this observation.   (WJEC80)

8. a) Explain what is meant by the *order of a chemical reaction*.
   b) For a first-order reaction

$$kt = \ln\frac{P}{P_t}$$

   where k is the reaction rate constant, P is the initial partial pressure and P_t is the partial pressure after time t.
   i) The following data refer to the reaction

$$2N_2O_5(g) \longrightarrow 4NO_2(g) + O_2(g) \quad \text{at } 338\,K$$

Partial pressure of N_2O_5 (Pa)	100	90	80	70	60	50	40	30	20
Time (s)	0	20	45	73	105	140	185	243	325

   Show graphically that these results are consistent with the decomposition of N_2O_5 being a first order reaction, and evaluate the rate constant for the decomposition at 338 K.

$$\ln P = 2.303 \log P$$

   ii) Give a reason why the equation

$$2N_2O_5(g) \longrightarrow 4NO_2(g) + O_2(g)$$

   does not give the overall order of the reaction.
   c) i) Explain what is meant by the *half-life* of a reaction.
      ii) Show that, for the first order reaction above, the half-life is independent of the initial partial pressure, and deduce a relationship between the rate constant k and the half-life of the reaction.
      iii) What is the half-life of this reaction?   (AEB79)

9. The following data refer to the first-order decomposition of X, at 300 K

$X/mol\,l^{-1}$	1.27	0.91	0.65	0.47	0.33	0.24	0.17	0.14	0.11
t/min	0	15	30	45	60	75	90	100	110

a) Plot a graph of the concentration against the time and explain the shape of the curve you obtain.

b) By means of tangents, measure the rates of decomposition at 5, 35 and 70 minutes. Calculate a value for the rate constant at each of these times, deducing the units of the constant. Comment on these results.

c) At 310 K, the rate is approximately double the corresponding one at 300 K. Explain this and state what you would expect the value of the rate constant to be at the higher temperature.

d) Explain the difference between order and molecularity. (JMB79, S)

11 Equilibria

CHEMICAL EQUILIBRIUM

An example of a reversible reaction between gases is the reaction between hydrogen and iodine to form hydrogen iodide:

$$H_2(g) + I_2(g) \rightleftharpoons 2HI(g)$$

If the reaction takes place in a closed vessel, the combination of hydrogen and iodine gradually slows down as the concentrations of these gases decrease. At first, there is very little decomposition of hydrogen iodide into hydrogen and iodine, but, as the concentration of hydrogen iodide increases, the rate of decomposition of hydrogen iodide into hydrogen and iodine increases until the rates of the forward and reverse reactions are equal, and the concentration of each species is constant.

An example of a reversible reaction which takes place in solution is the reaction between ethanoic acid and ethanol to form ethyl ethanoate:

$$CH_3CO_2H(l) + C_2H_5OH(l) \rightleftharpoons CH_3CO_2C_2H_5(l) + H_2O(l)$$

As the concentrations of ester and water increase, the reverse reaction — hydrolysis of the ester to form the acid and alcohol — speeds up. At equilibrium, the rate of the forward reaction is equal to the rate of the reverse reaction. Esterification is catalysed by inorganic acids. The presence of a catalyst speeds up the rate at which equilibrium is established.

THE EQUILIBRIUM LAW

If a reversible reaction is allowed to reach equilibrium, it is found that the product of the concentrations of the products divided by the product of the concentrations of the reactants has a constant value at a particular temperature. In the esterification reaction,

$$CH_3CO_2H(l) + C_2H_5OH(l) \rightleftharpoons CH_3CO_2C_2H_5(l) + H_2O(l)$$

it is found that

$$\frac{[CH_3CO_2C_2H_5][H_2O]}{[CH_3CO_2H][C_2H_5OH]} = K_c$$

where K_c is the equilibrium constant for the reaction in terms of concentration. In the reaction between hydrogen and iodine,

$$H_2(g) + I_2(g) \rightleftharpoons 2HI(g)$$

and

$$\frac{[HI]^2}{[H_2][I_2]} = K_c$$

Since this is a reaction between gases, the concentration of each gas can be expressed as a partial pressure. Then,

$$\frac{p_{HI}^2}{p_{H_2} \times p_{I_2}} = K_p$$

K_p is the equilibrium constant in terms of partial pressures. In the reaction between iron and steam,

$$3Fe(s) + 4H_2O(g) \rightleftharpoons Fe_3O_4(s) + 4H_2(g)$$

The equilibrium constant is given by

$$\frac{p_{H_2}^4}{p_{H_2O}^4} = K_p$$

The solids do not appear in the expression. Their vapour pressures are constant (at a constant temperature) as long as there is some of each solid present. These constant vapour pressures are incorporated into the value of the constant K_p.

Another type of reaction which reaches an equilibrium position is thermal dissociation. For example, when phosphorus(V) chloride is heated, it dissociates partially to form phosphorus(III) chloride and chlorine:

$$PCl_5(g) \rightleftharpoons PCl_3(g) + Cl_2(g)$$

As explained in Chapter 6 (pp. 86-8) the dissociation increases the number of moles of substance present and causes an increase in volume, or, if the volume is kept constant, an increase in pressure. The result is that the experimental determinations of molar mass give an unexpectedly low value. The degree of dissociation, α, can be obtained from the ratio,

$$\frac{\text{Molar mass calculated in the absence of dissociation}}{\text{Experimentally determined molar mass}} = 1 + \alpha$$

Inserting the value for α into the expression for the equilibrium constant, and putting c = initial concentration of PCl_5, we get

$$K_c = \frac{[Cl_2][PCl_3]}{[PCl_5]}$$

$$K_c = \frac{\alpha c \times \alpha c}{(1-\alpha)c} = \frac{\alpha^2 c}{1-\alpha}$$

If the total pressure $= P$, the partial pressures of PCl_3 and Cl_2 are $P\alpha/(1+\alpha)$, the partial pressure of PCl_5 is $P(1-\alpha)/(1+\alpha)$, and

$$K_p = \frac{P^2\alpha^2/(1+\alpha)^2}{P(1-\alpha)/(1+\alpha)} = \frac{\alpha^2 P}{1-\alpha^2}$$

EXAMPLE 1 1.00 mole of ethanoic acid was allowed to react with: a) 0.50 mole, b) 1.00 mole, c) 2.00 mole, and d) 4.00 mole of ethanol. At equilibrium, the amount of acid remaining was a) 0.58 mole, b) 0.33 mole, c) 0.15 mole and d) 0.07 mole. Calculate the equilibrium constant for the esterification reaction.

METHOD If the original amounts of acid and ethanol are a mol and b mol, then, at equilibrium, the amount of ester formed is x mol, and the amounts of acid and ethanol remaining are $(a - x)$ and $(b - x)$ mol.

$$CH_3CO_2H(l) + C_2H_5OH(l) \rightleftharpoons CH_3CO_2C_2H_5(l) + H_2O(l)$$
$$(a-x) \qquad\quad (b-x) \qquad\qquad\qquad x \qquad\qquad x$$

Since the equilibrium constant is given by

$$\frac{[CH_3CO_2C_2H_5]\,[H_2O]}{[CH_3CO_2H]\,[C_2H_5OH]} = K_c$$

then $\dfrac{(x/V)(x/V)}{[(a-x)/V]\,[(b-x)/V]} = K_c$ or $\dfrac{x^2}{(a-x)(b-x)} = K_c$

In reaction a)

$$(a - x) = 0.58; \quad x = 0.42; \quad (b - x) = 0.08$$

$$\therefore \qquad \frac{(0.42)^2}{0.58 \times 0.08} = K_c = 3.8$$

Substituting the other values of a, b and x in the equation gives the following values of K:

a	b	$a - x$	x	$b - x$	K_c
1.00	0.50	0.58	0.42	0.08	3.8
1.00	1.00	0.33	0.67	0.33	4.1
1.00	2.00	0.15	0.85	1.15	4.4
1.00	4.00	0.07	0.93	3.07	4.0

ANSWER The average value of the equilibrium constant is 4.1.

EXAMPLE 2 Calculate the amount of ethyl ethanoate formed when 1 mole of ethanoic acid and 3 moles of ethanol and 3 moles of water are allowed to come to equilibrium. The equilibrium constant for the reaction is 4.0.

METHOD Let the amount of ethyl ethanoate $= x$ mol
Then
Equilibrium amount of acid $= (1-x)$ mol
Equilibrium amount of ethanol $= (3 - x)$ mol
Equilibrium amount of water $= (3 + x)$ mol

Then, since $\dfrac{[CH_3CO_2C_2H_5]\,[H_2O]}{[CH_3CO_2H]\,[C_2H_5OH]} = K_c = 4.0$

$$\frac{x(3+x)}{(1-x)(3-x)} = 4.0$$

$$3x^2 - 19x + 12 = 0$$

Solving this quadratic equation (see p. 8) gives

$$x = 5.6 \quad \text{or} \quad 0.72$$

The value $x = 5.6$ can be excluded because it is higher than the number of moles of ethanoic acid present initially. The solution $x = 0.72$ must be the practical one.

ANSWER   The amount of ethyl ethanoate formed is 0.72 mol.

EXAMPLE 3 A mixture of iron and steam is allowed to come to equilibrium at 600 °C. The equilibrium pressures of hydrogen and steam are 3.2 kPa and 2.4 kPa. Calculate the equilibrium constant K_p for the reaction.

METHOD   The reaction is

$$3\,Fe(s) \;+\; 4H_2O(g) \;\rightleftharpoons\; 4H_2(g) \;+\; Fe_3O_4(s)$$

The equilibrium constant is given by

$$K_p = \frac{p_{H_2}{}^4}{p_{H_2O}{}^4}$$

Substituting in this equation gives

$$K_p = \left(\frac{3.2}{2.4}\right)^4$$

ANSWER $K_p = 3.1$

EXAMPLE 4 A molar mass determination on dinitrogen tetraoxide, N_2O_4, gave a value of $60\,\mathrm{g\,mol^{-1}}$ at 50 °C and 1.01×10^5 Pa. Find the equilibrium constant for the dissociation

$$N_2O_4(g) \;\rightleftharpoons\; 2NO_2(g)$$

METHOD   If the degree of dissociation is α, then a total of $1 + \alpha$ moles of particles are formed from 1 mole of N_2O_4. P is the total pressure.

$$\frac{\text{Molar mass}}{\text{Experimentally determined molar mass}} = \frac{92}{60} = 1 + \alpha$$

$$\alpha = 0.53$$

Since $K_p = \dfrac{p_{NO_2}{}^2}{p_{N_2O_4}}$

and $\quad p_{NO_2} = \dfrac{2\alpha}{1+\alpha} P \quad$ and $\quad p_{N_2O_4} = \dfrac{1-\alpha}{1+\alpha} P$

$$K_p = \frac{4\alpha^2}{1-\alpha^2} \times P = \frac{4(0.53)^2}{1-(0.53)^2} \times 1.01 \times 10^5 \, \text{Pa}$$

ANSWER $\qquad K_p = 1.58 \times 10^5 \, \text{Pa}$

EXERCISE 63   Problems on Equilibria

1. Write an expression for the equilibrium constant for the reaction,
   $A + B \rightleftharpoons C + D$:
   a) when A, B, C and D are gases
   b) when A, B, C and D are solutions
   c) when A and C are solids, B and D are gases
   d) when C is a solid, and A, B and D are solutions.

2. In the equilibrium
   $$N_2O_4(g) \rightleftharpoons 2NO_2(g)$$
   $3.20 \, \text{g}$ of dinitrogen tetraoxide occupy a volume $1.00 \, \text{dm}^3$ at $1.00 \times 10^5 \, \text{Pa}$ and $25 \, ^\circ\text{C}$. Calculate: a) the degree of dissociation, and b) the equilibrium constant.

3. Sulphur dichloride dioxide has an apparent molar mass of $75 \, \text{g mol}^{-1}$ at $400 \, ^\circ\text{C}$ and a pressure of $10^5 \, \text{N m}^{-2}$. Calculate the equilibrium constant for the reaction
   $$SO_2Cl_2(g) \rightleftharpoons SO_2(g) + Cl_2(g)$$
   at this temperature.

4. The equilibrium constant for the reaction
   $$CO(g) + H_2O(g) \rightleftharpoons CO_2(g) + H_2(g)$$
   is 4.00. If 1 mole of CO and 1 mole of H_2O are allowed to come to equilibrium, what fraction of the carbon monoxide will remain?

5. $1 \, \text{dm}^3$ of dinitrogen tetraoxide, N_2O_4, weighs $2.50 \, \text{g}$ at $60 \, ^\circ\text{C}$ and $1.01 \times 10^5 \, \text{N m}^{-2}$ pressure. Find a) the degree of dissociation into NO_2 and b) the value of K_p.

6. Equimolar amounts of hydrogen and iodine are allowed to reach equilibrium:
   $$H_2(g) + I_2(g) \rightleftharpoons 2HI(g)$$
   If 80% of the hydrogen can be converted to hydrogen iodide, what is the value of K_p at this temperature?

7. A mixture of 3 moles of hydrogen and 1 mole of nitrogen is allowed to reach equilibrium at a pressure of $5 \times 10^6 \, N \, m^{-2}$. The composition of the gaseous mixture is then 8% NH_3, 23% N_2, 69% H_2 by volume. Calculate K_p.

8. a) Hydrogen and iodine react to form hydrogen iodide. When the initial concentrations of both reactants are $0.11 \, mol \, dm^{-3}$, the initial rate of reaction is $2.42 \times 10^{-5} \, mol \, dm^{-3} \, s^{-1}$. What is the rate constant for this second-order reaction?

   b) When 1 mol of hydrogen and 1 mol of iodine are allowed to reach equilibrium at 600 K, the equilibrium mixture contains 1.6 mol hydrogen iodide. Calculate the equilibrium constant for the reaction

$$H_2(g) \; + \; I_2(g) \; \rightleftharpoons \; 2HI(g)$$

9. At a certain temperature the reaction

$$C_2H_5CO_2H(l) \; + \; C_2H_5OH(l) \; \rightleftharpoons \; C_2H_5CO_2C_2H_5(l) \; + \; H_2O(l)$$

   has an equilibrium constant of 6.

   a) Calculate the amount of ester (in moles) that will be formed at equilibrium from 1 mol acid and 1 mol alcohol.

   b) What mass of propanoic acid must react with 46 g ethanol in order to give 51 g ethylpropanoate at equilibrium?

10. The equilibrium, $H_2(g) + Cl_2(g) \rightleftharpoons 2HCl(g)$ has $K_p = 2.5 \times 10^4$ at 1500 K. What percentage of hydrogen is converted to hydrogen chloride at this temperature in a 1:1 mixture of hydrogen and chlorine?

11. The equilibrium constant for the reaction $H_2(g) + I_2(g) \rightleftharpoons 2HI(g)$ is 60 at 450 °C. The number of moles of hydrogen iodide in equilibrium with 2 mol of hydrogen and 0.3 mol of iodine at 450 °C is:

    a 1/100   b 1/10   c 6   d 36   e 3

12. The esterification reaction

$$RCO_2H \; + \; R'OH \; \rightleftharpoons \; RCO_2R' \; + \; H_2O$$

    has an equilibrium constant of 10.0 at 25 °C. If 1 mole of ester is dissolved in 5 moles of water, what are the equilibrium amounts of:
    a) acid, b) alcohol, c) ester and d) water?

13. A mixture of 1 mol nitrogen and 3 mol hydrogen is allowed to reach equilibrium at $1.0 \times 10^7 \, Pa$ and 500 °C. The equilibrium mixture contains 20% of ammonia by volume. Calculate the value of K_p for the reaction

$$N_2(g) \; + \; 3H_2(g) \; \rightleftharpoons \; 2NH_3(g)$$

    at 500 °C.

14. Sulphur dioxide and oxygen in the ratio 2 mol:1 mol are allowed to reach equilibrium in the presence of a catalyst, at a pressure of 5 atm. At equilibrium, $\frac{1}{3}$ of the SO_2 was converted to SO_3. Calculate the equilibrium constant for the reaction

$$2SO_2(g) + O_2(g) \rightleftharpoons 2SO_3(g)$$

15. The equilibrium constant K_p for the reaction

$$CO_2(g) + H_2(g) \rightleftharpoons CO(g) + H_2O(g)$$

is 0.72 at $1000\,°C$. Calculate the composition of the mixture which results when:

a) 0.5 mole CO_2 and 0.5 mole H_2 are mixed at a pressure of 1 atm and $1000\,°C$.

b) 5 moles CO_2 and 1 mole H_2 are mixed at a pressure of 1 atm and $1000\,°C$.

16. The oxidation of sulphur dioxide is a reversible process:

$$2SO_2(g) + O_2(g) \rightleftharpoons 2SO_3(g)$$

Calculate the value of the equilibrium constant, K_p, in terms of partial pressures from the following data, which were obtained at $1000\,K$:

Partial pressures/N m^{-2}		
p_{SO_2}	p_{O_2}	p_{SO_3}
10 000	68 800	80 100

EXERCISE 64   Questions from A-level Papers

1. a) Write an equation to show the relationship between the equilibrium constant K_p and the partial pressures $P_{N_2O_4}$ and P_{NO_2} of the reactants in the following gaseous equilibrium:

$$N_2O_4 \rightleftharpoons 2NO_2; \qquad \Delta H(298\ K) = 54\ kJ\ mol^{-1}$$

b) State the effect, if any, on the above equilibrium of:
   i) increasing the pressure,
   ii) raising the temperature.
   Give reasons for your answers.

c) It was found that $1.000\,dm^3$ of the gaseous mixture weighed $2.777\,g$ at $50.0\,°C$ and under a pressure of $1.01 \times 10^5\,N\,m^{-2}$ (= 1 atmosphere). Calculate:
   i) the fraction of the N_2O_4 that is dissociated;
   ii) the percentage of NO_2 molecules in the mixture;
   iii) the value of K_p.
   (One mole of a gas occupies $22.4\,dm^3$ at s.t.p.)   (O77)

2. a) Explain with the aid of vapour pressure/temperature diagrams how the allotropy of phosphorus differs from that of sulphur.

   b) The dissociation pressure of carbon dioxide in the system

   $$CaCO_3 \rightleftharpoons CaO + CO_2$$

   is $5.9 \times 10^3\,N\,m^{-2}$ at 993 K, and $27 \times 10^3\,N\,m^{-2}$ at 1073 K. What can you infer from this about the sign of the enthalpy change when calcium carbonate dissociates? Give reasons for your answer. Why is the dissociation pressure independent of the relative masses of the two solid phases so long as both are present?

   c) When a mixture consisting initially of one mole of ethanoic acid and one mole of ethanol is allowed to reach equilibrium at a certain temperature, the resulting solution is found to contain $\frac{2}{3}$ of a mole of ethyl ethanoate. Estimate what the amount of this substance in the equilibrium mixture would be if the initial mixture consisted of two moles of ethanol and one mole of ethanoic acid.   (O78)

3. a) Write the equations for two reversible gaseous reactions, in one of which the forward reaction is exothermic and in the other of which it is endothermic. In each case state how i) the rate of forward reaction, ii) the position of equilibrium changes with change in temperature.

   b) At 50 °C under a pressure of 13.32 kPa (100 mm Hg) N_2O_4 is 70% dissociated. What would be the equilibrium mixture produced if $200\,cm^3$ of N_2O_4 was allowed to dissociate at 50 °C and a pressure of 26.64 kPa (200 mm Hg)?   (AEB78)

4. This question concerns the reaction

   $$H_2(g) + I_2(g) \rightleftharpoons 2HI(g)$$

   a) Fig. 11.1 is an approximate energy profile for this reaction.

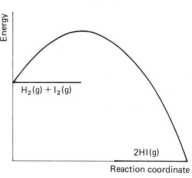

Fig. 11.1

Tracing Fig. 11.1 and on your tracing place labelled arrows to indicate:
   i) the activation energy of the reaction, and
   ii) the enthalpy change of the reaction.

b) The rate of the reaction is given by
$$\text{Rate} = k\,[H_2]\,[I_2]$$
   i) Name the constant k.
   ii) What is the order of the reaction with respect to hydrogen?
   iii) What is the reaction overall?
   iv) When 0.1 mol each of hydrogen and iodine were mixed at 700 K in a total volume of one litre, the initial rate of formation of hydrogen iodide was found to be $1.5 \times 10^{-5}\,\text{mol}\,l^{-1}\,s^{-1}$. Calculate the value of k at 700 K, stating the units.

c) The standard molar entropies of hydrogen, gaseous iodine and hydrogen iodide are 131, 261 and $207\,\text{J}\,\text{mol}^{-1}\,K^{-1}$ respectively. Calculate $\Delta S^{\ominus}$, the standard entropy change, for the reaction
$$H_2(g) \;+\; I_2(g) \;\rightleftharpoons\; 2HI(g)$$

d) A mixture of 1.90 mol of hydrogen and 1.90 mol of iodine was allowed to reach equilibrium at 710 K. The equilibrium mixture was found to contain 3.00 mol of hydrogen iodide.
   i) Calculate the equilibrium constant at 710 K for the reaction
$$H_2(g) \;+\; I_2(g) \;\rightleftharpoons\; 2HI(g)$$
   ii) What would be the effect on the position of equilibrium of increasing the pressure on the system? Give a reason for your answer. (L77)

5. Write a *short account* of the factors affecting the position of equilibrium of a balanced reaction, the rate at which equilibrium is attained and the value of the equilibrium constant.

a) Using partial pressures, show that for gaseous reactions of the type
$$XY(g) \;\rightleftharpoons\; X(g) \;+\; Y(g)$$
at a given temperature, the pressure at which XY is exactly *one-third* dissociated is numerically equal to *eight* times the equilibrium constant at that temperature.

b) When one mole of ethanoic acid (acetic acid) is maintained at 25 °C with 1 mole of ethanol, one-third of the ethanoic acid remains when equilibrium is attained. How much would have remained if one-half of a mole of *ethanol* had been used instead of one mole at the same temperature? (SUJB78)

6. This question is about the reversible reaction
$$2HI(g) \;\rightleftharpoons\; H_2(g) \;+\; I_2(g)$$
a) Using this system as an example, explain what is meant by:
i) equilibrium mixture, and ii) equilibrium constant K_c.

It can be shown that if some radioactive iodine is added to the equilibrium mixture, the iodine in the hydrogen iodide also becomes radioactive. What can be deduced from this observation?

b) In an experiment to determine K_c, 0.210 g of hydrogen iodide was heated at 800 K in a bulb of volume 100 cm³ until equilibrium was attained. The bulb was broken under potassium iodide solution and the iodine present found to be sufficient to react with 4.0 cm³ of sodium thiosulphate solution, concentration 0.1 mol dm⁻³. ($2S_2O_3^{2-} + I_2 = 2I^- + S_4O_6^{2-}$.)

   i) Why was it adequate to absorb the iodine at room temperature although its amount at the high temperature of the experiment was actually required?

   ii) Calculate:

the number of moles of hydrogen iodide in 0.210 g,

the number of moles of iodine (I_2) formed in the experiment.

Hence find:

the number of moles of hydrogen (H_2) also formed,

the number of moles of hydrogen iodide left unreacted.

   iii) Calculate K_c for the reaction at 800 K.

c) The bulb had a volume of 100 cm³. If a 200 cm³ bulb had been used, what values would have been obtained for the following?

   i) The number of moles of iodine present at equilibrium,

   ii) the equilibrium constant.

Explain your answer.   (L79)

12 Organic Chemistry

All the techniques you need to enable you to tackle problems in organic chemistry have been covered in Chapter 2 in the sections on empirical formulae, calculations based on chemical equations and reacting volumes of gases, and in Chapter 3 on volumetric analysis.

Numerical problems in organic chemistry give you some quantitative data and ask you to use it in conjunction with your knowledge of the reactions of organic compounds. There is no set pattern for tackling such problems. They are solved by a combination of calculation, familiarity with the reactions of the compounds involved and logic. The following examples and problems will show you what to expect.

EXAMPLE 1 When 0.2500 g of a hydrocarbon X burns in a stream of oxygen, it forms 0.7860 g of carbon dioxide and 0.3210 g of water. When 0.2500 g of X is vaporised, the volume which it occupies (corrected to s.t.p.) is 80.0 cm^3. Deduce the molecular formula of X.

METHOD X burns to form carbon dioxide and water.

$$\text{Mass of C in } 0.7860 \text{ g of } CO_2 = \frac{12.0}{44.0} \times 0.7860 = 0.2143 \text{ g}$$

$$\text{Mass of H in } 0.3210 \text{ g of } H_2O = \frac{2.02}{18.0} \times 0.3210 = 0.0360 \text{ g}$$

Therefore 0.2500 g of X contains 0.2143 g of C and 0.0360 g of H

These masses give the molar ratio for C:H of $\dfrac{0.2143}{12.0}$ to $\dfrac{0.0360}{1.01}$

$$= 0.0178 \text{ to } 0.0360 = 1 \text{ to } 2$$

Thus, the empirical formula is CH_2.

Since 80.0 cm^3 is the volume occupied by 0.2500 g of X,

$$22.4 \text{ dm}^3 \text{ is occupied by } \frac{22.4}{80.0 \times 10^{-3}} \times 0.2500 \text{ g of X} = 70.0 \text{ g of X}$$

The formula mass of CH_2 is 14. To give a molar mass of 70.0 g mol^{-1}, the empirical formula must be multiplied by 5. Therefore:

ANSWER The molecular formula is C_5H_{10}.

EXAMPLE 2 An organic liquid, P, contains 52.2% carbon, 13.0% hydrogen and 34.8% oxygen by mass. Mild oxidation converts P to Q, and, on further oxidation, R is formed. P and Q react together in the presence of anhydrous calcium chloride to form S, which has a molecular formula of $C_6H_{14}O_2$. P and R react to give T, which has a molecular formula of $C_4H_8O_2$. Identify compounds P to T, and explain the reactions involved.

METHOD   First, calculate the empirical formula of P. This comes to C_2H_6O. This must be the molecular formula also as P and R combine to form T, which has 4C in the molecule. Since P contains one oxygen atom, it is an alcohol, an aldehyde, a ketone or possibly an ether. Other classes of compounds are ruled out by the absence of nitrogen and halogens. Oxidation proceeds in two stages, evidence that P is probably an alcohol, being oxidised first to an aldehyde and then to an acid.

According to the formulae, P would be C_2H_5OH, and Q would be CH_3CHO, and R would be CH_3CO_2H. The reaction between P and Q fits in with this theory as

ANSWER $2C_2H_5OH(l) + CH_3CHO(l) \longrightarrow CH_3CH(OC_2H_5)_2(l) + H_2O(l)$

The reaction between P and R to form T is thus

ANSWER $CH_3CO_2H(l) + C_2H_5OH(l) \longrightarrow CH_3CO_2C_2H_5(l) + H_2O(l)$

The molecular formula $C_6H_{14}O_2$ for S fits $CH_3CH(OC_2H_5)_2$, and the molecular formula $C_4H_8O_2$ for T fits $CH_3CO_2C_2H_5$.

ANSWER   P = C_2H_5OH, ethanol; Q = CH_3CHO, ethanal; R = CH_3CO_2H, ethanoic acid; S = $CH_3CH(OC_2H_5)_2$, 1,1-diethoxyethane; T = $CH_3CO_2C_2H_5$, ethyl ethanoate.

EXAMPLE 3   X is a liquid containing 31.5% by mass of C, 5.3% H and 63.2% O. An aqueous solution of X liberates carbon dioxide from sodium carbonate, with the formation of a solution from which a substance Y of formula $C_2H_3O_3Na$ can be obtained. X reacts with phosphorus(V) chloride to give hydrogen chloride and a compound, Z, of molecular formula $C_2H_2OCl_2$. Identify X, Y and Z. Write equations for the two reactions, X $\longrightarrow$ Y and X $\longrightarrow$ Z.

METHOD   The empirical formula of X is easily shown to be $C_2H_4O_3$. As X liberates carbon dioxide from a carbonate, it must be an acid. Taking CO_2H from $C_2H_4O_3$ leaves CH_3O as the formula for the rest of the molecule. This could be CH_2OH, making X $HOCH_2CO_2H$.

In the reaction with PCl_5, $C_2H_4O_3$ is converted into $C_2H_2OCl_2$. This would fit in with two hydroxyl groups being replaced by two chlorine atoms. This would be the case for the reaction

ANSWER $HOCH_2CO_2H(l) + 2PCl_5(l) \longrightarrow ClCH_2COCl(l) + 2POCl_3(l) + 2HCl(g)$

If X is $HOCH_2CO_2H$, its reaction with sodium carbonate has an equation

ANSWER $2HOCH_2CO_2H(aq) + Na_2CO_3(s) \longrightarrow 2HOCH_2CO_2Na(aq) + CO_2(g) + H_2O(l)$

and Y is $HOCH_2CO_2Na$.

ANSWER   All the information agrees with $X = HOCH_2CO_2H$, hydroxyethanoic acid, $Y = HOCH_2CO_2Na$, sodium hydroxyethanoate; $Z = ClCH_2COCl$, chloroethanoyl chloride.

EXAMPLE 4 $1.220\,g$ of a dicarboxylic aliphatic acid is dissolved in water and made up to $250\,cm^3$. A $25.0\,cm^3$ portion of the solution requires $21.0\,cm^3$ of $0.100\,mol\,dm^{-3}$ sodium hydroxide solution for neutralisation. Deduce the molecular formula of the acid, and write a structural formula for it.

METHOD   The equation for the neutralisation is

$$R(CO_2H)_2(aq) + 2NaOH(aq) \longrightarrow R(CO_2Na)_2(aq) + 2H_2O(l)$$

No. of moles of NaOH $= 21.0 \times 10^{-3} \times 0.100 = 2.10 \times 10^{-3}\,mol$

From the equation, No. of moles of acid $= \frac{1}{2} \times$ No. moles of NaOH
$$= 1.05 \times 10^{-3}\,mol$$

Mass of acid $= \frac{1}{10} \times 1.220\,g = 0.1220\,g$

$\therefore$ $1.05 \times 10^{-3}\,mol = 0.1220\,g$

$$1\,mol = 116\,g$$

The molar mass is $116\,g\,mol^{-1}$. Subtracting 90 for $(CO_2H)_2$ leaves $26\,g\,mol^{-1}$ for the rest of the molecule. This is the mass of $(CH_2)_2$. The molecular formula is $C_4H_4O_4$, and the structural formula are

$$
\begin{array}{cc}
HCCO_2H & HO_2CCH \\
\| & \| \\
HCCO_2H & HCCO_2H
\end{array}
$$

for cis- and trans-butenedioic acid.

EXERCISE 65   Problems on Organic Chemistry

1. A compound X contains carbon and hydrogen only. $0.135\,g$ of X, on combustion in a stream of oxygen, gave $0.410\,g$ of carbon dioxide and $0.209\,g$ of water. Calculate the empirical formula of X.

   X is a gas at room temperature, and $0.29\,g$ of the gas occupy $120\,cm^3$ at room temperature and 1 atm. What is the molecular formula of X?

2. A monobasic organic acid C is dissolved in water and titrated with sodium hydroxide solution. $0.388\,g$ of C require $46.5\,cm^3$ of $0.095\,mol\,dm^{-3}$ sodium hydroxide for neutralisation. Calculate the molar mass of C and deduce its formula.

3. An organic compound has a composition by mass of $83.5\%\,C$; $6.4\%\,H$; $10.1\%\,O$. The molar mass is $158\,g\,mol^{-1}$. Find the molecular formula and suggest a structural formula for the compound.

4. A compound A contains 5.20% by mass of nitrogen. The other elements present are carbon, hydrogen and oxygen. Combustion of 0.0850 g of A in a stream of oxygen gave 0.224 g of carbon dioxide and 0.0372 g of water. Calculate the empirical formula of A.

5. An alkene A contains one double bond per molecule. 0.560 g of bromine is required to react completely with 0.294 g of A. When A is treated with ozone and the product is hydrolysed and then oxidised, B, which is a monobasic carboxylic acid is formed. 0.740 g of this acid require 100 cm^3 of 0.100 mol dm^{-3} sodium hydroxide for neutralisation. Deduce the formulae of A and B and the structural formula of A.

6. An organic compound, A, contains 70.6% carbon, 5.88% hydrogen and 25.5% oxygen, by mass. It has a molar mass of 136 g mol^{-1}. When A is refluxed with sodium hydroxide solution, and the resulting liquid is distilled, a liquid B distils over, and a solution of C remains. On addition of dilute hydrochloric acid to C, a white precipitate D forms. When this precipitate is mixed with soda lime and heated, the vapour burns with a smoky flame. B reacts with ethanoyl chloride to give a product with a fruity smell. B does not give a positive result in the iodoform test. Identify A, B and C.

7. A compound contains C, H, N and O. When 0.225 g of the compound was heated with sodium hydroxide solution, the ammonia evolved was passed into 25.0 cm^3 of sulphuric acid of concentration 0.100 mol dm^{-3}. The sulphuric acid that remained required 19.1 cm^3 of 0.100 mol dm^{-3}. sodium hydroxide for neutralisation.

A 0.195 g sample of the compound gave on complete oxidation 0.352 g of carbon dioxide and 0.168 g of water.

A solution of 9.12 g of the compound in 500 cm^3 of water froze at $-0.465\,°C$. The cryoscopic constant for water is 1.86 K kg mol^{-1}.

a) Find the molecular formula of the compound, and b) suggest its identity.

8. Phenol reacts with bromine to form a crystalline product. 25.0 cm^3 of a solution of phenol of concentration 0.100 mol dm^{-3} were added to 30.0 cm^3 of a solution of 0.100 mol dm^{-3} potassium bromate(V). An excess of potassium bromide and hydrochloric acid were added to liberate bromine. The excess bromine was estimated by adding potassium iodide and titrating the iodine displaced with a solution of sodium thiosulphate. 30.0 cm^3 of a 0.100 mol dm^{-3} solution of sodium thiosulphate were required.

a) Find the ratio of moles of Br$_2$: moles of phenol, and b) deduce the equation for the reaction.

9. Measurements on an alkene showed that 100 cm^3 of the gas weighed 0.233 g at 25 °C and 1 atm. 25.0 cm^3 of the alkene reacted with 25.0 cm^3 of hydrogen.

a) Find the molar mass of the alkene, and b) give its molecular formula.

Give the names and structural formulae of two alkenes which have this molecular formula.

10. An organic acid has the percentage composition by mass: C, 41.4%; H, 3.4%; O, 55.2%. A solution containing 0.250 g of the acid, which is dibasic, required 26.6 cm^3 of 0.200 mol dm^{-3} sodium hydroxide solution for neutralisation.

   Calculate: a) the empirical formula, and b) the molecular formula of the acid. c) Give its name and write its structural formula or formulae.

11. An organic liquid contains carbon, hydrogen and oxygen. On oxidation, 0.250 g of the liquid gave 0.595 g of carbon dioxide and 0.304 g of water. When vaporised, 0.250 g of the liquid occupied 131 cm^3 at 200 °C and 1 atm.

   Find: a) the empirical formula, and b) the molecular formula of the liquid. c) Write the structural formulae of compounds with this molecular formula.

12. A is an organic compound with the percentage composition by mass C, 71.1%; N, 10.4%; O, 11.8%; H, 6.7%, and a molar mass of 135 g mol^{-1}.

   On hydrolysis by aqueous sodium hydroxide, A gives an oily liquid, B. B has the percentage composition by mass: C, 77.1%; N, 15.1%; H, 7.5%, and a molar mass of 93 g mol^{-1}. B is basic and gives a precipitate with bromine water.

   Find the molecular formulae for A and B. From their reactions, deduce the identify of A and B.

EXERCISE 66   Questions from A-level Papers

1. An organic compound, W, on analysis, gave C = 40%, H = 8.5%, N = 23.7%, O = 27.1% by mass. Refluxing W with dilute hydrochloric acid produced a compound X which contained 40% carbon by mass and had the general formula $C_nH_{2n}O_2$.

   X was also prepared by the oxidation of a compound Y which had the general formula $C_nH_{2n}O$.

   a) Calculate the empirical formula of W.

   b) Determine the molecular formula of X and hence deduce the molecular formula of W.

   c) Write a balanced equation to represent the reaction which occurred when W was refluxed with dilute hydrochloric acid.

   d) Identify Y giving reasons for your answer.

   e) Give *two* chemical tests to distinguish between Y and butanone ($CH_3CH_2COCH_3$).   (AEB80)

2. An aromatic compound, A, with a relative molecular mass of 197, was found to contain 79.2% carbon, 5.57% hydrogen, 8.12% oxygen and 7.11% nitrogen by weight.

When A was refluxed with aqueous sodium hydroxide and then steam distilled, drops of an oily liquid B separated in the receiver, whilst an aqueous solution of a compound C was left in the flask. B reacted i) with hydrogen chloride to form a crystalline salt D, and ii) with bromine water to form a derivative E with a relative molecular mass of 330.

When the hot aqueous solution of C was acidified with hydrochloric acid, a colourless solid F crystallised out on cooling. On heating F with an excess of soda-lime in a test-tube, a vapour G was expelled and this burnt with a luminous, smoky flame when a lighted splint was applied to the mouth of the test tube.

Calculate the molecular formula of A and identify all the compounds A to G, giving your reasons. In addition, give the structural formula of each compound and write equations for three of the changes which have been described. (O76)

3. An aromatic compound A with a relative molecular mass of 250 contained 33.6% carbon, 2.40% hydrogen and 64.0% bromine by mass. On refluxing A with aqueous sodium hydroxide, half of the bromine was removed and the resulting organic compound, B, was converted into C by the action of a mild oxidising agent. C gave a crystalline derivative with 2,4-dinitrophenylhydrazine and was readily oxidised, even by exposure to the air, to a monobasic acid, D. The latter contained the same number of carbon atoms as C, and 2.01 g of it neutralised 100 cm³ of an aqueous solution of sodium hydroxide containing 0.100 moles per dm³.

On further investigation it was found that A gave two isomers E and F when a radical X was substituted into the aromatic ring in place of one hydrogen atom.

Deduce the structural formulae of the compounds A, B, C, D, E and F, explaining fully your reasoning. (O77)

4. An organic compound A, with a relative molecular mass of 178, contains 74.2% C, 7.9% H, and 17.9% O. Boiling A with aqueous alkali gave a volatile compound B which did not give a positive haloform test. Acidification of the alkaline solution gave C which was soluble in aqueous sodium carbonate solution. A 0.100 g sample of C neutralised 7.35 cm³ of aqueous sodium hydroxide solution (0.100 mol dm⁻³). Reduction of either A or C with lithium tetrahydridoaluminate (LiAlH₄) yielded D which reacted with ethanoic (acetic) anhydride to form E, $C_{10}H_{12}O_2$. Oxidation of D yields benzene-1,4-dicarboxylic acid.

Identify compounds A to E, giving your reasoning. (O & C79)

5. A compound A, whose composition is 80.0% C, 6.67% H and 13.33% O, has a relative molecular mass of about 120. When treated with hydrogen cyanide, it is converted into B (C_9H_9ON). B, on boiling with dilute sulphuric acid, gives C ($C_9H_{10}O_3$) and C, on oxidation, gives an acid D. When A reacts with iodine and sodium hydroxide, it gives tri-iodomethane and the sodium salt of D.

Identify A, B, C and D and give equations for the reactions mentioned. What intermediate compounds would be formed in the conversion of B into C? (C $=$ 12, H $=$ 1.0, O $=$ 16.)   (L76, S)

6. When the neutral compound A, $C_{10}H_{13}NO$, was refluxed with dilute acid it formed two products B, C_2H_7N and C.

On analysis, C was found to contain 70.59% carbon, 23.53% oxygen and 5.88% hydrogen by weight. The relative molecular mass of C was found to be 136.

On reaction with alkaline potassium manganate(VII) (permanganate) solution, C was oxidized to D, $C_8H_6O_4$.

D, which was acidific, was readily dehydrated to the neutral substance E, $C_8H_4O_3$.

B reacted with gaseous hydrogen chloride to form the ionic solid F, C_2H_8NCl. When F was dissolved in dilute hydrochloric acid and sodium nitrite solution added, a yellow oil, G, was formed and no effervescence occurred.

a) What is the empirical formula of C?
b) What is the molecular formula of C?
c) Write the structural formulae for substances A to G.
d) What are the names of substances B, C and F?   (SUJB80)

7. 2.20 g of an ester of a monoalkanoic acid and a monohydric alcohol were refluxed for one hour with 100 cm^3 of an aqueous solution of sodium hydroxide of concentration 1.00 mol/dm^3. After cooling, the volume was made up to 250 cm^3 with pure water. 25.00 cm^3 of the diluted solution reacted with 15.00 cm^3 of dilute hydrochloric acid of concentration 0.50 mol/dm^3, using phenolphthalein as indicator.

a) Assuming that hydrolysis was complete:
   i) determine the molecular formula of the ester;
   ii) write the names and structural formula of *all* of the esters having the molecular formula found in i);
   iii) write the names and structural formulae of the alcohols obtained by hydrolysing each of the esters given in ii).

b) Alcohols can be oxidized and the oxidation products can be distinguished by chemical tests. In this way alcohols can themselves be distinguished from one another.

   For the alcohols given in a) iii), name *one* suitable oxidizing agent which can be used to oxidize *each* of the alcohols; describe the oxidation and the test(s) you would use on the oxidation product in each case.   (AEB78)

8. When 25 cm^3 of a gaseous alkene X are completely burnt in an excess of oxygen, 100 cm^3 of carbon dioxide are produced, both volumes being measured under the same conditions of temperature and pressure. Ozonolysis of X gives methanal (formaldehyde) and Y (empirical formula C$_3$H$_6$O), which does not give a precipitate on boiling with Fehling's solution. Identify X and Y, explaining your reasoning, and write equations for the reactions involved.   (C79, p)

9. Four isomeric organic compounds A, B, C and D, containing carbon, hydrogen, nitrogen and oxygen, were analyzed with the following results.

a) Combustion in excess oxygen of 0.200 g gave 0.362 g of carbon dioxide and 0.173 g of water.

b) The nitrogen contained in 0.300 g, on suitable treatment, was converted into ammonia. This was absorbed in 25.0 cm^3 of sulphuric acid (concentration 0.50 mol l^{-1}) and the resulting solution neutralised by sodium hydroxide solution (concentration 1.0 mol l^{-1}) of which 20.9 cm^3 were required.

c) On hydrolysis, A gave a monocarboxylic acid, B gave a ketone, C gave an aldehyde and D a primary amine.

Assuming that the relative molecular mass of the isomers is less than 100, suggest a structure for each and identify the products of hydrolysis.

Starting with simple organic compounds containing no more than two carbon atoms per molecule, suggest a synthesis for *one* of the isomers, giving only reagents, reaction conditions and the structure of intermediate compounds. (C = 12.0, N = 14.0, O = 16.0, H = 1.00.)

(L78, S)

10. a) 1.73 g of a compound, A, which contains carbon, manganese, nitrogen and oxygen only, was dissolved in distilled water and the volume made up to 250 cm^3.

25 cm^3 of the solution was treated to convert the manganese in the dissolved compound A to manganate(VII) (permanganate). The resulting solution contained sufficient manganate(VII) to react with 50 cm^3 of a solution of iron(II) ammonium sulphate of concentration 0.1 mol/l.

Calculate the percentage of manganese in A.

b) 25 cm^3 of the solution of A, as used in a), was treated to convert the nitrogen in the dissolved compound, A, to ammonia. The ammonia was dissolved in water and the solution titrated with a solution of hydrochloric acid of concentration 0.1 mol/l; 30 cm^3 of the hydrochloric acid solution were required.

Calculate the percentage of nitrogen in A.

c) On heating 1.73 g of the compound, A, in air, all the carbon in it was converted to carbon dioxide. The volume of carbon dioxide was 224 cm³ at s.t.p.

Calculate the percentage of carbon in A.

d) From your calculated percentages of manganese, nitrogen and carbon, and given that its relative molecular mass is 173, determine the molecular formula of A.   (AEB77, S)

11. A compound A reacted with a mixture of sodium nitrite and dilute hydrochloric acid to give a compound B which on separation and purification was found to contain carbon, hydrogen and oxygen only. 1.00 g of B gave 2.24 g of carbon dioxide and 1.07 g of water on complete combustion. On prolonged oxidation with acidified potassium dichromate, B gave an acidic substance C. 0.10 g of the silver salt of C, which contains two silver ions per formula, decomposed on strong heating to give 0.06 g of silver.

Write possible formulae for A, B and C and say how you would expect A to react with a) dilute hydrochloric acid, and b) copper(II) sulphate solution.   (LN77, S)

12. A colourless liquid, of relative molecular mass 106.5, which fumes in moist air, reacts *readily* with water to produce a mixture of hydrochloric acid and a monobasic carboxylic acid. The compound contains 45.1% carbon and 6.6% hydrogen. It also contains oxygen and chlorine. 1.00 g of the compound under suitable conditions, gives 1.35 g of silver chloride. Give its possible structural formulae, and outline the method you would use to prepare *one* of the compounds so shown from the corresponding carboxylic acid.

Describe *briefly*, but giving essential experimental details, how you would prepare (chloromethyl) benzene (benzyl chloride) from methylbenzene (toluene).

Compare the reactivity of the halogen in chlorobenzene, (chloromethyl)benzene and benzene carbonyl chloride (benzoyl chloride) giving reasons for any differences in behaviour. [A_r (H) = 1; A_r (C) = 12; A_r (O) = 16; A_r (Cl) = 35.5; A_r (Ag) = 108.]   (WJEC78)

13. An organic compound, P, which contains carbon, hydrogen, nitrogen and oxygen was analysed with these results:

a) Combustion of 0.322 g of P in excess oxygen gave 0.770 g of carbon dioxide and 0.338 g of water.

b) Under suitable conditions the nitrogen in 0.430 g of P was converted into ammonia. This was absorbed in 25.00 cm³ of 0.5 M sulphuric acid and the solution titrated with 1.0 M sodium hydroxide, of which 21.65 cm³ were required.

When P was distilled with aqueous sodium hydroxide a compound, Q, of relative molecular mass 59, and a white solid, R, were formed. Decarboxylation of R with soda-lime gave a gas S. Combustion of 10 cm³ of S with excess of oxygen gave a mixture of gases whose volume was decreased by 30 cm³ on exposure to aqueous alkali. (All volumes were measured at the same temperature and pressure.)

Deduce from this information the formulae of Q, R and S, and of all the possible isomers of P, explaining your reasoning.

For one of the isomers of P, suggest a method of preparation from organic compounds containing not more than three atoms of carbon per molecule. (C = 12.0; H = 1.00; N = 14.0; O = 16.0.)

(L79, S)

14. a) An organic compound X has the following percentage composition by mass: C, 12.76%; H, 2.12%; Br, 85.11%. A certain volume of sulphur dioxide diffused through a porous plug in 2 minutes 40 seconds. The same volume of X diffused through the same plug, under the same conditions, in 4 minutes 34 seconds.

Calculate the molecular formula of X and write out the structural formulae for isomers of X.

b) Outline: i) how these isomers could be prepared in the laboratory from ethanol, ii) what chemical test you would carry out in order to distinguish between the isomers. (AEB78)

15. When a colourless volatile liquid, A, is treated with phosphorus pentachloride and the product allowed to react with an alcoholic solution of potassium cyanide, a compound, B, is obtained.

On hydrolysis with a dilute acid, B yields a compound C.

A reacts with a small amount of concentrated sulphuric acid to give D, which contains 34.80 per cent oxygen.

On refluxing C for several hours with phosphorus(V) oxide (phosphorus pentoxide) a product, E, is obtained which contains 47.06 per cent oxygen.

A and C react together in the presence of concentrated sulphuric to give F which has the following percentage composition: C = 48.65; H = 8.11; O = 43.24.

F contains two oxygen atoms per molecule. Identify A, B, C, D, E and F, giving equations for the reactions involved. (AEB77, S)

Answers to Exercises

CHAPTER 1

Exercise 1

1. a) 2.3678×10^4 b) 4.376×10^2 c) 1.69×10^{-2} d) 3.45×10^{-4}
   e) 6.72891×10^5
2. a) 5.85×10^4 b) 2.66×10^6 c) 6.35×10^7 d) 1.21×10
   e) 1.34×10^2
3. a) 3.32×10^6 b) 2.72×10^5 c) 1.86×10^{-4} d) 6.44×10^{-5}
   e) 6.11×10^{-3}
4. a) 2.001×10^4 b) 5.648×10^3 c) 1.29×10^5 d) -1.12×10^{-3}
   e) 6.252×10^4
5. a) 4×10^{10} b) 2×10^2 c) 5×10^8 d) 1×10^3
   e) 2×10^{10}
6. a) 3.6753 b) 3.7052 c) -2.8771 d) 1.0033
   e) -5.6356
7. a) 2.862×10^3 b) 1.135 c) 6.969×10^7 d) 3.3791×10^{-7}
   e) 8.7680×10^{-3}
8. a) 4.264×10^{-3} b) 2.867×10^{-7} c) 4.037×10^2 d) 2.055×10^{-10}
   e) 3.781×10^4
9. a) $x = 7$ or -13 b) $x = 4$ or -0.4 c) $y = 5$ d) $z = 14$ or 2
   e) $x = -\frac{1}{2}$ or $-4\frac{1}{2}$

CHAPTER 2

Exercise 2

1. a) 14.3% b) 43.5% c) 29.2% d) 47.4%
   e) 5.70%
2. a) MgO b) $CaCl_2$ c) $FeCl_3$ d) CuS
   e) LiH
3. a) FeO b) Fe_2O_3 c) Fe_3O_4 d) K_2CrO_4
   e) $K_2Cr_2O_7$ f) CH g) C_3H_8
4. a) $a = 5$ b) $b = 6$ c) $c = 2$ d) $d = 3$
   e) $e = 6$ f) $f = 12$
5. a) $C_5H_{10}O$ b) $C_5H_{10}O$ 6. a) C_2H_4O b) C_2H_4O
7. a) $C_5H_{10}O_2$ b) $C_5H_{10}O_2$ 8. $C_9H_{10}O_2$

Exercise 3

1. $9.78\,kg$ 2. 522.7 tonnes   3. $0.1435\,g$ 4. $15.63\,kg$
5. $2.808\,g$ 6. a) 7.00 tonnes   b) 7.24 tonnes   7. $63.1\,g$
8. $36.46\,g$ 9. $4.481\,g$
10. $50.00\,g\ Ca_3(PO_4)_2$ $29.03\,g\ SiO_2$ $4.839\,g\ C$
11. $40\%\ CaCO_3$ $60\%\ MgCO_3$ 12. $3.5000\,g\ NaHCO_3$ $6.5000\,g\ Na_2CO_3$

Exercise 4

1. a) $2\,dm^3$ b) $750\,cm^3$ c) $625\,cm^3$ d) $937.5\,cm^3$ e) $2\,dm^3$
2. $500\,cm^3\ SO_2$ 3. 50% 4. C_3H_8 5. C_4H_6
6. $a = 30\,cm^3$ $b = 40\,cm^3$ 7. d

Exercise 5

1. $267 \, dm^3$ 2. $3.50 \, dm^3$ 3. $1.107 \, g$ 4. $2.388 \, g$
5. $3.646 \, g$ 6. $11.5 \, dm^3$ 7. $2460 \, dm^3$

Exercise 6

1. 92.9%   2. 90.5%   3. 89.0%   4. 91.0%
5. 99.2%

Exercise 7

1. 436 tonnes   2. 46 kg   3. 2.7 kg   4. 304 kg
5. 93.5%

Exercise 8

1. $S_2O_8^{2-}(aq) + 2I^-(aq) \longrightarrow I_2(aq) + 2SO_4^{2-}(aq)$
2. $H_2NSO_3^-(aq) + OH^-(aq) \longrightarrow NH_3(g) + SO_4^{2-}(aq)$
3. $Na_2S_2O_3(aq) + AgCl(s) \longrightarrow NaCl(aq) + NaAgS_2O_3(aq)$
4. $C_6H_8 + 2Br_2 \longrightarrow C_6H_8Br_4$
5. $C_6H_5NH_2 + 3Br_2 \longrightarrow C_6H_2Br_3NH_2 + 3HBr$

Exercise 9

1. a) $m = 2$ $n = 6$ b) $Mg(OH)Cl$ 2. 18   3. 9
4. $A = Ge(C_2H_5)_4$; $B = Ge(C_2H_5)_3Br$; Structure is tetrahedral   5. a
6. c) Sn   d) $7.84 \, g$
7. $x = y = 40 \, cm^3$
8. a) i) $C_2H_6(g) + 3\frac{1}{2}O_2(g) \longrightarrow 2CO_2(g) + H_2O(l)$
   ii) $C_3H_8(g) + 5O_2(g) \longrightarrow 3CO_2(g) + 4H_2O(l)$
   b) i) $-2\frac{1}{2}$ vol   ii) -3 vol
   c) $70 \, cm^3$ ethane + $30 \, cm^3$ propane
9. C_3H_6
10. a) $280 \, cm^3$ b) $360 \, cm^3$ c) C_7H_8 d) Avogadro's law
11. CO 44%; CO_2 20%; H_2 36%   12. b   13. b

CHAPTER 3

Exercise 10

1. a) $5.90 \, g$ b) $5.30 \, g$ c) $9.45 \, g$ d) $42.0 \, g$
   e) $19.1 \, g$
2. a) $0.500 \, dm^3$ b) $0.0750 \, dm^3$ or $75.0 \, cm^3$ c) $11.4 \, cm^3$
   d) $192 \, cm^3$ e) $192 \, cm^3$
3. a) i) $0.325 \, g$ ii) $0.560 \, g$ iii) $0.618 \, g$ iv) $1.43 \, g$
   b) i) $50.0 \, cm^3$ ii) $100 \, cm^3$ iii) $10.0 \, cm^3$ iv) $40.0 \, cm^3$
4. $34.0 \, cm^3$ 5. $4.76 \times 10^{-2} \, mol \, dm^{-3}$ 6. $133 \, cm^3$
7. $400 \, cm^3$ 8. 95.0%   9. 87.6%   10. 90.6%
11. $[Na_2CO_3] = [NaHCO_3] = 0.120 \, mol \, dm^{-3}$ 12. 18.0%   13. 92.2%
14. 3   15. 68.2%   16. 10

Exercise 11

1. a) $+2$ b) $+2$ c) 0   d) $+1$
   e) $+3$ f) $+1$ g) $+3$ h) $+5$
   i) $+5$ j) $+4$ k) 0   l) $+7$
   m) $+2$ n) -1 o) $+3$ p) $+6$
   q) $+5$ r) -1 s) $+6$ t) $+4$
   u) -1 v) $+6$ w) $+4$ x) $+1$
   y) $+4$ z) $+5$

2. a) Sn is oxidised from $+2$ to $+4$; Pb is reduced from $+4$ to $+2$.

b) Mn is oxidised from $+2$ to $+7$; Bi is reduced from $+5$ to $+3$.

c) As is oxidised from $+3$ to $+7$; Mn is reduced from $+7$ to $+2$.

3. a) F is reduced from 0 to -1.

b) Cl disproportionates from 0 to $+5$ and -1.

c) N disproportionates from -3 and $+5$ to -1.

d) Cr is reduced from $+6$ to $+3$.

e) C is oxidised from $+3$ to $+4$.

4. a) -2 b) $+2$

5. a) $IO_4^- + 7I^- + 8H^+ \longrightarrow 4I_2 + 4H_2O$

b) $BrO_3^- + 6I^- + 6H^+ \longrightarrow Br^- + 3I_2 + 3H_2O$

c) $2V^{3+} + H_2O_2 \longrightarrow 2VO^{2+} + 2H^+$

d) $SO_2 + 2H_2O + Br_2 \longrightarrow 4H^+ + SO_4^{2-} + 2Br^-$

e) $4NH_3 + 3O_2 \longrightarrow 2N_2 + 6H_2O$

f) $2NH_4 + 2O_2 \longrightarrow N_2O + 3H_2O$

g) $4NH_3 + 5O_2 \longrightarrow 4NO + 6H_2O$

h) $Fe^{2+}C_2O_4^{2-} + 3Ce^{3+} \longrightarrow 2CO_2 + 3Ce^{2+} + Fe^{3+}$

6. $Cr_2O_7^{2-} + 6I^- + 14H^+ \longrightarrow 3I_2 + 2Cr^{3+} + 7H_2O$

Exercise 12

1. a) $NO_2^- + H_2O \longrightarrow NO_3^- + 2H^+ + 2e^-$

b) $AsO_3^{3-} + H_2O \longrightarrow AsO_4^{3-} + 2H^+ + 2e^-$

c) $Hg_2^{2+} \longrightarrow 2Hg^{2+} + 2e^-$

d) $H_2O_2 \longrightarrow 2H^+ + 2e^- + O_2$

e) $V^{3+} + H_2O \longrightarrow VO^{2+} + 2H^+ + e^-$

2. a) $NO_3^- + 2H^+ + e^- \longrightarrow NO_2 + H_2O$

b) $NO_3^- + 4H^+ + 3e^- \longrightarrow NO + 2H_2O$

c) $NO_3^- + 10H^+ + 8e^- \longrightarrow NH_4^+ + 3H_2O$

d) $2BrO_3^- + 12H^+ + 10e^- \longrightarrow Br_2 + 6H_2O$

e) $PbO_2 + 4H^+ + 2e^- \longrightarrow Pb^{2+} + 2H_2O$

3. a) $2MnO_4^-(aq) + 5H_2O_2(aq) + 6H^+(aq) \longrightarrow 5O_2(g) + 2Mn^{2+}(aq) + 8H_2O(l)$

b) $MnO_2(s) + 4H^+(aq) + 2Cl^-(aq) \longrightarrow Mn^{2+}(aq) + Cl_2(g) + 2H_2O(l)$

c) $2MnO_4^-(aq) + 5C_2O_4^{2-}(aq) + 16H^+(aq) \longrightarrow 2Mn^{2+}(aq) + 10CO_2(g) + 8H_2O(l)$

d) $Cr_2O_7^{2-}(aq) + 3C_2O_4^{2-}(aq) + 14H^+(aq) \longrightarrow 2Cr^{3+}(aq) + 6CO_2(g) + 7H_2O(l)$

e) $Cr_2O_7^{2-}(aq) + 6I^-(aq) + 14H^+(aq) \longrightarrow 2Cr^{3+}(aq) + 3I_2(aq) + 7H_2O(l)$

f) $H_2O_2(aq) + NO_2^-(aq) \longrightarrow NO_3^-(aq) + H_2O(l)$

4. a) 1.5×10^{-2} mol   b) 7.5×10^{-3} mol   c) 7.5×10^{-3} mol   d) 7.5×10^{-3} mol

e) 1.5×10^{-2} mol

5. a) 6.0×10^{-4} mol   b) 3.0×10^{-4} mol   c) 6.0×10^{-4} mol   d) 3.0×10^{-4} mol

e) 3.0×10^{-4} mol

6. a) 4.0×10^{-3} mol   b) 2.0×10^{-3} mol   c) 2.0×10^{-3} mol   d) 4.0×10^{-3} mol

e) 6.7×10^{-4} mol

7. a) $62.5 \, cm^3$ b) $250 \, cm^3$ c) $5.00 \, cm^3$ d) $12.5 \, cm^3$

e) $8.3 \, cm^3$

8. a) $45.0 \, cm^3$ b) $12.0 \, cm^3$ c) $7.2 \, cm^3$ d) $4.50 \, cm^3$

e) $9.0 \, cm^3$

9. $0.090 \, mol \, dm^{-3}$ **10.** $0.0195 \, mol \, dm^{-3}$ **11.** $0.0447 \, mol \, dm^{-3}$ **12.** 99.5%

13. 90.6%   **14.** $1.64 \times 10^{-2} \, mol \, dm^{-3}$ **15.** $0.103 \, mol \, dm^{-3}$

16. (a) $[Fe^{2+}] = 0.0600 \, mol \, dm^{-3}$ b) $[Fe^{3+}] = 0.0160 \, mol \, dm^{-3}$

17. a) $20.0 \, cm^3$ b) $22.4 \, cm^3$ **18.** $7.63 \times 10^{-2} \, mol \, dm^{-3}$

19. $+4$ **20.** $7.2 \times 10^{-3} \, mol \, dm^{-3}$

21. a) $1NH_2OH : 2Fe^{3+}$ b) -1 c) -1 d) $+1$

e) N_2O f) $2NH_2OH + 4Fe^{3+} \longrightarrow 4Fe^{2+} + N_2O + H_2O + 4H^+$

22. 82.3%   **23.** b) $0.74 \, mol \, dm^{-3}$

Exercise 13

1. $9.37 \times 10^{-3} \, mol \, dm^{-3}$ **2.** 78 ppm

3. 60.0% CaO   40.0% MgO   **4.** 18

Exercise 14

1. $1.48 \times 10^{-2} \, mol \, dm^{-3}$ 2. 49.7%
3. 34.6% NaCl   65.4% NaBr   4. $2.77 \times 10^{-2} \, mol \, dm^{-3}$
5. a) $1.58 \times 10^{-2} \, mol \, dm^{-3}$ b) $4.97 \times 10^{-2} \, mol \, dm^{-3}$
6. 55.3%

Exercise 15

1. $2.24 \, cm^3$ 2. a) 2nd and 3rd   b) 52.3%   3. d
4. $ICl_x + 3I^- \longrightarrow 2I_2 + xCl^-; \; +3$
5. a) i) 0.010   ii) 0.025   iii) 2.5
   b) i) $+2 \longrightarrow 0$ ii) $0 \longrightarrow +5$ iii) 3
   c) $2P + 6H_2O + 5CuSO_4 \longrightarrow 5Cu + 2HPO_3 + 5H_2SO_4$
6. a) $UOC_2O_4 \cdot 4H_2O$ b) U is oxidised to $+8$ 7. a
8. a) $2Fe^{3+} : 1NH_3OH^+$ b) $4Fe^{3+} + 2NH_3OH^+ \longrightarrow 4Fe^{2+} + N_2O + H_2O + 6H^+$
9. b) i) $Cr(OH)_3(s) \longrightarrow CrO_4^-(aq) \longrightarrow Cr_2O_7^{2-}(aq)$
      ii) $+1$ iii) $[Cr(H_2O)_4Cl_2]^+$
10. A, $2.10 \times 10^{-2} \, mol \, dm^{-3}$ B, $1.76 \times 10^{-2} \, mol \, dm^{-3}$
    $6Fe^{2+} + ClO_3^- + 6H^+ \longrightarrow 6Fe^{3+} + Cl^- + 3H_2O$

CHAPTER 4

Exercise 16

1. $85.6 \, m_u$ 2. $69.8 \, m_u$ 3. 24.3
4. $3 \, ^{35}Cl : 1 \, ^{37}Cl; \; 35.5 \, m_u$ 5. 6.93
6. $1, ^1H$ $2, ^2H$ $3, ^1H^2H$ $4, ^2H_2$ $17 \, ^{16}O^1H$ $18, ^{16}O^2H$ and $^1H_2 \, ^{16}O$ $19, ^1H^2H^{16}O$
   $20, ^2H_2 \, ^{16}O$
7. $39.1 \, m_u$
8. ^{63}Cu ^{65}Cu ^{63}CuO ^{65}CuO $^{63}CuNO_3$ $^{65}CuNO_3$ $^{63}Cu(NO_3)_2$ $^{65}Cu(NO_3)_2$

Exercise 17

1. a) Mean $A_r = 80$ b) isotopes are 79, 81, in relative abundance $1:1$
2. 35.5
3. b) 63.6
4. $Ne = 20.2$ The peaks at 32 and multiples of 32 are due to $S, S_2, S_3, S_4, S_5, S_6, S_7$ and S_8
   There is no evidence here for any isotope other than ^{32}S

Exercise 18

1. a) $a = 1, b = 0$ b) $a = 17, b = 8$ c) $a = 4, b = 2$ d) $a = 210, b = 83$
   e) $a = 4, b = 2$ f) $a = 4, b = 2$ g) $a = 14, b = 6$ h) $a = 16, b = 7$
   i) $a = 207, b = 83$ j) $a = 4, b = 11$ k) $q = 0, p = 1, r = 35$
   l) $c = 3, d = 1$

Exercise 19

1. $a = 24$ $b = 11$ $Z = Na$
2. a) $^{27}_{13}Al + {}^1_1H \longrightarrow {}^{24}_{12}Mg + {}^4_2He$ b) $^{32}_{16}S + {}^1_0n \longrightarrow {}^1_1H + {}^{32}_{15}P$
3. b) (1) is $_{-1}^0 e$ (2) is ^4_2He Pb is in Gp 4, X in Gp 5, Y in Gp 3 and Z in Gp 4
4. a) $P = {}^{230}_{90}Th$ $Q = {}^{239}_{93}Np$ $R = {}^{235}_{92}U$ $S = {}^1_1H$ b) $X = {}_{-1}^0 e$

CHAPTER 5

Exercise 20

1. a) $187 \, cm^3$ b) $387 \, cm^3$ c) $6.23 \, dm^3$ d) $132 \, cm^3$
   e) $2.43 \, dm^3$

2. $1.75 \times 10^5 \, \mathrm{N\,m^{-2}}$ **3.** $0.943 \, \mathrm{dm^3}$ **4.** 484 K   **5.** $586 \, \mathrm{cm^3}$
6. a) $185 \, \mathrm{cm^3}$ b) $36.7 \, \mathrm{cm^3}$ c) $6.46 \, \mathrm{dm^3}$ d) $3.83 \, \mathrm{dm^3}$
e) $436 \, \mathrm{cm^3}$

Exercise 21

1. a) 46   b) NO_2 **2.** 59 cm   **3.** 160
4. 4   **5.** $A = H_2$ $B = O_3$ **6.** 48 s
7. 65.3   59% of the NO_2 molecules are dimerised   **8.** 100.5 s
9. $16.3 \, \mathrm{cm^3\,min^{-1}}$ **10.** $24.9 \, \mathrm{cm^3}$ **11.** $44 \, \mathrm{g\,mol^{-1}}$
12. CO, 25% CO_2, 75%   **13.** NO_2, 43.7% N_2O_4, 56.3%

Exercise 22

1. $44.1 \, \mathrm{g\,mol^{-1}}$ **2.** $39.9 \, \mathrm{g\,mol^{-1}}$ **3.** $6.01 \, \mathrm{dm^3}$ **4.** $83.8 \, \mathrm{g\,mol^{-1}}$
5. 0.583 mol   **6.** $44.0 \, \mathrm{g\,mol^{-1}}$ **7.** $6.18 \, \mathrm{dm^3}$ **8.** 7.54×10^{-2} mol

Exercise 23

1. a) $176.5 \, \mathrm{cm^3}$ b) $207 \, \mathrm{cm^3}$ c) $26.3 \, \mathrm{dm^3}$
2. $2.73 \times 10^5 \, \mathrm{N\,m^{-2}}$ **3.** $2.53 \times 10^4 \, \mathrm{N\,m^{-2}}$ **4.** $4.50 \times 10^5 \, \mathrm{N\,m^{-2}}$
5. a) $p(N_2) = 3.00 \times 10^4 \, \mathrm{N\,m^{-2}}$ $p(O_2) = 2.63 \times 10^4 \, \mathrm{N\,m^{-2}}$, $p(CO_2) = 1.88 \times 10^4 \, \mathrm{N\,m^{-2}}$
b) $p(N_2) = 3.00 \times 10^4 \, \mathrm{N\,m^{-2}}$ $p(O_2) = 2.63 \times 10^4 \, \mathrm{N\,m^{-2}}$
6. a) $p(NH_3) = 6.00 \times 10^4 \, \mathrm{N\,m^{-2}}$ $p(H_2) = 3.75 \times 10^4 \, \mathrm{N\,m^{-2}}$ $p(N_2) = 5.25 \times 10^4 \, \mathrm{N\,m^{-2}}$
b) $p(H_2) = 3.75 \times 10^4 \, \mathrm{N\,m^{-2}}$ $p(N_2) = 5.25 \times 10^4 \, \mathrm{N\,m^{-2}}$

Exercise 24

1. a) $67.25 \, \mathrm{cm^3}$ b) $16.7\% \, N_2$ $8.9\% \, O_2$ $74.3\% \, CO_2$
2. a) $48.12 \, \mathrm{cm^3}$ b) $64.3\% \, N_2$ $34.0\% \, O_2$ $1.63\% \, CO_2$
3. $3.7\% \, CO$ $96.3\% \, CO_2$ **4.** $95\% \, Cl_2$ $5\% \, O_2$

Exercise 25

1. $84.4 \, \mathrm{g\,mol^{-1}}$ **2.** $1840 \, \mathrm{m\,s^{-1}}$ **3.** $3410 \, \mathrm{J\,mol^{-1}}$ **4.** $413 \, \mathrm{m\,s^{-1}}$
5. $44.1 \, \mathrm{g\,mol^{-1}}$ **6.** 2.02   **7.** $242 \, \mathrm{m\,s^{-1}}$
8. a) 4.47   b) 5460 K

Exercise 26

1. b) i) $1.9 \times 10^{-6} \, \mathrm{mol\,year^{-1}}$ ii) $6.1 \times 10^{23} \, \mathrm{mol^{-1}}$ c) $^{224}_{86}$ Rn
2. H_2, 40 kPa   He, 60 kPa
3. b) i) Y   ii) Cl_2 c)

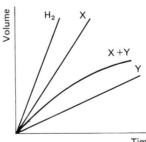

4. b) i) $1.22 \times 10^5 \, \mathrm{N\,m^{-2}}$ ii) I_2 is 40.2% dissociated   **5.** $750 \, \mathrm{cm^3\,dm^{-3}}$
6. $19.3\% \, O_2$ $80.0\% \, N_2$ $0.73\% \, CO_2$ **7.** a) 89   b) 759 K
8. a) 0.70 atm or 70.9 kPa   b) ii) 1 volume N_2 2 volumes H_2 4 volumes O_2

CHAPTER 6

Exercise 27

1. $343\,\mathrm{g\,mol^{-1}}$ 2. PF_5 3. $90\,\mathrm{g\,mol^{-1}}$ 4. $64.6\,\mathrm{g\,mol^{-1}}$
5. $156\,\mathrm{g\,mol^{-1}}$ 6. $85\,\mathrm{g\,mol^{-1}}$ 7. $134\,\mathrm{g\,mol^{-1}}$
8. a) $46\,\mathrm{g\,mol^{-1}}$ b) $58\,\mathrm{g\,mol^{-1}}$ c) $74\,\mathrm{g\,mol^{-1}}$

Exercise 28

1. 0.400   2. 0.500   3. 73.9%   4. 0.300
5. 0.68

Exercise 29

1. $3.46\,\mathrm{g}$ 2. $1.98\,\mathrm{g}$ 3. $1.62 \times 10^{17}\,\mathrm{cm^{-3}}$ 4. $3.17\,\mathrm{kPa}$

Exercise 30

1. $20\,690\,\mathrm{N\,m^{-2}}$ 2. a) $40\,000\,\mathrm{N\,m^{-2}}$ b) 0.300
3. a) $1.60 \times 10^3\,\mathrm{Pa}$ b) $6.38 \times 10^2\,\mathrm{Pa}$ 4. a) $38.0\,\mathrm{kPa}$ b) $33.6\,\mathrm{kPa}$

Exercise 31

1. $156\,\mathrm{g\,mol^{-1}}$ 2. 26.9%   3. $182\,\mathrm{g\,mol^{-1}}$ 4. 27.8%
5. $32.3\,\mathrm{g}$ 6. $123\,\mathrm{g\,mol^{-1}}$

Exercise 32

1. $13.6\,\mathrm{g}$ 2. $4.76\,\mathrm{g}$ 3. 20.0   4. $0.0514\,\mathrm{mol}$
5. a) $4.74\,\mathrm{g}$ b) $4.95\,\mathrm{g}$ 6. 25.0   7. 1.18×10^{-2}
8. a) $3.0\,\mathrm{g}$ b) $3.5\,\mathrm{g}$ 9. a) $4.75\,\mathrm{g}$ b) $4.95\,\mathrm{g}$
10. $Cu(NH_3)_4^{2+}$

Exercise 33

1. b) 0.375   2. b) 0.400   3. $340.6\,\mathrm{Pa}$
4. 22% ethanol   78% methanol   5. a) 1.67×10^{25} b) 4.31×10^{20}
6. b) i) 0.59   ii) $258\,\mathrm{cm^3\,N_2O_4}$ $742\,\mathrm{cm^3\,NO_2}$ iii) $p(N_2O_4) = 26.1\,\mathrm{kPa}$
    iii) $p(N_2O_4) = 26.1\,\mathrm{kPa}$ $p(NO_2) = 75.2\,\mathrm{kPa}$
7. $M_r = 325$ (dimerised)
8. b) i) 200   ii) $4\,\mathrm{mol\,AlCl_3} : 1\,\mathrm{mol\,Al_2Cl_6}$
    iii) M_r (apparent) will decrease with an increase in temperature and increase with an
    increase in pressure
9. c) 5×10^{13} 10. b) 20.1%   11. c) 0.20   12. b) 4
13. c) i) $4.80\,\mathrm{g}$ ii) $5.33\,\mathrm{g}$ 14. $1.23 \times 10^{-3}\,\mathrm{mol\,dm^{-3}}$
15. $103\,\mathrm{mg}$

CHAPTER 7

Exercise 34

1. a) $2.33 \times 10^3\,\mathrm{N\,m^{-2}}$ b) $2.30 \times 10^3\,\mathrm{N\,m^{-2}}$ c) $2.29 \times 10^3\,\mathrm{N\,m^{-2}}$
    d) $1.80 \times 10^3\,\mathrm{N\,m^{-2}}$
2. e   3. $63.7\,\mathrm{g\,mol^{-1}}$ 4. $3.09 \times 10^3\,\mathrm{Pa}$ 5. $59.8\,\mathrm{g\,mol^{-1}}$
6. $73.5\,\mathrm{g\,mol^{-1}}$ 7. $0.342\,\mathrm{mol\,dm^{-3}}$

Exercise 35

1. a) $100.21\,^\circ\mathrm{C}$ b) $100.31\,^\circ\mathrm{C}$ c) $100.59\,^\circ\mathrm{C}$ d) $100.12\,^\circ\mathrm{C}$
    e) $100.35\,^\circ\mathrm{C}$ f) 100.53
2. a) $60.0\,\mathrm{g\,mol^{-1}}$ b) $342\,\mathrm{g\,mol^{-1}}$ c) $73\,\mathrm{g\,mol^{-1}}$ d) $90\,\mathrm{g\,mol^{-1}}$
    e) $180\,\mathrm{g\,mol^{-1}}$

Exercise 36

1. a) $52.5\,\mathrm{g\,mol^{-1}}$ b) $80.0\,\mathrm{g\,mol^{-1}}$ c) $59.9\,\mathrm{g\,mol^{-1}}$ d) $180\,\mathrm{g\,mol^{-1}}$
    e) $342\,\mathrm{g\,mol^{-1}}$
2. $180\,\mathrm{g\,mol^{-1}}$

Exercise 37

1. a) $0.36\,^{\circ}\mathrm{C}$ b) $82.6\,^{\circ}\mathrm{C}$ c) $1.23 \times 10^{4}\,\mathrm{N\,m^{-2}}$
2. a) $100.13\,^{\circ}\mathrm{C}$ b) $-0.47\,^{\circ}\mathrm{C}$ c) $2.33 \times 10^{3}\,\mathrm{N\,m^{-3}}$
3. $146\,\mathrm{g\,mol^{-1}}$
4. a) $-1.86\,^{\circ}\mathrm{C}$ $100.52\,^{\circ}\mathrm{C}$ b) $-0.37\,^{\circ}\mathrm{C}$ $100.10\,^{\circ}\mathrm{C}$
    c) $-0.37\,^{\circ}\mathrm{C}$ $100.10\,^{\circ}\mathrm{C}$ d) $-2.07\,^{\circ}\mathrm{C}$ $100.58\,^{\circ}\mathrm{C}$
    e) $-4.04\,^{\circ}\mathrm{C}$ $101.13\,^{\circ}\mathrm{C}$ 5. $4.00\,\mathrm{kg}$

Exercise 38

1. a) $5.19 \times 10^{5}\,\mathrm{N\,m^{-2}}$ b) $1.98 \times 10^{6}\,\mathrm{N\,m^{-2}}$ c) $7.87 \times 10^{5}\,\mathrm{N\,m^{-2}}$
    d) $1.41 \times 10^{6}\,\mathrm{N\,m^{-2}}$ e) $2.44 \times 10^{6}\,\mathrm{N\,m^{-2}}$
2. a) $68.6\,\mathrm{g\,mol^{-1}}$ b) $166\,\mathrm{g\,mol^{-1}}$ c) $259\,\mathrm{g\,mol^{-1}}$
3. $-0.58\,^{\circ}\mathrm{C}$ 4. 50 5. $2\,400$
6. $1.66 \times 10^{4}\,\mathrm{g\,mol^{-1}}$ 7. a) $2.01 \times 10^{5}\,\mathrm{Pa}$ b) $5.00\,\mathrm{g\,dm^{-3}}$ 8. $8.38 \times 10^{5}\,\mathrm{N\,m^{-2}}$

Exercise 39

1. $-0.818\,^{\circ}\mathrm{C}$ $100.229\,^{\circ}\mathrm{C}$ 2. $0.578\,\mathrm{mol\,dm^{-3}}$ 3. $0.439\,\mathrm{mol\,dm^{-3}}$
4. $3.70\,\mathrm{g}$ 5. $101.793\,^{\circ}\mathrm{C}$ 6. 0.0215 7. 0.785
8. 0.779 9. $-0.0772\,^{\circ}\mathrm{C}$ 10. $C_6H_4Br_2$ 11. 0.763
12. $-0.389\,^{\circ}\mathrm{C}$ 13. $C_7H_6O_2$ in water $C_{14}H_{12}O_4$ in benzene
14. $0.333\,\mathrm{kg}$ per kg water   15. $3.16 \times 10^{3}\,\mathrm{Pa}$
16. $M_r = 74$ in water, 146 in benzene   The acid is almost completely dimerised in benzene
    Degree of dimerisation $= 0.99$
17. a) $80\,\mathrm{g\,mol^{-1}}$ b) $5.66 \times 10^{5}\,\mathrm{Pa}$ 18. $1.95 \times 10^{5}\,\mathrm{N\,m^{-2}}$

Exercise 40

1. b) 122 from neutralisation 244 from b.p. of solution in benzene 122 from b.p. of
    aqueous solution   The acid is dimerised in benzene, and is only slightly ionised in
    water
2. c) Benzoic acid is 88% dimerised in solution in benzene   3. $M_r = 124\,(P_4)$
4. a) $AlCl_3$ is covalent   NaCl is ionic
    b) $AlCl_3$ is 95% dimerised in benzene
    c)

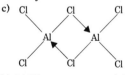

5. b) 14.9% c) 51 6. $1.24 \times 10^{3}\,\mathrm{N\,m^{-2}}$
7. a) i) 148 ii) $73\,\mathrm{g\,mol^{-1}}$ b) $C_2H_5CO_2H$ c) 0.014

CHAPTER 8

Exercise 41

1. $0.265\,\mathrm{g}$
2. $0.403\,\mathrm{g}$ a) doubled   b) doubled   c) unchanged
3. $1.24\,\mathrm{g\,Ca}$ $2.21\,\mathrm{g\,Cl_2}$ 4. $0.0560\,\mathrm{g}$
5. a) $0.0672\,\mathrm{A}$ b) $23.1\,\mathrm{cm^3}$ c) $0.195\,\mathrm{g}$ 6. 268 minutes
7. $0.454\,\mathrm{A}$ 8. $1.77\,\mathrm{mol\,dm^{-3}}$ 9. 2482 hours
10. $2.14\,\mathrm{dm^3\,O_2}$ $4.28\,\mathrm{dm^3\,H_2}$ 11. $1.84 \times 10^{4}\,\mathrm{C}$ 12. $1050\,\mathrm{s}$
13. $0.292\,\mathrm{dm^3}$

Exercise 42

1. a) $0.100 \, \Omega \, m$ or $10.0 \, \Omega \, cm$ b) $10.0 \, \Omega^{-1} m^{-1}$ $0.100 \, \Omega^{-1} cm^{-1}$
2. a) $0.060 \, \Omega \, m$ or $6.00 \, \Omega \, cm$ b) $16.0 \, \Omega^{-1} m^{-1}$ $0.160 \, \Omega^{-1} cm^{-1}$
3. $1.77 \, m^{-1}$ or $1.77 \times 10^{-2} cm^{-1}$
4. $1.63 \times 10^{-2} \, \Omega^{-1} m^2 mol^{-1}$ or $163 \, \Omega^{-1} cm^2 mol^{-1}$
5. $0.210 \, \Omega^{-1} m^2 mol^{-1}$

Exercise 43

1. $1.75 \times 10^{-5} \, mol \, dm^{-3}$
2. a) 0.0271 b) $6.78 \times 10^{-4} \, mol \, dm^{-3}$ c) $1.89 \times 10^{-5} \, mol \, dm^{-3}$
3. a) 0.230 b) $1.37 \times 10^{-3} \, mol \, dm^{-3}$
4. $1.75 \times 10^{-4} \, mol \, dm^{-3}$
5. a) 0.0256 b) $2.02 \times 10^{-5} \, mol \, dm^{-3}$

Exercise 44

1.

	pH	pOH		pH	pOH		pH	pOH		pH	pOH
a)	8	6	b)	4	10	c)	7	7	d)	2.2	11.8
e)	4.5	9.5	f)	1.5	12.5	g)	0.60	13.4	h)	8.3	5.7
i)	6.2	7.8	j)	1.0	13.0						

2. a) 12 b) 11 c) 6.0 d) 12.7
   e) 11 f) 12.9 g) 12 h) 9.7
   i) 6.8 j) 4.6

3. In $mol \, dm^{-3}$, the values are:
   a) 1.00 b) 5.01×10^{-5} c) 4.47×10^{-3} d) 0.0132
   e) 7.08×10^{-5} f) 1.45×10^{-8} g) 6.17×10^{-10} h) 2.00×10^{-14}
   i) 3.16×10^{-1} j) 2.34×10^{-3}

4. a) 0.784 b) 1.05 c) 13.3 d) 12.7
   e) 13.4

5. 2.52 6. a) $1.00 \times 10^{-6} \, mol \, dm^{-3}$ b) 6.00

7. 9.92

8. a) 2.04×10^{-4} b) 1.77×10^{-5} c) 3.96×10^{-10} d) 1.84×10^{-4}
   e) 3.37×10^{-2} (all in $mol \, dm^{-3}$)

9. a) 1.81×10^{-5} b) 3.97×10^{-10} c) 1.38×10^{-3} d) 1.43×10^{-5}
   e) 2.00×10^{-9} (all in $mol \, dm^{-3}$)

10. a) 2.28×10^{-11} b) 5.62×10^{-10} c) 1.86×10^{-11} d) 1.44×10^{-11}
    e) 4.24×10^{-10} (all in $mol \, dm^{-3}$)

11. a) 4.46×10^{-4} b) 1.89×10^{-2} c) 2.41×10^{-2} d) 1.89×10^{-2}
    e) 7.93×10^{-2} (all in $mol \, dm^{-3}$)

12. a) 2.00 b) 12.0 c) 2.30 13. 0.0110%

14. 11.1 15. a) 10^{-6} b) 6

16. pH $= 3.0$ a) $9.0 \, cm^3$ b) $0.90 \, cm^3$

Exercise 45

1. a   2. 1.00 mole   3. a) 3.34 b) 3.94
4. a) 4.73 b) 0.117 mol   5. 2.86
6. a) A; $4.75 \longrightarrow 4.66$ B: $4.45 \longrightarrow 4.31$
        C: $4.05 \longrightarrow 3.71$ D: $4.75 \longrightarrow 4.71$
   b) Buffering capacity decreases as [Salt]/[Acid] decreases, and increases as concentration increases
7. a) 2.40 b) 3.49 c) 3.75
8. a) 2.72 b) 4.44 c) 4.38

Exercise 46

1. a) 5.13 b) 6.32 c) 11.2 d) 8.38
   e) 8.38

2. a) 8.72   b) 4.93   c) 11.35   d) 8.32
    e) 5.08   f) 0.569 mol dm^{-3}
3. a) 8.88   b) 8.03   c) 10.81   d) 5.91
    e) 3.72   f) 2.96

Exercise 47

1. 3.0   **2.** 4.1 × 10^{-17} mol dm^{-3}
3. 1.1 × 10^{-3} mol dm^{-3} **4.** 2.8 × 10^{-7} mol^4 dm^{-12}
5. 4.7 × 10^{-13} mol^4 dm^{-12}

Exercise 48

1. a) 1.69 × 10^{-28} mol^2 dm^{-6} b) 3.99 × 10^{-19} mol^2 dm^{-6}
    c) 5.93 × 10^{-51} mol^3 dm^{-9} d) 1.08 × 10^{-93} mol^5 dm^{-15}
    e) 5.00 × 10^{-16} mol^3 dm^{-9}
2. a) 2.52 × 10^{-27} mol^2 dm^{-6} b) 8.29 × 10^{-17} mol^2 dm^{-6}
    c) 1.07 × 10^{-10} mol^2 dm^{-6} d) 2.68 × 10^{-11} mol^3 dm^{-6}
    e) 1.25 × 10^{-16} mol^2 dm^{-6}
3. a) 6.3 × 10^{-34} mol dm^{-3} b) 6.3 × 10^{-43} mol dm^{-3}
    c) 3.2 × 10^{-14} mol dm^{-3} d) 2.2 × 10^{-30} mol dm^{-3}
    e) 1.6 × 10^{-46} mol dm^{-3} f) 6.3 × 10^{-12} mol dm^{-3}
4. a) Yes   b) No   c) Yes   d) No
    e) Yes   f) No
5. CdS and NiS
6. K_{sp}(AgCl) = 1.96 × 10^{-10} mol^2 dm^{-6}
    K_{sp}(Ag$_2$CrO$_4$) = 3.61 × 10^{-12} mol^3 dm^{-9}
    a) [Ag$^+$] = 1.96 × 10^{-9} mol dm^{-3} b) [Ag$^+$] = 2.69 × 10^{-5} mol dm^{-3}
7. CaCO$_3$ 9.99 g   **8.** 0.3 ion dm^{-3} **9.** PbI$_2$ 45.6 g
10. a) 8.12 × 10^{-3} g dm^{-3} b) 6.38 × 10^{-8} g dm^{-3} c) 9.63 × 10^{-4} g dm^{-3}
11. a) 6.3 × 10^{-4} mol dm^{-3} b) 4.0 × 10^{-6} mol dm^{-3}
12. a) 1.2 × 10^{-3} b) 1.8 × 10^{-7} mol dm^{-3}
13. a) 2.15 × 10^{-4} mol dm^{-3} b) 4.00 × 10^{-5} mol dm^{-3}
    c) 7.12 × 10^{-8} mol dm^{-3}

Exercise 49

1. Ag, I$^-$ **2.** I$_2$, Fe^{3+}
3. a) +0.40 V   b) +0.26 V   c) −0.27 V
4. a) +0.94 V   b) +0.44 V   c) −0.67 V   d) +0.78 V
    e) +0.63 V
5. Fe(s) + Fe^{3+}(aq) ⟶ 2Fe^{2+}(aq)
6. Cr$_2$O$_7$$^{2-}$(aq) + 14H$^+$(aq) + 6Fe^{3+}(aq) ⟶ 2Cr^{3+}(aq) + 6Fe^{3+}(aq) + 14H$_2$O
7. a) 2Fe^{3+}(aq) + 2I$^-$(aq) ⟶ 2Fe^{2+}(aq) + I$_2$(aq)
    b) 2Ag$^+$(aq) + Cu(s) ⟶ 2Ag(s) + Cu^{2+}(aq)
    c) No reaction   d) No reaction
    e) Br$_2$(aq) + 2Fe^{2+}(aq) ⟶ 2Br$^-$(aq) + 2Fe^{3+}(aq)
        i) Cl$_2$ and Br$_2$ ii) Cl$_2$, Br$_2$ and I$_2$

Exercise 50

1. a) 0.0790 g   b) 13.9 cm^3 **2.** b) 5 min   d) 1.60 × 10^{-19} C
3. d) i) 390.6 ohm^{-1} cm^2 mol^{-1} ii) α = 0.134, K = 2.07 × 10^{-5}, pH = 3.87
    f) from 1:10 to 10:1 over pH 3.68 to 5.68
4. b) i) 151 Ω^{-1} cm^2 mol^{-1} ii) 0.0159 mol dm^{-3}
5. a) 5.45 × 10^{-4} mol dm^{-3} 11.7   **6.** b) 4.71 × 10^{-6} mol dm^{-3}
7. b) 6.5   c) 8.72
8. 7.2 cm^3 Na$_2$HPO$_4$ 2.8 cm^3 KH$_2$PO$_4$ pH = 7.22

9. c) i) Negative   ii) Yes   iii) No   iv) Zn negative, e.m.f. $= -0.54$ V
$$Zn(s) + Ni^{2+}(aq) \longrightarrow Zn^{2+}(aq) + Ni(s) \qquad\qquad\qquad d) 6.30 \, g \, s^{-1}$$
10. a) 5   b) i) $2 \times 10^{-12} \, mol \, dm^{-3}$ ii) Yes
11. b) i) $4.5 \times 10^{-3} \, mol \, l^{-1}$ ii) $2.0 \times 10^{-4} \, mol \, l^{-1}$ iii) $6.7 \times 10^{-5} \, mol \, l^{-1}$
12. b) i) $1.59 \times 10^{-2} \, mol \, l^{-1}$ ii) $4 \times 10^{-3} \, mol$ d) iii) $1.23 \times 10^{-15} \, mol \, l^{-1}$
13. b) $1.72 \times 10^{-10} \, mol^2 \, dm^{-6}$
14. a) $1.91 \times 10^{-6} \, mol^3 \, dm^{-9}$ b) $7.46 \times 10^{-2} \, g \, dm^{-3}$

CHAPTER 9

Exercise 51

1. $-52.9 \, kJ \, mol^{-1}$ 2. $-58.4 \, kJ \, mol^{-1}$ 3. $-49.3 \, kJ \, mol^{-1}$ 4. $-56.4 \, kJ \, mol^{-1}$
5. $-59.1 \, kJ \, mol^{-1}$

Exercise 52

1. a) $-3.23 \, MJ \, mol^{-1}$ b) $-3.24 \, MJ \, mol^{-1}$ 2. a) $-9.91 \, kJ \, mol^{-1}$ b) $-3520 \, kJ \, mol^{-1}$
3. a) $1.24 \, kJ \, mol^{-1}$ b) $\Delta U^{\ominus} - \Delta H^{\ominus} = 1.24 \, kJ \, mol^{-1}$
4. a) $-4210 \, kJ \, mol^{-1}$ b) $-2220 \, kJ \, mol^{-1}$

Exercise 53

1. a) $-484 \, kJ \, mol^{-1}$ b) $108 \, kJ \, mol^{-1}$ c) -75 and $+33 \, kJ \, mol^{-1}$
   d) $-118 \, kJ \, mol^{-1}$ e) $+11 \, kJ \, mol^{-1}$ f) -106 and $+504 \, kJ \, mol^{-1}$
   g) $-251, -246, -471$ and $-152 \, kJ \, mol^{-1}$
2. $-78.2 \, kJ \, mol^{-1}$ 3. a) -297 b) -394 c) $-262 \, kJ \, mol^{-1}$
4. $-64 \, kJ \, mol^{-1}$ 5. $-847 \, kJ \, mol^{-1}$ exothermic   6. $-0.30 \, kJ \, mol^{-1}$
7. $184 \, kJ \, mol^{-1}$ 8. $-106 \, kJ \, mol^{-1}$ Mg reduces Al_2O_3
9. a) $-890 \, kJ \, mol^{-1}$ b) $+ 317 \, kJ \, mol^{-1}$ c) $-55 \, kJ \, mol^{-1}$
   d) i) $-1.43 \times 10^5 \, kJ$ ii) $-2.97 \times 10^4 \, kJ$ iii) $-4.84 \times 10^4 \, kJ$
10. $416 \, kJ \, mol^{-1}$
11. a) $C_2H_6, -76; C_2H_4, +48 \, kJ \, mol^{-1}$ b) $-95 \, kJ \, mol^{-1}$ c) $-200 \, kJ \, mol^{-1}$
    d) $-343 \, kJ \, mol^{-1}$
    e) $+264 \, kJ \, mol^{-1}$ The difference is the 'bond delocalisation energy' of benzene
    f) $+136 \, kJ \, mol^{-1}$ $28 \, kJ \, mol^{-1}$ higher than the value from combustion   Butadiene
    is more stable than it is calculated to be because it is stabilised by bond delocalisation
    g) i) $+74 \, kJ \, mol^{-1}$ ii) $-20 \, kJ \, mol^{-1}$ Reaction ii) will occur
12. $-362 \, kJ \, mol^{-1}$ 13. $-440 \, kJ \, mol^{-1}$ 14. $-372 \, kJ \, mol^{-1}$ 15. b) $-380 \, kJ \, mol^{-1}$
16. $-775 \, kJ \, mol^{-1}$
17. a) $-387 \, kJ \, mol^{-1}$ b) $-220 \, kJ \, mol^{-1}$
    $\Delta H_f^{\ominus}(CaCl_2)$ has a larger negative value than $\Delta H_f^{\ominus}(CaCl)$
18. a) Solubility (LiCl) > Solubility (NaCl) since $(\Delta H_{lattice}^{\ominus} + \Delta H_{hydration}^{\ominus})$ has a more
    negative value for LiCl than for NaCl.
    b) Solubility (NaCl) > Solubility (NaF) since $(\Delta H_{lattice}^{\ominus} + \Delta H_{hydration}^{\ominus})$ has a positive
    value for NaF and a negative value for NaCl.
19. $-367 \, kJ \, mol^{-1}$
20. $\Delta H_c^{\ominus}$ is $< 3\Delta H_a^{\ominus}$ because bond delocalisation makes benzene more stable than
    calculated from bond energy terms, and less enthalpy than expected is released when
    benzene $\longrightarrow$ cyclohexane
    The hydrogenation of the first double bond destroys the bond delocalisation in
    benzene, and ΔH_b is therefore positive, showing that the enthalpy content of the system
    has increased

Exercise 54

1. $87.8 \, J \, K^{-1} \, mol^{-1}$ 2. $8.5 \, J \, K^{-1} \, mol^{-1}$
3. HF, $25.6 \, J \, K^{-1} \, mol^{-1}$ HCl, $86.2 \, J \, K^{-1} \, mol^{-1}$ HBr, $85.4 \, J \, K^{-1} \, mol^{-1}$
   HI, $83.2 \, J \, K^{-1} \, mol^{-1}$

4. 36.6 J K^{-1} mol^{-1} 5. 124 J K^{-1} mol^{-1}

6. In J K^{-1} mol^{-1}:   a) +20.0   b) −199   c) −163.5
                                       d) −121   e) +176   f) −90.1
                                       g) +285   h) +681

7. a) +   b) +   c) −   d) −
   e) +   f) −

8. a) 97.0 J K^{-1} mol^{-1} b) 145 J K^{-1} mol^{-1} c) 97.3 J K^{-1} mol^{-1}

9. a) +0.202 kJ mol^{-1} at 25 °C   +0.127 kJ mol^{-1} at 100°C   above 227 °C
   b) +2.56 kJ mol^{-1} at 25 °C   above 491°C

10. a) −2.91 kJ mol^{-1} b) +2.91 kJ mol^{-1} trans

Exercise 55

1. b) −139.6 kJ mol^{-1}
2. b) −645 kJ mol^{-1}
3. b) i) −600 kJ mol^{-1} ii) −630 kJ mol^{-1}
   iii) Bond energy is not the same in different compounds   iv) bond delocalisation energy
4. −511. kJ mol^{-1} 5. −382 kJ mol^{-1}
6. b) i) 90 kJ mol^{-1} ii) −16 kJ mol^{-1} f) 60%
7. b) i) −3 444 kJ mol^{-1} ii) −3 770 kJ mol^{-1} c) −133 kJ mol^{-1}
8. b) ii) −4.3 kJ   iii) −516 kJ mol^{-1} c) −1 104 kJ mol^{-1}
   d) $\Delta H^{\ominus}_{\text{ionisation}}$ (Ca) < $\Delta H^{\ominus}_{\text{ionisation}}$ (Mg) as Ca > Mg in size
   $\Delta H^{\ominus}_{\text{lattice}}$ (CaCO$_3$) < $\Delta H^{\ominus}_{\text{lattice}}$ (MgCO$_3$)
   $\Delta H^{\ominus}_{\text{formation}}$ (CaCO$_3$) and $\Delta H^{\ominus}_{\text{formation}}$ (MgCO$_3$) are therefore very similar in value
9. 104 J K^{-1} mol^{-1} 10. d) i) 0.600 mol dm^{-3} ii) 52.1 kJ mol^{-1}
11. b) i) decrease   ii) decrease   iii) increase   c) ΔG negative
    d) i) −121 J K^{-1} mol^{-1} ii) −199 J K^{-1} mol^{-1}
    e) i) $\Delta G^{\ominus} = -191$ kJ mol^{-1} Yes   ii) the presence of a catalyst
12. c) −441.0 kJ mol^{-1}

CHAPTER 10

Exercise 56

1. 2   2. 1 w.r.t. A   2 w.r.t. B   3. d

4. $\dfrac{d[X]}{dt} = k[X]^0$ 0   5. 10.0 mol^{-1} dm^3 s^{-1}

6. $\dfrac{d[P]}{dt} = k[A][B]$ 2.0 × 10^{-3} dm^3 mol^{-1} s^{-1}

7. a) 1   b) 2   1.67 × 10^5 mol^{-2} dm^6 h^{-1} or 46.4 mol^{-2} dm^6 s^{-1}
8. 1   1.92 × 10^{-4} s^{-1}

9. $\dfrac{d[N_2O_5]}{dt} = k[N_2O_5]$ a) 0.150 mol dm^{-3} s^{-1} b) 1.80 mol dm^{-3} s^{-1}

Exercise 57

1. 1.54 × 10^{-4} s^{-1} 2. 1.10 × 10^{-5} s^{-1}
3. Gradients/mol dm^{-3} min^{-1}:   a) −0.842   b) −0.386
   c) −0.200   d) 0   0.365 min^{-1} or 6.10 × 10^{-3} s^{-1} 1
4. a) Initial rates/mol dm^{-3} min^{-1}:   1) 6.30 × 10^{-3} 2) 1.27 × 10^{-2} 3) 1.90 × 10^{-2}
   b) 1   c) 3.75 min   7.50 min   1
   d) $\dfrac{d[C]}{dt} = k[A][B]$ e) 1.06 × 10^{-2} dm^3 mol^{-1} s^{-1}

Exercise 58

1. 50.0 dm^3 mol^{-1} s^{-1} 2. 1.40 × 10^{-4} dm^3 mol^{-1} s^{-1}
3. 1 w.r.t. A   2 w.r.t. B   Rate = $k[A][B]^2$ 4.0 dm^6 mol^{-2} s^{-1}

Exercise 59

1. 170 min   2. 64.5 h
3. An isotope with $t_{1/2} = 0.85$ h decays to form an isotope with $t_{1/2} = 36$ h
4. 14 days   5. a) 0.03125 ($= 1/2^5$)   b) 9.3×10^{-10} ($= 1/2^{30}$)
6. 832 years   7. 1.00×10^{-5} g   8. 4.20 h
9. 43.1 min   10. a) 75.5%   b) 5.9%   11. 3276 years

Exercise 60

1. 3.85×10^{-4} s^{-1} 70 min   2. 4.16×10^{-6} s^{-1} 3. 5.86×10^{-4} s^{-1}
4. a) 6.17×10^{-4} mol^{-1} dm^3 s^{-1} b) 1.62×10^4 s (4.5 h)
   c) 1.79×10^4 s (5.0 h)
5. a) 1.6×10^{-6} s^{-1} b) 20 days   6. 2.95×10^{-8} mol dm^{-3} s^{-1}
7. 1.36×10^{-7} mol dm^{-3} s^{-1}
8. a) 2   b) 1.0×10^{-7} m^2 N^{-1} s^{-1}

9. Experiment 1: $\dfrac{d[A]}{dt} = k[A]^2[B]$ $k = 0.417$

   Experiment 2: $\dfrac{d[A]}{dt} = k[A]^2$ $k = 0.05$ B is adsorbed on the surface

10. a) 1   b) 7.48×10^{-4} s^{-1}
11. a) 2   b) 5.96×10^{-4} kN m^{-2} s^{-1}

Exercise 61

1. 29.0 kJ mol^{-1} 2. 25.5 kJ mol^{-1} 3. 2.25   4. 78.3 kJ mol^{-1}
5. 13.6 kJ mol^{-1}

Exercise 62

1. b) 2   d) 2.04×10^{-3} dm^3 mol^{-1} s^{-1} 2. 6.13×10^{-4} s^{-1}
3. c) 107 kJ mol^{-1}
4. c) 2.82×10^{-2} min^{-1} or 4.70×10^{-4} s^{-1} (at all times)   d) i) 1
5. c) 5.8×10^{-4} s^{-1} 6. 1 w.r.t. [H$^+$]   0 w.r.t. [I$_2$]
7. 1 w.r.t. [ester]   1 w.r.t. [H$_3$O$^+$] $K_c = 3.45$
8. b) 4.95×10^{-3} s^{-1} d) ii) 140 s
9. b) Gradients/(mol l^{-1} s^{-1}) $\times 10^3$: 0.440   0.190   0.097
   Rate constants/(s^{-1}) $\times 10^4$: 3.9   3.3   3.6
   c) 6.6×10^{-4} s^{-1} $E = 23.0$ kJ mol^{-1}

CHAPTER 11

Exercise 63

1. a) $K_p = \dfrac{p_C \times p_D}{p_A \times p_B}$ b) $K_c = \dfrac{[C][D]}{[A][B]}$ c) $K_{Het} = \dfrac{p_D}{p_B}$ d) $K_{Het} = \dfrac{[D]}{[A][B]}$
2. a) 0.176   b) 1.28×10^4 Pa   3. 1.78×10^5 N m^{-2} 4. $\frac{1}{3}$
5. a) 0.35   b) 8.1×10^4 N m^{-2}
6. 64   7. 3.39×10^{-15} N^{-2} m^4
8. a) 2.00×10^{-3} dm^3 mol^{-1} s^{-1} b) 64.0
9. a) 0.71 mol   b) 43 g   10. 99.8%   11. c
12. a) 0.483 mol   b) 0.483 mol   c) 0.517 mol   d) 4.517 mol
13. 9.26×10^{-15} Pa^{-2} 14. 0.2 atm^{-1}
15. a) CO$_2$, H$_2$: 0.27 mol   CO, H$_2$O: 0.23 mol
    b) CO$_2$: 4.2 mol   H$_2$: 0.21 mol   CO, H$_2$O: 0.79 mol
16. 9.33×10^{-4} N^{-1} m^2

Exercise 64

1. c) i) 0.25   ii) 40% NO_2 iii) $2.70 \times 10^4 \, N \, m^{-2}$
2. b) +   c) 0.845 mol   3. b) 145 cm^3 NO_2 55 cm^3 N_2O_4
4. b) ii) 1   iii) 2   iv) $1.5 \times 10^{-3} \, l \, mol^{-1} \, s^{-1}$
   c) $22.0 \, J \, mol^{-1} \, K^{-1}$ d) i) 56.25   ii) none
5. a) $\dfrac{\left(\frac{1}{3}/\frac{4}{3}\right)^2 p^2}{\left(\frac{2}{3}/\frac{4}{3}\right)p} = K_p \quad \therefore \; p = 8K_p$ b) 0.577 mol
6. b) ii) 1.64×10^{-3} mol HI 2.00×10^{-4} mol I_2 2.00×10^{-4} mol H_2
      1.24×10^{-3} mol HI left.   iii) 0.026
   c) i) and ii) no change since the concentration units cancel and K_c is dimensionless

CHAPTER 12

Exercise 65

1. C_2H_5 C_4H_{10} 2. $88 \, g \, mol^{-1}$ $C_3H_7CO_2H$ 3. $C_{11}H_{10}O$
4. $C_{16}H_{13}O_3N$ 5. A is $C_6H_{12} = CH_3CH_2CH{=}CHCH_2CH_3$ B is $C_2H_5CO_2H$
6. A is $C_6H_5CO_2CH_3$ B is CH_3OH C is $C_6H_5CO_2Na$ D is $C_6H_5CO_2H$
7. a) C_3H_7ON b) propanamide, $C_2H_5CONH_2$
8. a) 3 mol Br_2 : 1 mol C_6H_5OH
   b) $C_6H_5OH + 3Br_2 \longrightarrow C_6H_2Br_3OH + 3HBr$
9. a) $55.9 \, g \, mol^{-1}$ b) C_4H_8
     but-1-ene, $CH_3CH_2CH{=}CH_2$ and but-2-ene, $CH_3CH{=}CHCH_3$
10. a) CHO   b) $C_4H_4O_4$
   c) butenedioic acid, cis $HCCO_2H$ and trans HO_2CH
                $\|$ $\|$
           $HCCO_2H$ $HCCO_2H$
11. a) $C_4H_{10}O$ b) $C_4H_{10}O$ c) $C_2H_5OC_2H_5$ $CH_3OCH_2CH_2CH_3$
     $CH_3OCH(CH_3)_2$ $CH_3CH_2CH_2CH_2OH$ $CH_3CH_2C(OH)CH_3$
     $CH_3CH(CH_3)CH_2OH$ $CH_3CH_2CH(CH_3)OH$
12. A is $C_8H_9ON = C_6H_5NHCOCH_3$ B is $C_6H_7N = C_6H_5NH_2$

Exercise 66

1. a) C_2H_5NO b) X is $C_2H_4O_2$ W is CH_3CONH_2
   c) $CH_3CONH_2(aq) + HCl(aq) + H_2O(l) \longrightarrow CH_3CO_2H(aq) + NH_4Cl(aq)$
   d) Y is CH_3CHO
2. A is $C_{13}H_{11}NO = C_6H_5NHCOCH{=}CHC_6H_5$
   B is $C_6H_5NH_2$ C is $C_6H_5CH{=}CHCO_2Na$ D is $C_6H_5NH_3{}^+Cl^-$
   E is $C_6H_2Br_3NH_2$ F is $C_6H_5CH{=}CHCO_2H$ G is $C_6H_5CH{=}CH_2$
3. A is $BrC_6H_4CH_2Br$ B is $BrC_6H_4CH_2OH$ C is BrC_6H_4CHO D is $BrC_6H_4CO_2H$
   E is 2-X-4-bromobenzoic acid   F is 3-X-4-bromobenzoic acid
4. A is $CH_3C_6H_4CO_2C_3H_7$ B is $CH_3CH_2CH_2OH$ C is $CH_3C_6H_4CO_2H$
   D is $CH_3C_6H_4CH_2OH$ E is $CH_3C_6H_4CH_2O_2CCH_3$
5. A is $C_6H_5COCH_3$ B is $C_6H_5C(OH)(CN)CH_3$ C is $C_6H_5C(OH)(CO_2H)CH_3$
   D is $C_6H_5CO_2H$
6. a) C_4H_4O b) $C_8H_8O_2$
   c) A is $CH_3C_6H_4CON(CH_3)_2$ B is $(CH_3)_2NH$ C is $CH_3C_6H_4CO_2H$
     D is $C_6H_5CO_2H$ E is $C_6H_4\!\!\begin{array}{c}C{=}O\\ \diagdown O\\ C{=}O\end{array}$ F is $(CH_3)_2NH_2{}^+Cl^-$
                                                 G is $(CH_3)_2N{-}N{=}O$

7. a) i) $C_4H_8O_2$
          ii) $HCO_2CH_2CH_2CH_3$ propyl methanoate
              $HCO_2CH(CH_3)_2$ 1-methylethylmethanoate
              $CH_3CO_2C_2H_5$ ethylethanoate
              $C_2H_5CO_2CH_3$ methylpropanoate

8. X is $(CH_3)_2C{=}CH_2$ Y is $(CH_3)_2CO$

9. A is $C_2H_5CONH_2$ B is $(CH_3)_2C{=}NOH$ C is $C_2H_5CH(OH)NH_2$
D is $CH_3NHCOCH_3$

10. a) 31.8%   b) 24.3%   c) 6.9%   d) $MnCO_4N_3$

11. A is $H_2N(CH_2)_6NH_2$ B is $HO(CH_2)_6OH$ C is $HO_2C(CH_2)_6CO_2H$

12. $CH_3CH_2CH_2COCl$ and $(CH_3)_2CHCOCl$

13. Q is $C_3H_7NH_2$ R is $C_3H_7CO_2H$ S is C_3H_8
P is $CH_3CH_2CH_2CONHCH_2CH_2CH_3$ or $(CH_3)_2CHCONHCH_2CH_2CH_3$
or $CH_2CH_2CH_2CONHCH(CH_3)_2$ or $(CH_3)_2CHCONHCH(CH_3)_2$

14. a) $C_2H_4Br_2 = BrCH_2CH_2Br$ and CH_3CHBr_2

15. A is CH_3OH B is CH_3CN C is CH_3CO_2H D is CH_3OCH_3 E is CH_3CO

F is $CH_2CO_2CH_3$

Table of Relative Atomic Masses

Element	Symbol	Atomic number	Relative atomic mass	Element	Symbol	Atomic number	Relative atomic mass
Aluminium	Al	13	27.0	Lead	Pb	82	207
Antimony	Sb	51	122	Lithium	Li	3	6.94
Argon	Ar	18	40.0	Magnesium	Mg	12	24.3
Arsenic	As	33	75.0	Manganese	Mn	25	54.9
Barium	Ba	56	137	Mercury	Hg	80	200
Beryllium	Be	4	9.0	Neon	Ne	10	20.2
Bismuth	Bi	83	209	Nickel	Ni	28	58.7
Boron	B	5	10.8	Nitrogen	N	7	14.0
Bromine	Br	35	80.0	Oxygen	O	8	16.0
Cadmium	Cd	48	112.5	Phosphorus	P	15	31.0
Calcium	Ca	20	40.1	Platinum	Pt	78	195
Carbon	C	6	12.0	Potassium	K	19	39.1
Chlorine	Cl	17	35.5	Selenium	Se	34	79.0
Chromium	Cr	24	52.0	Silicon	Si	14	28.1
Cobalt	Co	27	59.0	Silver	Ag	47	108
Copper	Cu	29	63.5	Sodium	Na	11	23.0
Fluorine	F	9	19.0	Strontium	Sr	38	87.6
Germanium	Ge	32	72.5	Sulphur	S	16	32.1
Gold	Au	79	197	Tin	Sn	50	119
Helium	He	2	4.00	Titanium	Ti	22	47.9
Hydrogen	H	1	1.01	Vanadium	V	23	50.9
Iodine	I	53	127	Xenon	Xe	54	131
Iron	Fe	26	55.8	Zinc	Zn	30	65.4
Krypton	Kr	36	83.8				

Periodic Table of the Elements

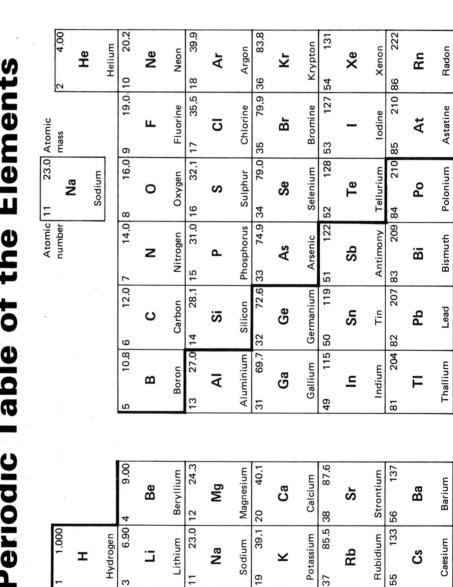

Atomic number		Atomic mass
11		23.0
	Na	
	Sodium	

1	1.000
	H
Hydrogen	

3	6.90	4	9.00
	Li		Be
Lithium		Beryllium	

11	23.0	12	24.3
	Na		Mg
Sodium		Magnesium	

19	39.1	20	40.1
	K		Ca
Potassium		Calcium	

37	85.5	38	87.6
	Rb		Sr
Rubidium		Strontium	

55	133	56	137
	Cs		Ba
Caesium		Barium	

		2	4.00
			He
			Helium

5	10.8	6	12.0	7	14.0	8	16.0	9	19.0	10	20.2
	B		C		N		O		F		Ne
Boron		Carbon		Nitrogen		Oxygen		Fluorine		Neon	

13	27.0	14	28.1	15	31.0	16	32.1	17	35.5	18	39.9
	Al		Si		P		S		Cl		Ar
Aluminium		Silicon		Phosphorus		Sulphur		Chlorine		Argon	

31	69.7	32	72.6	33	74.9	34	79.0	35	79.9	36	83.8
	Ga		Ge		As		Se		Br		Kr
Gallium		Germanium		Arsenic		Selenium		Bromine		Krypton	

49	115	50	119	51	122	52	128	53	127	54	131
	In		Sn		Sb		Te		I		Xe
Indium		Tin		Antimony		Tellurium		Iodine		Xenon	

81	204	82	207	83	209	84	210	85	210	86	222
	Tl		Pb		Bi		Po		At		Rn
Thallium		Lead		Bismuth		Polonium		Astatine		Radon	

21	45.0 Sc Scandium	22	47.9 Ti Titanium	23	50.9 V Vanadium	24	52.0 Cr Chromium	25	54.9 Mn Manganese	26	55.8 Fe Iron	27	58.9 Co Cobalt	28	58.7 Ni Nickel	29	63.5 Cu Copper	30	65.4 Zn Zinc
39	88.9 Y Yttrium	40	91.2 Zr Zirconium	41	92.9 Nb Niobium	42	95.9 Mo Molybdenum	43	99 Tc Technetium	44	101 Ru Ruthenium	45	103 Rh Rhodium	46	106 Pd Palladium	47	108 Ag Silver	48	112 Cd Cadmium
57	139 La Lanthanum	72	178 Hf Hafnium	73	181 Ta Tantalum	74	184 W Tungsten	75	186 Re Rhenium	76	190 Os Osmium	77	192 Ir Iridium	78	195 Pt Platinum	79	197 Au Gold	80	201 Hg Mercury

Logarithms

LOGARITHMS

	0	1	2	3	4	5	6	7	8	9	1	2	3	4	5	6	7	8	9
10	0000	0043	0086	0128	0170						4	8	13	17	21	25	30	34	38
						0212	0253	0294	0334	0374	4	8	12	16	20	24	28	32	36
11	0414	0453	0492	0531	0569						4	8	12	15	19	23	27	31	35
						0607	0645	0682	0719	0755	4	7	11	15	18	22	26	30	33
12	0792	0828	0864	0899	0934						4	7	11	14	18	21	25	28	32
						0969	1004	1038	1072	1106	3	7	10	14	17	20	24	27	31
13	1139	1173	1206	1239	1271						3	7	10	13	16	20	23	26	30
						1303	1335	1367	1399	1430	3	6	9	13	16	19	22	25	28
14	1461	1492	1523	1553	1584						3	6	9	12	15	18	21	24	27
						1614	1644	1673	1703	1732	3	6	9	12	15	18	21	24	27
15	1761	1790	1818	1847	1875						3	6	9	11	14	17	20	23	26
						1903	1931	1959	1987	2014	3	6	8	11	14	17	19	22	25
16	2041	2068	2095	2122	2148						3	5	8	11	13	16	19	21	24
						2175	2201	2227	2253	2279	3	5	8	10	13	16	18	21	23
17	2304	2330	2355	2380	2405						3	5	8	10	13	15	18	20	23
						2430	2455	2480	2504	2529	2	5	7	10	12	15	17	20	22
18	2553	2577	2601	2625	2648						2	5	7	10	12	14	17	19	21
						2672	2695	2718	2742	2765	2	5	7	9	12	14	16	19	21
19	2788	2810	2833	2856	2878						2	5	7	9	11	14	16	18	20
						2900	2923	2945	2967	2989	2	4	7	9	11	13	15	18	20
20	3010	3032	3054	3075	3096	3118	3139	3160	3181	3201	2	4	6	8	11	13	15	17	19
21	3222	3243	3263	3284	3304	3324	3345	3365	3385	3404	2	4	6	8	10	12	14	16	18
22	3424	3444	3464	3483	3502	3522	3541	3560	3579	3598	2	4	6	8	10	12	14	15	17
23	3617	3636	3655	3674	3692	3711	3729	3747	3766	3784	2	4	6	7	9	11	13	15	17
24	3802	3820	3838	3856	3874	3892	3909	3927	3945	3962	2	4	5	7	9	11	12	14	16
25	3979	3997	4014	4031	4048	4065	4082	4099	4116	4133	2	3	5	7	9	10	12	14	15
26	4150	4166	4183	4200	4216	4232	4249	4265	4281	4298	2	3	5	7	8	10	11	13	15
27	4314	4330	4346	4362	4378	4393	4409	4425	4440	4456	2	3	5	6	8	9	11	13	14
28	4472	4487	4502	4518	4533	4548	4564	4579	4594	4609	2	3	5	6	8	9	11	12	14
29	4624	4639	4654	4669	4683	4698	4713	4728	4742	4757	1	3	4	6	7	9	10	12	13
30	4771	4786	4800	4814	4829	4843	4857	4871	4886	4900	1	3	4	6	7	9	10	11	13
31	4914	4928	4942	4955	4969	4983	4997	5011	5024	5038	1	3	4	6	7	8	10	11	12
32	5051	5065	5079	5092	5105	5119	5132	5145	5159	5172	1	3	4	5	7	8	9	11	12
33	5185	5198	5211	5224	5237	5250	5263	5276	5289	5302	1	3	4	5	6	8	9	10	12
34	5315	5328	5340	5353	5366	5378	5391	5403	5416	5428	1	3	4	5	6	8	9	10	11
35	5441	5453	5465	5478	5490	5502	5514	5527	5539	5551	1	2	4	5	6	7	9	10	11
36	5563	5575	5587	5599	5611	5623	5635	5647	5658	5670	1	2	4	5	6	7	8	10	11
37	5682	5694	5705	5717	5729	5740	5752	5763	5775	5786	1	2	3	5	6	7	8	9	10
38	5798	5809	5821	5832	5843	5855	5866	5877	5888	5899	1	2	3	5	6	7	8	9	10
39	5911	5922	5933	5944	5955	5966	5977	5988	5999	6010	1	2	3	4	5	7	8	9	10
40	6021	6031	6042	6053	6064	6075	6085	6096	6107	6117	1	2	3	4	5	6	8	9	10
41	6128	6138	6149	6160	6170	6180	6191	6201	6212	6222	1	2	3	4	5	6	7	8	9
42	6232	6243	6253	6263	6274	6284	6294	6304	6314	6325	1	2	3	4	5	6	7	8	9
43	6335	6345	6355	6365	6375	6385	6395	6405	6415	6425	1	2	3	4	5	6	7	8	9
44	6435	6444	6454	6464	6474	6484	6493	6503	6513	6522	1	2	3	4	5	6	7	8	9
45	6532	6542	6551	6561	6571	6580	6590	6599	6609	6618	1	2	3	4	5	6	7	8	9
46	6628	6637	6646	6656	6665	6675	6684	6693	6702	6712	1	2	3	4	5	6	7	7	8
47	6721	6730	6739	6749	6758	6767	6776	6785	6794	6803	1	2	3	4	5	5	6	7	8
48	6812	6821	6830	6839	6848	6857	6866	6875	6884	6893	1	2	3	4	4	5	6	7	8
49	6902	6911	6920	6928	6937	6946	6955	6964	6972	6981	1	2	3	4	4	5	6	7	8

	0	1	2	3	4	5	6	7	8	9	1	2	3	4	5	6	7	8	9
50	6990	6998	7007	7016	7024	7033	7042	7050	7059	7067	1	2	3	3	4	5	6	7	8
51	7076	7084	7093	7101	7110	7118	7126	7135	7143	7152	1	2	3	3	4	5	6	7	8
52	7160	7168	7177	7185	7193	7202	7210	7218	7226	7235	1	2	2	3	4	5	6	7	7
53	7243	7251	7259	7267	7275	7284	7292	7300	7308	7316	1	2	2	3	4	5	6	6	7
54	7324	7332	7340	7348	7356	7364	7372	7380	7388	7396	1	2	2	3	4	5	6	6	7
55	7404	7412	7419	7427	7435	7443	7451	7459	7466	7474	1	2	2	3	4	5	5	6	7
56	7482	7490	7497	7505	7513	7520	7528	7536	7543	7551	1	2	2	3	4	5	5	6	7
57	7559	7566	7574	7582	7589	7597	7604	7612	7619	7627	1	2	2	3	4	5	5	6	7
58	7634	7642	7649	7657	7664	7672	7679	7686	7694	7701	1	1	2	3	4	4	5	6	7
59	7709	7716	7723	7731	7738	7745	7752	7760	7767	7774	1	1	2	3	4	4	5	6	7
60	7782	7789	7796	7803	7810	7818	7825	7832	7839	7846	1	1	2	3	4	4	5	6	6
61	7853	7860	7868	7875	7882	7889	7896	7903	7910	7917	1	1	2	3	4	4	5	6	6
62	7924	7931	7938	7945	7952	7959	7966	7973	7980	7987	1	1	2	3	3	4	5	6	6
63	7993	8000	8007	8014	8021	8028	8035	8041	8048	8055	1	1	2	3	3	4	5	5	6
64	8062	8069	8075	8082	8089	8096	8102	8109	8116	8122	1	1	2	3	3	4	5	5	6
65	8129	8136	8142	8149	8156	8162	8169	8176	8182	8189	1	1	2	3	3	4	5	5	6
66	8195	8202	8209	8215	8222	8228	8235	8241	8248	8254	1	1	2	3	3	4	5	5	6
67	8261	8267	8274	8280	8287	8293	8299	8306	8312	8319	1	1	2	3	3	4	5	5	6
68	8325	8331	8338	8344	8351	8357	8363	8370	8376	8382	1	1	2	3	3	4	4	5	6
69	8388	8395	8401	8407	8414	8420	8426	8432	8439	8445	1	1	2	2	3	4	4	5	6
70	8451	8457	8463	8470	8476	8482	8488	8494	8500	8506	1	1	2	2	3	4	4	5	6
71	8513	8519	8525	8531	8537	8543	8549	8555	8561	8567	1	1	2	2	3	4	4	5	5
72	8573	8579	8585	8591	8597	8603	8609	8615	8621	8627	1	1	2	2	3	4	4	5	5
73	8633	8639	8645	8651	8657	8663	8669	8675	8681	8686	1	1	2	2	3	4	4	5	5
74	8692	8698	8704	8710	8716	8722	8727	8733	8739	8745	1	1	2	2	3	3	4	5	5
75	8751	8756	8762	8768	8774	8779	8785	8791	8797	8802	1	1	2	2	3	3	4	5	5
76	8808	8814	8820	8825	8831	8837	8842	8848	8854	8859	1	1	2	2	3	3	4	5	5
77	8865	8871	8876	8882	8887	8893	8899	8904	8910	8915	1	1	2	2	3	3	4	4	5
78	8921	8927	8932	8938	8943	8949	8954	8960	8965	8971	1	1	2	2	3	3	4	4	5
79	8976	8982	8987	8993	8998	9004	9009	9015	9020	9025	1	1	2	2	3	3	4	4	5
80	9031	9036	9042	9047	9053	9058	9063	9069	9074	9079	1	1	2	2	3	3	4	4	5
81	9085	9090	9096	9101	9106	9112	9117	9122	9128	9133	1	1	2	2	3	3	4	4	5
82	9138	9143	9149	9154	9159	9165	9170	9175	9180	9186	1	1	2	2	3	3	4	4	5
83	9191	9196	9201	9206	9212	9217	9222	9227	9232	9238	1	1	2	2	3	3	4	4	5
84	9243	9248	9253	9258	9263	9269	9274	9279	9284	9289	1	1	2	2	3	3	4	4	5
85	9294	9299	9304	9309	9315	9320	9325	9330	9335	9340	1	1	2	2	3	3	4	4	5
86	9345	9350	9355	9360	9365	9370	9375	9380	9385	9390	1	1	2	2	3	3	4	4	5
87	9395	9400	9405	9410	9415	9420	9425	9430	9435	9440	0	1	1	2	2	3	3	4	4
88	9445	9450	9455	9460	9465	9469	9474	9479	9484	9489	0	1	1	2	2	3	3	4	4
89	9494	9499	9504	9509	9513	9518	9523	9528	9533	9538	0	1	1	2	2	3	3	4	4
90	9542	9547	9552	9557	9562	9566	9571	9576	9581	9586	0	1	1	2	2	3	3	4	4
91	9590	9595	9600	9605	9609	9614	9619	9624	9628	9633	0	1	1	2	2	3	3	4	4
92	9638	9643	9647	9652	9657	9661	9666	9671	9675	9680	0	1	1	2	2	3	3	4	4
93	9685	9689	9694	9699	9703	9708	9713	9717	9722	9727	0	1	1	2	2	3	3	4	4
94	9731	9736	9741	9745	9750	9754	9759	9763	9768	9773	0	1	1	2	2	3	3	4	4
95	9777	9782	9786	9791	9795	9800	9805	9809	9814	9818	0	1	1	2	2	3	3	4	4
96	9823	9827	9832	9836	9841	9845	9850	9854	9859	9863	0	1	1	2	2	3	3	4	4
97	9868	9872	9877	9881	9886	9890	9894	9899	9903	9908	0	1	1	2	2	3	3	4	4
98	9912	9917	9921	9926	9930	9934	9939	9943	9948	9952	0	1	1	2	2	3	3	4	4
99	9956	9961	9965	9969	9974	9978	9983	9987	9991	9996	0	1	1	2	2	3	3	3	4

Antilogarithms

ANTILOGARITHMS

	0	1	2	3	4	5	6	7	8	9	1	2	3	4	5	6	7	8	9
0.00	1000	1002	1005	1007	1009	1012	1014	1016	1019	1021	0	0	1	1	1	1	2	2	2
0.01	1023	1026	1028	1030	1033	1035	1038	1040	1042	1045	0	0	1	1	1	1	2	2	2
0.02	1047	1050	1052	1054	1057	1059	1062	1064	1067	1069	0	0	1	1	1	1	2	2	2
0.03	1072	1074	1076	1079	1081	1084	1086	1089	1091	1094	0	0	1	1	1	1	2	2	2
0.04	1096	1099	1102	1104	1107	1109	1112	1114	1117	1119	0	1	1	1	1	2	2	2	2
0.05	1122	1125	1127	1130	1132	1135	1138	1140	1143	1146	0	1	1	1	1	2	2	2	2
0.06	1148	1151	1153	1156	1159	1161	1164	1167	1169	1172	0	1	1	1	1	2	2	2	2
0.07	1175	1178	1180	1183	1186	1189	1191	1194	1197	1199	0	1	1	1	1	2	2	2	2
0.08	1202	1205	1208	1211	1213	1216	1219	1222	1225	1227	0	1	1	1	1	2	2	2	3
0.09	1230	1233	1236	1239	1242	1245	1247	1250	1253	1256	0	1	1	1	1	2	2	2	3
0.10	1259	1262	1265	1268	1271	1274	1276	1279	1282	1285	0	1	1	1	1	2	2	2	3
0.11	1288	1291	1294	1297	1300	1303	1306	1309	1312	1315	0	1	1		2	2	2	2	3
0.12	1318	1321	1324	1327	1330	1334	1337	1340	1343	1346	0	1	1	1	2	2	2	3	3
0.13	1349	1352	1355	1358	1361	1365	1368	1371	1374	1377	0	1	1	1	2	2	2	3	3
0.14	1380	1384	1387	1390	1393	1396	1400	1403	1406	1409	0	1	1	1	2	2	2	3	3
0.15	1413	1416	1419	1422	1426	1429	1432	1435	1439	1442	0	1	1	1	2	2	2	3	3
0.16	1445	1449	1452	1455	1459	1462	1466	1469	1472	1476	0	1	1	1	2	2	2	3	3
0.17	1479	1483	1486	1489	1493	1496	1500	1503	1507	1510	0	1	1	1	2	2	2	3	3
0.18	1514	1517	1521	1524	1528	1531	1535	1538	1542	1545	0	1	1	1	2	2	2	3	3
0.19	1549	1552	1556	1560	1563	1567	1570	1574	1578	1581	0	1	1	1	2	2	3	3	3
0.20	1585	1589	1592	1596	1600	1603	1607	1611	1614	1618	0	1	1	1	2	2	3	3	3
0.21	1622	1626	1629	1633	1637	1641	1644	1648	1652	1656	0	1	1	2	2	2	3	3	3
0.22	1660	1663	1667	1671	1675	1679	1683	1687	1690	1694	0	1	1	2	2	2	3	3	3
0.23	1698	1702	1706	1710	1714	1718	1722	1726	1730	1734	0	1	1	2	2	2	3	3	4
0.24	1738	1742	1746	1750	1754	1758	1762	1766	1770	1774	0	1	1	2	2	2	3	3	4
0.25	1778	1782	1786	1791	1795	1799	1803	1807	1811	1816	0	1	1	2	2	2	3	3	4
0.26	1820	1824	1828	1832	1837	1841	1845	1849	1854	1858	0	1	1	2	2	3	3	3	4
0.27	1862	1866	1871	1875	1879	1884	1888	1892	1897	1901	0	1	1	2	2	3	3	4	4
0.28	1905	1910	1914	1919	1923	1928	1932	1936	1941	1945	0	1	1	2	2	3	3	4	4
0.29	1950	1954	1959	1963	1968	1972	1977	1982	1986	1991	0	1	1	2	2	3	3	4	4
0.30	1995	2000	2004	2009	2014	2018	2023	2028	2032	2037	0	1	1	2	2	3	3	4	4
0.31	2042	2046	2051	2056	2061	2065	2070	2075	2080	2084	0	1	1	2	2	3	3	4	4
0.32	2089	2094	2099	2104	2109	2113	2118	2123	2128	2133	0	1	1	2	2	3	3	4	4
0.33	2138	2143	2148	2153	2158	2163	2168	2173	2178	2183	0	1	1	2	2	3	3	4	4
0.34	2188	2193	2198	2203	2208	2213	2218	2223	2228	2234	1	1	2	2	3	3	4	4	5
0.35	2239	2244	2249	2254	2259	2265	2270	2275	2280	2286	1	1	2	2	3	3	4	4	5
0.36	2291	2296	2301	2307	2312	2317	2323	2328	2333	2339	1	1	2	2	3	3	4	4	5
0.37	2344	2350	2355	2360	2366	2371	2377	2382	2388	2393	1	1	2	2	3	3	4	4	5
0.38	2399	2404	2410	2415	2421	2427	2432	2438	2443	2449	1	1	2	2	3	3	4	4	5
0.39	2455	2460	2466	2472	2477	2483	2489	2495	2500	2506	1	1	2	2	3	3	4	5	5
0.40	2512	2518	2523	2529	2535	2541	2547	2553	2559	2564	1	1	2	2	3	4	4	5	5
0.41	2570	2576	2582	2588	2594	2600	2606	2612	2618	2624	1	1	2	2	3	4	4	5	5
0.42	2630	2636	2642	2649	2655	2661	2667	2673	2679	2685	1	1	2	2	3	4	4	5	6
0.43	2692	2698	2704	2710	2716	2723	2729	2735	2742	2748	1	1	2	3	3	4	4	5	6
0.44	2754	2761	2767	2773	2780	2786	2793	2799	2805	2812	1	1	2	3	3	4	4	5	6
0.45	2818	2825	2831	2838	2844	2851	2858	2864	2871	2877	1	1	2	3	3	4	5	5	6
0.46	2884	2891	2897	2904	2911	2917	2924	2931	2938	2944	1	1	2	3	3	4	5	5	6
0.47	2951	2958	2965	2972	2979	2985	2992	2999	3006	3013	1	1	2	3	3	4	5	5	6
0.48	3020	3027	3034	3041	3048	3055	3062	3069	3076	3083	1	1	2	3	4	4	5	6	6
0.49	3090	3097	3105	3112	3119	3126	3133	3141	3148	3155	1	1	2	3	4	4	5	6	6

	0	1	2	3	4	5	6	7	8	9	1	2	3	4	5	6	7	8	9
0.50	3162	3170	3177	3184	3192	3199	3206	3214	3221	3228	1	1	2	3	4	4	5	6	7
0.51	3236	3243	3251	3258	3266	3273	3281	3289	3296	3304	1	2	2	3	4	5	5	6	7
0.52	3311	3319	3327	3334	3342	3350	3357	3365	3373	3381	1	2	2	3	4	5	5	6	7
0.53	3388	3396	3404	3412	3420	3428	3436	3443	3451	3459	1	2	2	3	4	5	6	6	7
0.54	3467	3475	3483	3491	3499	3508	3516	3524	3532	3540	1	2	2	3	4	5	6	6	7
0.55	3548	3556	3565	3573	3581	3589	3597	3606	3614	3622	1	2	2	3	4	5	6	7	7
0.56	3631	3639	3648	3656	3664	3673	3681	3690	3698	3707	1	2	3	3	4	5	6	7	8
0.57	3715	3724	3733	3741	3750	3758	3767	3776	3784	3793	1	2	3	3	4	5	6	7	8
0.58	3802	3811	3819	3828	3837	3846	3855	3864	3873	3882	1	2	3	4	4	5	6	7	8
0.59	3890	3899	3908	3917	3926	3936	3945	3954	3963	3972	1	2	3	4	5	5	6	7	8
0.60	3981	3990	3999	4009	4018	4027	4036	4046	4055	4064	1	2	3	4	5	6	6	7	8
0.61	4074	4083	4093	4102	4111	4121	4130	4140	4150	4159	1	2	3	4	5	6	7	8	9
0.62	4169	4178	4188	4198	4207	4217	4227	4236	4246	4256	1	2	3	4	5	6	7	8	9
0.63	4266	4276	4285	4295	4305	4315	4325	4335	4345	4355	1	2	3	4	5	6	7	8	9
0.64	4365	4375	4385	4395	4406	4416	4426	4436	4446	4457	1	2	3	4	5	6	7	8	9
0.65	4467	4477	4487	4498	4508	4519	4529	4539	4550	4560	1	2	3	4	5	6	7	8	9
0.66	4571	4581	4592	4603	4613	4624	4634	4645	4656	4667	1	2	3	4	5	6	7	9	10
0.67	4677	4688	4699	4710	4721	4732	4742	4753	4764	4775	1	2	3	4	5	7	8	9	10
0.68	4786	4797	4808	4819	4831	4842	4853	4864	4875	4887	1	2	3	4	6	7	8	9	10
0.69	4893	4909	4920	4932	4943	4955	4966	4977	4989	5000	1	2	3	5	6	7	8	9	10
0.70	5012	5023	5035	5047	5058	5070	5082	5093	5105	5117	1	2	4	5	6	7	8	9	11
0.71	5129	5140	5152	5164	5176	5188	5200	5212	5224	5236	1	2	4	5	6	7	8	10	11
0.72	5248	5260	5272	5284	5297	5309	5321	5333	5336	5358	1	2	4	5	6	7	9	10	11
0.73	5370	5383	5395	5408	5420	5433	5445	5458	5470	5483	1	3	4	5	6	8	9	10	11
0.74	5495	5508	5521	5534	5546	5559	5572	5585	5598	5610	1	3	4	5	6	8	9	10	12
0.75	5623	5636	5649	5662	5675	5689	5702	5715	5728	5741	1	3	4	5	7	8	9	10	12
0.76	5754	5768	5781	5794	5808	5821	5834	5848	5861	5875	1	3	4	5	7	8	9	11	12
0.77	5888	5902	5916	5929	5943	5957	5970	5984	5998	6012	1	3	4	5	7	8	10	11	12
0.78	6026	6039	6053	6067	6081	6095	6109	6124	6138	6152	1	3	4	6	7	8	10	11	13
0.79	6166	6180	6194	6209	6223	6237	6252	6266	6281	6295	1	3	4	6	7	9	10	11	13
0.80	6310	6324	6339	6353	6368	6383	6397	6412	6427	6442	1	3	4	6	7	9	10	12	13
0.81	6457	6471	6486	6501	6516	6531	6546	6561	6577	6592	2	3	5	6	8	9	11	12	14
0.82	6607	6622	6637	6653	6668	6683	6699	6714	6730	6745	2	3	5	6	8	9	11	12	14
0.83	6761	6776	6792	6808	6823	6839	6855	6871	6887	6902	2	3	5	6	8	9	11	13	14
0.84	6918	6934	6950	6966	6982	6998	7015	7031	7047	7063	2	3	5	6	8	10	11	13	15
0.85	7079	7096	7112	7129	7145	7161	7178	7194	7211	7228	2	3	5	7	8	10	12	13	15
0.86	7244	7261	7278	7295	7311	7328	7345	7362	7379	7396	2	3	5	7	8	10	12	13	15
0.87	7413	7430	7447	7464	7482	7499	7516	7534	7551	7568	2	3	5	7	9	10	12	14	16
0.88	7586	7603	7621	7638	7656	7674	7691	7709	7727	7745	2	4	5	7	9	11	12	14	16
0.89	7762	7780	7798	7816	7834	7852	7870	7889	7907	7925	2	4	5	7	9	11	13	14	16
0.90	7943	7962	7980	7998	8017	8035	8054	8072	8091	8110	2	4	6	7	9	11	13	15	17
0.91	8128	8147	8166	8185	8204	8222	8241	8260	8279	8299	2	4	6	8	9	11	13	15	17
0.92	8318	8337	8356	8375	8395	8414	8433	8453	8472	8492	2	4	6	8	10	12	14	15	17
0.93	8511	8531	8551	8570	8590	8610	8630	8650	8670	8690	2	4	6	8	10	12	14	16	18
0.94	8710	8730	8750	8770	8790	8810	8831	8851	8872	8892	2	4	6	8	10	12	14	16	18
0.95	8913	8933	8954	8974	8995	9016	9036	9057	9078	9099	2	4	6	8	10	12	15	17	19
0.96	9120	9141	9162	9183	9204	9226	9247	9268	9290	9311	2	4	6	8	11	13	15	17	19
0.97	9333	9354	9376	9397	9419	9441	9462	9484	9506	9528	2	4	7	9	11	13	15	17	20
0.98	9550	9572	9594	9616	9638	9661	9683	9705	9727	9750	2	4	7	9	11	13	16	18	20
0.99	9772	9795	9817	9840	9863	9886	9908	9931	9954	9977	2	5	7	9	11	14	16	18	20

Index